Thymus Transcriptome and Cell Biology

Geraldo A. Passos

Editor

Thymus Transcriptome and Cell Biology

 Springer

Editor
Geraldo A. Passos
Laboratory of Genetics and Molecular Biology
Department of Basic and Oral Biology
School of Dentistry of Ribeirão Preto
University of São Paulo
Ribeirão Preto, SP, Brazil

Molecular Immunogenetics Group
Department of Genetics
Ribeirão Preto Medical School
University of São Paulo
Ribeirão Preto, SP, Brazil

ISBN 978-3-030-12039-9   ISBN 978-3-030-12040-5   (eBook)
https://doi.org/10.1007/978-3-030-12040-5

Library of Congress Control Number: 2019935516

This Springer imprint is published by the registered company Springer Nature Switzerland AG
The registered company address is: Gewerbestrasse 11, 6330 Cham, Switzerland

Foreword

The importance of the thymus as the primary lymphoid organ responsible for the generation and selection of T lymphocytes is now obvious. Nevertheless, the thymus has long been a mysterious organ. It was not until 1961 that J. F. Miller showed in seminal studies that "the thymus at birth can be essential to life," quickly followed by its role in immunological tolerance by skin grafting experiments in mice. It is always surprising that this key discovery for any immunologist, physician, or biologist did not happen sooner. The skepticism surrounding this discovery, from such eminent immunologists as Burnet, Medawar, or Mitchison, may also be astonishing, Sir Peter Medawar even going so far as to declare in 1963 "we will come to look at the presence of lymphocytes in the thymus as an evolutionary accident with little significance." This controversy remains exemplary and instructive for our ways to get out of accepted dogmas. This was a "golden age" of immunology, the 1960s being particularly remarkable with the discovery of the major histocompatibility complex (MHC), the H-2 system in mice and HLA in humans, by Jean Dausset (1980 Nobel Prize with Baruj Benacerraf and George Snell), Jon van Rood, and many others. These two major discoveries paved the way for the demonstration of thymic selection, a major physiological function of the thymus in shaping the T-cell adaptive immunity. They also provided the basis of our current understanding of how the immune system works. With time, T-lymphocyte subpopulations, T- and B-cell cooperation, mechanisms of allogeneic MHC restriction, T-cell receptor structure and T-cell selection mechanisms, and identification of regulatory T cells have been gradually described. Each of these steps leads us back to thymopoiesis, ranging from the identification of factors required for the entry of hematopoietic progenitors into the T-lymphocyte development program to the factors regulating the expression of "tissue-restricted antigens" within the thymic epithelium. Those are key in establishing the central tolerance, among which *AIRE* and *Fezf2* are the best known, others still being to be described. This has been remarkably studied in murine models, and several chapters of this book are devoted to this central issue.

Although the concepts are very similar, the data concerning human thymopoiesis still need to be further developed. However, they are progressing rapidly thanks to methodological approaches and to large-scale studies. It is now clear that, contrary to conventional wisdom still present in textbooks, the thymus remains functional in adults. This is important to better understand the parameters that govern the thymic function under physiological conditions, during aging, or in pathologies, especially lymphopenic situations after hematopoietic cell grafts, during HIV disease, or in autoimmunity. We are thus evolving from studies purely focusing on thymocytes to a broader view considering the thymus as an organ in its complexity. Excellent chapters of this book deal with what is called the "cross-talk" between thymocytes and cell populations housed in the thymus whose heterogeneity and complexity are gradually being uncovered. They include of course the different subpopulations of cortical and medullary thymic epithelial cells but also dendritic cells, macrophages, and so-called innate lymphoid cells, important in the process of thymic regeneration, without putting aside endothelial cells, key in the entry of lymphoid progenitors and also in the egress of recently generated naive T cells or "recent thymic emigrants." All this global knowledge will enable to consider future thymic regeneration strategies that will be personalized according to age, gender, and clinical contexts. This is to say the importance of this book and its timely content. The thymus still has a lot to teach!

Université Paris Diderot, INSERM U.1160,   Antoine Toubert
Hôpital Saint-Louis, APHP, Paris, France

Preface

If I had to choose an organ that in humans and mice symbolizes the intersection between immunology, endocrinology, molecular biology, and genetics, this organ would certainly be the thymus gland.

Galen in Greece had anatomically described this organ more than twenty centuries ago but only had its function scientifically attributed in the second half of the twentieth century by Jacques Miller. In his experiments of neonatal thymectomy of mice, Miller observed that the operated pups were suffering from susceptibility to infections with concomitant lymphopenia. These experiments were instrumental in finally assigning to the thymus its function in the immune system.

The fact that the thymus was the last of the major organs of the body to have its function assigned promptly incites exclamation.

Even intersected as said, the main function of the thymus is immunological and strongly associated to the development and positive/negative selection of T cells and induction of central immune tolerance.

One of its major cellular components, the thymocytes, undergoes a maturation process that is dependent on the random recombination of DNA segments (V[D]J recombination) of T-cell receptor (TCR) leading to the generation of the diversity of T-cell clones. The experimental demonstration of V(D)J recombination by Susumu Tonegawa in the mid-1970s, initially involving immunoglobulin gene segments in B cells and later TCR in T cells, has opened a unique perspective, i.e., understanding the molecular basis of diversity of lymphocyte repertoire. For genetics, this represented a great impact, since it was demonstrated that the genome is dynamic and can recombine somatically.

Another very intriguing aspect about the thymus is the functioning of its stroma. Thymic stroma is not merely a connective tissue or a supporting structure. The thymic epithelial cells (TECs), which are part of the stroma, establish a close physical contact with developing thymocytes through TECs-thymocytes adhesion, which is crucial, both for the thymus as a whole and for the thymocytes themselves. This property is termed thymic crosstalk during which the medullary TECs (mTECs) present to the developing thymocytes a vast amount of self-peptides that represent virtually all the organs and tissues of the body.

This showed the intersection of the thymus with the molecular biology of large-scale gene expression or transcriptomics. The mTEC cells have become a very intriguing cell type because they express almost the entire functional genome without losing their characteristics. The meaning of this property is immunological, relative to self-representation and induction of central immune tolerance. Due to the enormous diversity of self-peptide antigens expressed by mTECs, this property was termed *promiscuous gene expression* (PGE), which is controlled by *Aire* and *Fezf2* genes.

The demonstration that the thymic stromal cells express the functional oxytocin hormone made possible an important relation of this organ with endocrinology.

The thymus is also closely associated with human genetics since mutations in *Aire* cause the APECED (APS-1) autoimmune syndrome, which is linked to chromosome 21q22.3, the exact physical location of the *Aire* gene in humans.

One curiosity that, perhaps, some researchers still do not know: the mouse genome project and science of transcriptomics have in the past benefited greatly from the thymus. To begin massive sequencing of mouse DNA in the early 1990s, researchers in fact made cDNA libraries from total RNA extracted from a mouse thymus. Briefly, thousands of expressed sequence tags (ESTs) were then generated and later positioned along the genome, which in turn was being assembled. The sequenced mouse EST libraries were then used in microarray technology that emerged in 1995 making the science of transcriptomics possible!

As we can see, the thymus is a fascinating organ. It is crucial for the maintenance of immune homeostasis and is the place where the self-non-self distinction occurs. Even so, it is still a neglected organ in immunological research. Immunology is one of the branches of biological sciences that has progressed the most in the last decades, and most of the works are directed to the peripheral effector cells, i.e., B and T cells, NK cells, dendritic cells, macrophages, etc. However, much still has to be better known about thymus maturation, its ontogeny and differentiation, the origin of the TEC cells, and the control of the thymus gene expression in health and autoimmune diseases.

A promising field of research that is still in its early stages is about the control of gene expression of "second" thymus (cervical thymus discovered in 2006) that is present in about 50% of humans and mice.

It is for these reasons that this book was organized, i.e., to review the main aspects of cell biology, gene expression, and clinical intervention of the thymus. It was conceived during the realization of the Third Meeting on Thymus Transcriptome and Cell Biology, held at the University of São Paulo, Ribeirão Preto Medical School, Ribeirão Preto, Brazil, on 21–22 November 2017. Many of the authors of this book participated in this fruitful meeting. The purpose of this book is also to attempt to motivate young scientists to research the thymus.

I am grateful to all the researchers who have dedicated a part of their time to writing their chapters and the Springer Nature Publishing to welcome us and give full support for this book to be published.

Finally, I would like to pay a homage to Dr. Bruno Kyewski (1950–2018) who was an excellent scientist and devoted much of his career at the German Cancer Research Center in Heidelberg, Germany, studying cell biology and promiscuous gene expression in the thymus.

Ribeirão Preto, Brazil   Geraldo A. Passos
November, 2018

Contents

Contributors

Nuno L. Alves Instituto de Investigação e Inovação em Saúde, Universidade do Porto, Porto, Portugal

Thymus Development and Function Laboratory, Instituto de Biologia Molecular e Celular, Porto, Portugal

Lucas C. M. Arruda Department of Clinical Science, Intervention and Technology, Karolinska Institutet, Stockholm, Sweden

Amanda F. Assis Molecular Immunogenetics Group, Department of Genetics, Ribeirão Preto Medical School, University of São Paulo, Ribeirão Preto, SP, Brazil

Silvia Yumi Bando Department of Pediatrics, Faculdade de Medicina da Universidade de São Paulo, São Paulo, SP, Brazil

Fernanda Bernardi Bertonha Department of Pediatrics, Faculdade de Medicina da Universidade de São Paulo, São Paulo, SP, Brazil

Magda Carneiro-Sampaio Department of Pediatrics, Faculdade de Medicina da Universidade de São Paulo, São Paulo, SP, Brazil

Edward L. Y. Chen Department of Immunology, Sunnybrook Research Institute, University of Toronto, Toronto, ON, Canada

Ann P. Chidgey Department of Anatomy and Developmental Biology, Biomedicine Discovery Institute, Monash University, Melbourne, VIC, Australia

Vinicius Cotta-de-Almeida Laboratory on Thymus Research, Oswaldo Cruz Institute, Oswaldo Cruz Foundation, Rio de Janeiro, RJ, Brazil

Estácio de Sá University (UNESA), Rio de Janeiro, RJ, Brazil

Maria Carolina de Oliveira Center for Cell-based Therapy, Regional Hemotherapy Center of Ribeirão Preto Medical School, University of São Paulo, Ribeirão Preto, SP, Brazil

Division of Clinical Immunology, Department of Internal Medicine, Ribeirão Preto Medical School, University of São Paulo, Ribeirão Preto, SP, Brazil

Basic and Applied Immunology Program, Ribeirão Preto Medical School, University of São Paulo, Ribeirão Preto, SP, Brazil

Eduardo A. Donadi Department of Clinical Medicine, Ribeirão Preto Medical School, University of São Paulo, Ribeirão Preto, SP, Brazil

Max J. Duarte Molecular Immunogenetics Group, Department of Genetics, Ribeirão Preto Medical School, University of São Paulo, Ribeirão Preto, SP, Brazil

Jarrod Dudakov Program in Immunology, Clinical Research Division, and Immunotherapy Integrated Research Center, Fred Hutchinson Cancer Research Center, Seattle, WA, USA

Rafaella Ferreira-Reis Laboratory on Thymus Research, Oswaldo Cruz Institute, Oswaldo Cruz Foundation, Rio de Janeiro, RJ, Brazil

Vincent Geenen GIGA-I^3 Immunoendocrinology, GIGA Institute, University of Liège, Liège, Belgium

Adriana B. Genari Molecular Immunogenetics Group, Department of Genetics, Ribeirão Preto Medical School, University of São Paulo, Ribeirão Preto, SP, Brazil

Matthieu Giraud Centre de Recherche en Transplantation et Immunologie, UMR 1064, Institut National de la Santé et de la Recherche Médicale (INSERM), Université de Nantes, Nantes, France

Georg Holländer Department of Paediatrics and the Weatherall Institute of Molecular Medicine, University of Oxford, Oxford, UK

MRC Functional Genomics Unit, Department of Physiology, Anatomy and Genetics, University of Oxford, Oxford, UK

Michael L. Hun Department of Anatomy and Developmental Biology, Biomedicine Discovery Institute, Monash University, Melbourne, VIC, Australia

Magali Irla Centre d'Immunologie de Marseille-Luminy (CIML), INSERM U1104, CNRS UMR7280, Aix-Marseille Université UM2, Marseille, France

Arnon Dias Jurberg Laboratory on Thymus Research, Oswaldo Cruz Institute, Oswaldo Cruz Foundation, Pavilhão Leônidas Deane, Rio de Janeiro, RJ, Brazil

Julia Pereira Lemos Laboratory on Thymus Research, Oswaldo Cruz Institute, Oswaldo Cruz Foundation, Rio de Janeiro, RJ, Brazil

National Institute of Science and Technology on Neuroimmunomodulation (INCT-NIM), Rio de Janeiro, RJ, Brazil

João R. Lima-Júnior Biosciences and Biotechnology Program, School of Pharmaceutical Sciences of Ribeirão Preto, University of São Paulo, Ribeirão Preto, SP, Brazil

Center for Cell-based Therapy, Regional Hemotherapy Center of Ribeirão Preto Medical School, University of São Paulo, Ribeirão Preto, SP, Brazil

Mayara V. Machado Molecular Immunogenetics Group, Department of Genetics, Ribeirão Preto Medical School, University of São Paulo, Ribeirão Preto, SP, Brazil

Kelen C. R. Malmegrim Biosciences and Biotechnology Program, School of Pharmaceutical Sciences of Ribeirão Preto, University of São Paulo, Ribeirão Preto, SP, Brazil

Center for Cell-based Therapy, Regional Hemotherapy Center of Ribeirão Preto Medical School, University of São Paulo, Ribeirão Preto, SP, Brazil

Department of Clinical Analysis, Toxicology and Food Sciences, School of Pharmaceutical Sciences of Ribeirão Preto, University of São Paulo, Ribeirão Preto, SP, Brazil

Romário Mascarenhas Molecular Immunogenetics Group, Department of Genetics, Ribeirão Preto Medical School, University of São Paulo, Ribeirão Preto, SP, Brazil

Minoru Matsumoto Division of Molecular Immunology, Institute for Enzyme Research, Tokushima University, Tokushima, Japan

Department of Molecular and Environmental Pathology, Institute of Biomedical Sciences, The University of Tokushima Graduate School, Tokushima, Japan

Mitsuru Matsumoto Division of Molecular Immunology, Institute for Enzyme Research, Tokushima University, Tokushima, Japan

Daniella Arêas Mendes-da-Cruz Laboratory on Thymus Research, Oswaldo Cruz Institute, Oswaldo Cruz Foundation, Rio de Janeiro, RJ, Brazil

National Institute of Science and Technology on Neuroimmunomodulation (INCT-NIM), Rio de Janeiro, RJ, Brazil

Carolina Valença Messias Laboratory on Thymus Research, Oswaldo Cruz Institute, Oswaldo Cruz Foundation, Rio de Janeiro, RJ, Brazil

National Institute of Science and Technology on Neuroimmunomodulation (INCT-NIM), Rio de Janeiro, RJ, Brazil

Ana C. Monteleone-Cassiano Molecular Immunogenetics Group, Department of Genetics, Ribeirão Preto Medical School, University of São Paulo, Ribeirão Preto, SP, Brazil

Carlos Alberto Moreira-Filho Department of Pediatrics, Faculdade de Medicina da Universidade de São Paulo, São Paulo, SP, Brazil

J. J. Muñoz Center for Cytometry and Fluorescence Microscopy, Complutense University, Madrid, Spain

Bergithe E. Oftedal KG Jebsen Center for Autoimmune Disorders, Department of Clinical Science, University of Bergen, Bergen, Norway

Department of Clinical Science, Haukeland University Hospital, University of Bergen, Bergen, Norway

Ernna H. Oliveira Molecular Immunogenetics Group, Department of Genetics, Ribeirão Preto Medical School, University of São Paulo, Ribeirão Preto, SP, Brazil

João Ramalho Ortigão-Farias Laboratory on Thymus Research, Oswaldo Cruz Institute, Oswaldo Cruz Foundation, Rio de Janeiro, RJ, Brazil

Geraldo A. Passos Laboratory of Genetics and Molecular Biology, Department of Basic and Oral Biology, School of Dentistry of Ribeirão Preto, University of São Paulo, Ribeirão Preto, SP, Brazil

Molecular Immunogenetics Group, Department of Genetics, Ribeirão Preto Medical School, University of São Paulo, Ribeirão Preto, SP, Brazil

Pärt Peterson Molecular Pathology, Institute of Biomedicine and Translational Medicine, University of Tartu, Tartu, Estonia

Pedro M. Rodrigues Instituto de Investigação e Inovação em Saúde, Universidade do Porto, Porto, Portugal

Thymus Development and Function Laboratory, Instituto de Biologia Molecular e Celular, Porto, Portugal

Wilson Savino Laboratory on Thymus Research, Oswaldo Cruz Institute, Oswaldo Cruz Foundation, Rio de Janeiro, RJ, Brazil

National Institute of Science and Technology on Neuroimmunomodulation (INCT-NIM), Rio de Janeiro, RJ, Brazil

Brazilian National Institute of Science and Technology on Neuroimmunomodulation, Oswaldo Cruz Institute, Oswaldo Cruz Foundation, Rio de Janeiro, RJ, Brazil

Jastaranpreet Singh Department of Immunology, Sunnybrook Research Institute, University of Toronto, Toronto, ON, Canada

Laura Sousa Instituto de Investigação e Inovação em Saúde, Universidade do Porto, Porto, Portugal

Pedro P. Tanaka Molecular Immunogenetics Group, Department of Genetics, Ribeirão Preto Medical School, University of São Paulo, Ribeirão Preto, SP, Brazil

Antoine Toubert Université Paris Diderot, INSERM U 1160, Hôpital Saint-Louis, APHP, Paris, France

Koichi Tsuneyama Department of Molecular and Environmental Pathology, Institute of Biomedical Sciences, The University of Tokushima Graduate School, Tokushima, Japan

Larissa Vasconcelos-Fontes Laboratory on Thymus Research, Oswaldo Cruz Institute, Oswaldo Cruz Foundation, Rio de Janeiro, RJ, Brazil

Anette S. B. Wolff KG Jebsen Center for Autoimmune Disorders, Department of Clinical Science, University of Bergen, Bergen, Norway

Department of Clinical Science, Haukeland University Hospital, University of Bergen, Bergen, Norway

Kahlia Wong Department of Anatomy and Developmental Biology, Biomedicine Discovery Institute, Monash University, Melbourne, VIC, Australia

Kogulan Yoganathan Department of Immunology, Sunnybrook Research Institute, University of Toronto, Toronto, ON, Canada

Agustín G. Zapata Center for Cytometry and Fluorescence Microscopy, Complutense University, Madrid, Spain

Department of Cell Biology, Complutense University, Madrid, Spain

Juan Carlos Zúñiga-Pflücker Department of Immunology, Sunnybrook Research Institute, University of Toronto, Toronto, ON, Canada

Chapter 1
History of the Thymus: From a Vestigial Organ to the Programming of Immunological Self-Tolerance

Vincent Geenen and Wilson Savino

Abstract This introductory chapter presents the most important disruptions of concepts concerning the thymus since its discovery in Antique Greece. For centuries, the thymus was considered as a vestigial organ, and its role in T-cell differentiation was proposed only in the 1960s. Most recent studies attribute to the thymus an essential and unique role in programming central immunological self-tolerance. The basic mechanism implicated in this function is the transcription in the thymic epithelium of genes encoding precursors of neuroendocrine-related and tissue-restricted self-peptides. Their processing leads to the presentation of self-antigens by the major histocompatibility complex (MHC) machinery expressed by thymic epithelial and dendritic cells. Already during foetal life, this presentation promotes negative selection of T lymphocytes harbouring a receptor with high affinity for MHC/self-peptide complexes. Mainly after birth, this presentation also drives the generation of regulatory T cells specific for these complexes. Numerous studies, as well as the identification of *Aire* and *Fezf2* genes, have shown that a thymus defect plays a crucial role in the development of autoimmunity. The discovery of the central tolerogenic action of the thymus revolutionized the whole field of immunology, and such knowledge will pave the way for innovative tolerogenic therapies against autoimmunity, the so heavy tribute paid by mankind for the extreme diversity and efficiency of adaptive immunity.

V. Geenen (✉)
GIGA-I[3] Immunoendocrinology, GIGA Institute, University of Liège, Liège, Belgium
e-mail: vgeenen@uliege.be

W. Savino
Laboratory on Thymus Research, Oswaldo Cruz Institute, Oswaldo Cruz Foundation, Rio de Janeiro, RJ, Brazil

Brazilian National Institute of Science and Technology on Neuroimmunomodulation, Oswaldo Cruz Institute, Oswaldo Cruz Foundation, Rio de Janeiro, Brazil

© Springer Nature Switzerland AG 2019
G. A. Passos (ed.), *Thymus Transcriptome and Cell Biology*,
https://doi.org/10.1007/978-3-030-12040-5_1

1.1   Historical Summary from Antique Greece
to the Twentieth Century

The name 'thymus' first appeared in Galen's manuscripts (± 160 AD) and was so named because of its morphological analogy with the leaf of *Thymus cunula*. Galen considered the thymus as an excrescence, the function of which was to serve as a cushion between the sternum and basal blood vessels. He also observed that the thymus was larger in young animals and that its volume decreased with age.

Jacopo Berengario de Carpi (1460–1530) was the first anatomist to describe the human thymus by dissecting cadavers at the University of Bologna, one of the great European centres of anatomy at that time, with Padua and Paris. He precisely depicted both vascularisation and innervation of the thymus.

With the help of the students in Titian's school in Venice, the Belgian Andre Vesalius (1514–1564), the great anatomist of the Renaissance period, published in Padua the first anatomical board with a thymus in *De Humani Corporis Fabrica*: it was a small multi-lobed organ just behind the sternum that Vesalius also considered as a cushion protecting blood vessels in the superior mediastinum. Bartolomeo Eustachi (1510–1574), who was the first to describe adrenal glands, drew also the thymus showing its anatomical relationships in the anterior mediastinum. At the same time, the French surgeon Ambroise Paré (1510–1590) mentioned the thymus as a very soft and spongy gland. Felix Plater (1536–1614), a Swiss physician, reported a case of a child death by asphyxia secondary to a tracheal obstruction by an internal mass developed from the thymus.

During the seventeenth century, the English physician Francis Glisson (1599–1677), who first described rickets, speculated that the thymus could produce a fluid devoted to feeding and growth of the baby. William Hensson (1739–1774), an English surgeon known for his discovery of fibrin, published the first textbook on the thymus. He also reported the variations of thymus size during aging, its involution during some acute or chronic diseases, and noted that the thymus was full of the same 'particles' than those present in blood and lymph. He deducted that the thymus only exists at the beginning of life when these particles seem to be most necessary.

In 1832, Astley Cooper (1768–1841) published *The Anatomy of the Thymus Gland* enriched with precise illustrations of this organ. He also described malignant thymoma, a cancer of the thymus. Later, an essay on the physiology of the thymus gland (1845) brought to the English surgeon John Simon (1816–1904) his nomination to the Royal Society of London. Arthur Hill Hassal (1817–1894), physician-chemist, compared the histology of the thymus and other lymphoid organs (spleen, lymph nodes) and described Hassal's corpuscles in the thymic medulla. The Scottish embryologist John Beard (1858–1924) considered that the thymus could be the sources of all lymphocytes in the body. He also suggested an analogy of invasive power between cancer and placental trophoblast, saying that cancer would be an 'irresponsible trophoblast'. In 1902, he published in *The Lancet* a paper entitled 'Embryological aspects and aetiology of carcinoma', which prefigures the current concept of cancer stem cells. He was nominated in 1906 for the Nobel prize of physiology or medicine with the following argumentation: 'For the discovery of the presence in early vertebrates of a nervous structure that develops and is functional only during first embryological stages, the discovery of the real nature of the thymus gland, and the demonstration of a direct morphological continuation of germinal cells in all vertebrates'. That year however, the Nobel Prize was attributed to Camillo Golgi and Santiago Ramon y Cajal for their work on the cellular structure of the nervous system.

In France, Jona Salkind showed in his PhD thesis a first and complete comparative study on the histology of the thymus, from fish to human. In this work he even reported experiments with thymectomy in adult fish but did not find any significant changes in this animal (Salkind 1915).

1.2   The Thymus at the Crossroad Between Endocrinology and Immunology

At the beginning of the twentieth century, the thymus was still some 'enigma'; it could be an epithelial gland infiltrated by many small lymphocytes at the second month of the embryonic life in humans. These thymic lymphocytes (thymocytes) actively divide and thus, contrary to the epithelial framework, are very sensitive to the X-rays identified in 1895 by Wilhelm Röntgen. The scientific community then considered that the thymus was a vestigial and transitory organ, which prematurely declined and ceased to function very early in life. However, already in 1890, the German anatomist Wilhelm Waldeyer (1836–1921) had noticed that, even in elderly people, islets of thymic tissue could be observed in adipose thymus. Jan-August Hammar (Sweden, 1861–1946) confirmed these findings and detailed that, although the maximal development of the thymus is reached at puberty, normal thymic tissue persists until advanced age. He also showed that animal castration before puberty maintains an important volume of the thymus and that an involution of this organ accompanies pregnancy, undernourishment, as well as some infectious diseases Hammar (1921). Inversely, thymus hyperplasia is associated with autoimmune Graves' thyroid disease, Addison's adrenal deficiency, myasthenia and acromegaly. Consequently, the thymus was considered as another glandular component of the endocrine system. Actually, it was directly linked to the hypothalamus-pituitary-adrenal (HPA) axis by the Hungarian investigator Hans Selye who showed in 1946 that stress conditions, stimulating the HPA axis, simultaneously caused thymic atrophy (Selye 1946). Such findings led him to create the concept of an HPA-thymus axis.

Despite brilliant works made before 1950, the immunological function of the thymus and thymic lymphocytes remained completely unknown: indeed, similar to what had been reported for fish, adult mice that were thymectomized did not present any immunological problems. However, Jacques Miller (Australia, born in 1931) made a key observation, showing that thymectomy in mice, performed immediately after birth, provokes their premature death: the presence of a thymus at birth would thus be essential for survival. Further studies by Miller showed that mice thymecto-mized at the first day of life (and not later than 1 week) are very susceptible to infec-tions. He observed an important lymphopenia in blood, spleen and lymph nodes of these animals, which were also unable to reject a foreign skin graft, an essential immune response. In 1961, he concluded that the thymus is the organ responsible for the development of immunocompetent cells that constitute a specific cell popu-lation, thymus-dependent (T) lymphocytes (Miller 1961, 1964). Not without irony, he will write later that the success of his experiments was due to the fact that mice had been bred in a non germ-free environment. Despite the rightness of Miller's experiments and conclusions, Sir Peter Medawar still wrote in 1963: '*We shall come to regard the presence of lymphocytes in the thymus as an evolutionary accident of no very great significance*' (Medawar 1963).

In the following of studies conducted by Donald Metcalf (1929–2014), in respect to the identification of hematopoietic growth factors, Jacques Miller advanced the hypothesis of one or several soluble thymic factors that would be responsible for driving T-cell differentiation (Osoba and Miller 1963). Innumerable studies will try to characterize such factor(s) but it will be never possible to establish the reality of such thymus-specific growth factor(s) and to apply the endocrine model to the com-munication between epithelial cells and lymphocytes in the thymus. The demonstra-tion of a crucial role of the thymus during embryonic and foetal life, as well as the absence of any pathogenicity resulting from thymectomy a few days after birth, reinforced the idea that the thymus is, if not a vestigial organ, at least an organ that quickly becomes useless, this being verified in human clinics: the Di George congenital syndrome, which is the most common form of genetic micro-deletion, associates, besides other defects, the absence or hypoplasia of the thymus and a severe immunodeficiency. On the other hand, children who have been thymecto-mized during surgical correction of a congenital cardiac defect do not present any patent immune deficiency further in life. This latter question, however, would deserve to be further investigated in careful longitudinal studies.[1]

1.3 Interactions Between the Immune and Neuroendocrine Systems

Pioneer work published in 1970s by the Argentinean researchers (working in Switzerland) Hugo Besedovsky and Adriana del Rey revealed that neonatal thymec-tomy also promoted hypoplasia in the development of secondary reproductive organs (Besedovsky and Sorkin 1974; Besedovsky et al. 1985).

---

[1] In parallel to the launching of the FP6 European Integrated Project *Euro-Thymaide*, the European Commission published a directive prohibiting total thyroidectomy in children needing corrective cardiac surgery.

Using experimental psychology approaches, and reproducing old studies carried out during the 1950s at the Pavlov Institute (Saint Petersburg, Russia), the psychologist Robert Ader, showed early in the 1980s that typical Pavlovian reflex could be seen in the immune response. For such discoveries, he is presently considered the father of psychoneuroimmunology (Ader 1983).

Later, in October 1983, the Cardiologic Princess Lilian Foundation organized in Brussels the international symposium *Neural Modulation of Immunity* (Guillemin et al. 1985), chaired by the neuroendocrinologist Roger Guillemin, Nobel Prize of physiology or medicine in 1977, and two eminent immunologists, Melvin Cohn and Theodor Melnechuk. Invited speakers of this symposium revealed a completely new field of research, immune-(neuro)endocrinology, as well as several original approaches for a better understanding of integrated physiology.

One year later, Herbert Novera Spector (1919–2017) organized the First International Workshop on Neuroimmunomodulation, at the National Institutes of Health (Bethesda, USA) devoted to the same subject. In particular, during this symposium, there was a nice discussion on the putative production of endogenous opioids, including met-enkephalin, by the human thymic epithelium.

The stimulation of uterine contractions and galactokinesis of thymus extracts had already been reported at the beginning of the twentieth century by English authors (Ott and Scott 1909). Today, we know that such action is specific of the neurohypophysial peptide oxytocin but this first non-steroid hormone was then unknown since Vincent du Vigneaud will characterize it, only in 1953 in New York (du Vigneaud et al. 1953).

The hypothesis of an oxytocin synthesis in the thymus immediately raised and, indeed, specific radioimmunoassays revealed important quantities of oxytocin and neurophysin[2] in the human thymus. In this organ, coexistence in equimolar concentrations of those peptides already argued for a local synthesis. Human thymus extracts were then sent to Françoise Acézat, a great specialist at that time of biological assays of oxytocin, who observed that such extracts were able to contract myometrium of female rats. Furthermore, quantification of their biological activity was in close concordance with data obtained by radioimmunoassay (Geenen et al. 1986). Oxytocin-synthesizing cells in the thymus are cortical and medullary thymic epithelial cells (TECs), but not thymocytes (Geenen et al. 1987). Oxytocin is also synthesized in the epithelial component of thymic 'nurse' cells (TNCs), but not in T cells engulfed within those cellular complexes (Geenen et al. 1988). Thus, TNCs became a remarkable example of an intimate association between to cell populations derived from the distinct neuroendocrine and immune systems (Geenen et al. 1987). Specific neurohypophysial receptors are expressed by distinct thymic T-cell subsets and, after binding to these receptors, oxytocin promotes phosphorylation of tyrosine-kinases implicated in focal adhesion (Martens et al. 1998; Hansenne et al. 2004) (Fig. 1.1).

---

[2] Neurophysin is a 10-kDa transport and binding protein of oxytocin, which is encoded by the same gene (pro-oxyphysin).

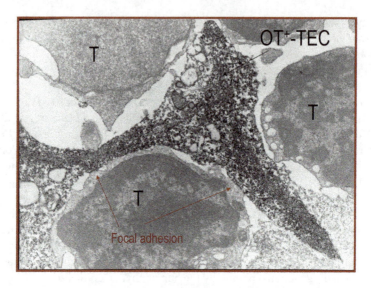

Fig. 1.1 The thymus microenvironment. Extension of a murine TEC (labelled with anti-oxytocin (OT) antibody) is surrounded by thymic T cells (T). No classical secretory granules could be observed in TEC cytoplasm, contrary to the situation in hypothalamic nerve endings of the posterior pituitary. A series of focal adhesion points between TEC and T cells are visualized and correspond to immunological 'synapses' between the two types of cells

In 1986, an international symposium of Neuroimmunomodulation was organized at the University of Liège under the chairmanship of Joseph Wybran and Jean-Jacques Vanderhaeghen, respectively professor of immunology and neuropathology at the Free University of Brussels (ULB). Starting in 1988, interactions between the neuroendocrine and immune systems were thoroughly investigated in the excellence research network *Neuroimmunomodulation* (1988–1992) under the auspices of the European Science Foundation (ESF).

A specific journal, also entitled *Neuroimmunomodulation*, was launched by Karger Editors in 1994, having as first editors the late brilliant researchers Samuel M. McCann (1925–2006) and James M. Lipton (1938–2016). This journal actually publishes for two times special issues devoted to the thymus, namely: Neuroendocrine Control of The Thymus (1999) and Neuroendocrine Immunology of The Thymus (2011).

1.4 Immunological Self-Tolerance

As soon as 1900, the father of immunology, Paul Ehrlich (1854–1915), proposed the formula '*horror autoxicus*' to claim the impossibility that one organism could be aggressed in normal conditions by its own cells in charge for its defence. Ehrlich thought then that either structures or mechanisms should exist to avoid autotoxicity, and this should be of the highest importance for individual health and species survival (Ehrlich 1900). In the continuation of his revolutionary theory of clonal selection, the virologist and immunologist Frank Macfarlane Burnet (1899–1985) introduced the term '*tolerance*' to characterize one of the cardinal properties of the adaptive immune system with diversity, specificity and memory. During a conference at the University of London in 1962, he declared: '*If, as I think, the thymus is the site where occur proliferation of lymphocytes in clones with precise immunological functions, we have also to consider another function: elimination or inhibition of clones with reactivity to self*'. Burnet also named 'forbidden' clones lymphocytes having escaped clonal negative selection (Burnet and Mackay 1962).

In 1976, Susumu Tonegawa provided the scientific basis with the explanation for the extreme diversity of the adaptive immune response against infectious pathogens through the elucidation of the molecular mechanisms responsible for the random recombination of gene segments encoding variable domains of the immunoglobulin B cell receptor for antigen (BCR) (Tonegawa 1976). In 1984, Tak Mak, Mark Davis and other authors showed that an analogous mechanism takes place in the thymus for the generation of TCR diversity, which recognizes the MHC/antigen complex (Malissen et al. 1984; Toyonaga et al. 1984; Davis et al. 1984). There is a close homology between TCR and BCR, and a more distant one with MHC, which suggests mechanisms of replication and diversification from ancestral genes throughout evolution.

Billions of BCR and TCR combinations result in the fantastic lottery acting in the generation of their diversity and a majority of these combinations are able to recognize molecular structures of the host (the '*self*'). In normal conditions however, the adaptive immune system does not aggress self and, for long, immunology has been defined as the science of self-nonself discrimination, almost evoking some philosophical question. Burnet was readily speaking about immunology as the science of self and nonself; however, lymphocytes are not intelligent cells able to discriminate between self and nonself. In 1987 and 1988, the research groups of Nicole Le Douarin (Nogent-sur-Marne, France) (Ohki et al. 1987), John Kappler and Philippa Marrack (Denver, USA) (Kappler et al. 1987), Hugh Robson Macdonald (Epalinges, Switzerland) (MacDonald et al. 1988), and Harald von Boehmer (Basel, Switzerland) (Kisielow et al. 1988) scientifically demonstrated the theory of thymic clonal selection proposed by Burnet so many years before. The study published in Nature by the group of the Basel Institute for Immunology (Kisielow et al. 1988) was particularly elegant after generating transgenic mice with lymphocytes expressing a unique TCR specific of the HY antigen. The thymus of male mice was extremely poor in living thymocytes contrary to the female thymus fulfilled with thymocytes. This could only be explained by the deletion of transgenic lymphocytes due to the presence of the HY antigen in the male thymus, a situation of course impossible in the female thymus. So, the thymus appeared first as a cemetery for early T cells expressing a TCR specific of self-antigens.

It is interesting to note that the various interactions leading to death versus survival fate of developing thymocytes occur simultaneously during the highly organised and the oriented migration throughout the thymic lobules. During the

1990s key contributions were provided stating that both soluble and insoluble moieties (respectively chemokines and extracellular matrix) are involved in migration of developing T-cells into the thymus (Savino et al. 1993, 2002). According to one hypothesis, cell migration itself into the thymus would result from several simultaneous interactions in a sort of multivectorial migration, each interaction being one individual vector (Mendez-da-Cruz et al. 2008). We presently think that thymocytes, in a given moment of their differentiation, can be simultaneously exposed to 30 or more cell migration-related molecular interactions. The future will tell us more about this particular issue.

By the same time, scientific community was facing a fundamental problem of basic cell biology. Oxytocin is the basis for the model of hormonal secretion by neurons (neurosecretion). However, primary cultures of human TECs, even after diverse stimuli, did not secrete any oxytocin, which was in contradiction not only with the scientific truth established for oxytocin, but also with works showing oxytocin synthesis in TECs from different species and expression of specific oxytocin receptor by thymic T cells. Furthermore, Martin Wiemann published a paper that confirmed by electronic microscopy the presence of oxytocin in TEC cytosol but not in classical secretory granules issued from the Golgi apparatus (Fig. 1.1) (Wiemann and Ehret 1993). A crucial problem thus raised: if oxytocin is not secreted by TECs, how can it bind to specific receptors expressed by thymocytes? Could this be some 'illegitimate' transcription of the oxytocin gene in thymus epithelium? All hypotheses, even that of a new gene member of the neurohypophysial family, have been investigated then ruled out. Several observations however led to understand the signification of oxytocin synthesis in the thymus. In 1990, the Australian expert in adrenal physiology,

John Funder, proposed a new model of cell-to-cell signalling, the cryptocrine communication (from the Greek word *cryptos*, hidden). According to Funder, who has worked before with Burnet on the expression of steroid receptors by TECs, two examples of cryptocrine communication exist in the body: on the one hand, in the testis, between nursing Sertoli cells and differentiating spermatozoids and, on the other hand, in the thymus, between TECs/TNCs, and developing T lymphocytes (Funder 1990). At the same time, Hans-Georg Rammensee (Max-Planck Institute, Tubingen, Germany) deciphered the biochemical mechanisms that were responsible of antigen presentation by MHC proteins (Rammensee et al. 1993). On this basis was advanced the hypothesis that thymic oxytocin could behave not as the classical secreted neurohormone but rather as the self-peptide of the neurohypophysial hormone family that would be presented to T cells during their education to self-tolerance in the thymus. This hypothesis was shown to be true after different experiments (Geenen et al. 1993a, Martens et al. 1996a). Oxytocin is indeed presented by MHC proteins at the outer surface of the TEC plasma membrane and this presentation is responsible for programming immunological self-tolerance to neurohypophysial functions assumed by oxytocin and vasopressin. Contrary to vasopressin, the antidiuretic hormone, oxytocin is highly expressed in the thymus and, consequently, tolerance to oxytocin is stronger than tolerance to vasopressin. This also explained why, in our hands, rabbits immunized with oxytocin reacted so

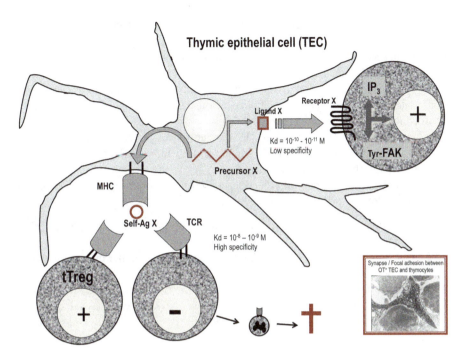

Fig. 1.2 The different roles of neuroendocrine precursors in T-cell differentiation. Following its transcription under Aire (or Fezf2) control in TEC nucleus, a neuroendocrine precursor X is processed according two distinct pathways. On the one hand, it is the source of a cryptocrine ligand X that is able to bind to a neuroendocrine receptor expressed by thymic T cells, to mobilize second messengers such as IP3, and to phosphorylate focal adhesion-related kinases (such as p125^{Fak} and p130^{Cas}) for thymic oxytocin. This constitutes a positive accessory signal during T-cell development. On the other hand, the same precursor X is also processed as a self-antigen/peptide X that is presented by MHC proteins of TEC or thymic dendritic cells. During foetal life, self-presentation induces the negative selection of T-cells, which are randomly bearing a TCR specific of this CMH/self-Ag X. Mainly after birth, self-presentation also promotes Treg cells specific of the same complex

weakly than after immunization with vasopressin. Tolerance to oxytocin is so high that it is extremely difficult to break contrary to tolerance to vasopressin.

On the basis of these studies, a theoretical model was proposed, which transposes at the molecular level the multiple roles of the thymus in T cell differentiation, both in thymopoiesis and in the programming of central self-tolerance (Martens et al. 1996b). This model relies on the two types of behaviour that oxytocin is able to display in the thymus (Fig. 1.2). As a cryptocrine signal targeted at the TEC outer surface, oxytocin can bind to a specific neuroendocrine receptor expressed by early T cells after their migration into the thymus. This binding promotes mobilization of second messengers (inositol triphosphate, IP3) within thymocytes, as well as phosphorylation of focal adhesion-related kinases. By this way, thymic oxytocin could be able to stimulate the formation of immunological '*synapses*' between TECs and T cells, a structure playing a major role in T-cell development. In the same time, thymic oxytocin also behaves as the self-peptide of the neurohypophysial family, which is presented by thymic MHC proteins and is responsible for programming central tolerance to neurohypophysial functions. Oxytocin presentation by TECs may induce either negative selection of T cells expressing a TCR specific of MHC/oxytocin, or generation thymic Treg cells with the same specificity. Thereafter, this model has been successfully applied to other dominant members of different neuroendocrine families synthetized in TECs such as neurokinin A for the tachykinins, neuropeptide Y (Ericsson et al. 1990), neurotensin (Vanneste et al. 1997), and insulin-like growth factor 2 (IGF-2) for the insulin family (Geenen et al. 1993b, Geenen and Lefebvre 1998) (Fig. 1.2).

The biochemical nature of neuroendocrine self may therefore be defined according to the following principles:

1. One dominant member per neuroendocrine gene family is expressed in TECs from different species.
2. Because of close homology, this dominant gene/protein in the thymus mediates cross-tolerance to all members of the family.
3. Most importantly, thymic neuroendocrine precursors are not processed according to the model of neurosecretion. They are actually processed as self-peptides that are presented by MHC proteins expressed by TECs and thymic dendritic cells.

Through this new paradigm of 'neuroendocrine self-peptides', very specific of the thymus, an integrated and harmonious was ensured between the neuroendocrine and immune systems when the genes RAG1 and RAG2 activating recombination of BCR and TCR appeared in cartilaginous fishes some 450 million years ago (Geenen 2012).

Thereafter, the elegant studies performed by Bruno Kyewski and colleagues showed that TECs, essentially in thymic medulla, are also the site for the promiscuous expression of genes encoding many tissue-specific antigens (Derbinski et al. 2001). Their publication in *Annual Reviews of Immunology* has definitively installed the central role of the thymus in central self-tolerance (Kyewski and Klein 2006). However, contrary to neuroendocrine self-peptides, most of tissue-restricted antigens do not exert any accessory signalling during T-cell development in the thymus.

In 1972, Richard Gershon (1932–1983) identified in Yale immunosuppressive cells regulating immunocompetent lymphocytes (Gershon et al. 1972). Quite ironically, he evoked this suppression as the 'second law of thymodynamics'. After his premature death, his studies were followed by Shimon Sakagushi who, in 1995, identified another thymus-dependent major tolerogenic mechanism (Sakagushi et al. 1995). Mainly after birth, the thymus is indeed the source of a new population of regulatory T (tTreg) cells that are able to inhibit in periphery self-reactive T cells having escaped central thymic negative selection. The generation/selection of tTreg cells also depends on presentation of self-peptides by thymic MHC proteins. How the same mechanism of MHC-mediated self-peptide presentation promotes two so distinct T cell fates (negative selection and tTreg cell generation) is still a matter of scientific discussion.

The programming of an intrathymic tolerance to neuroendocrine proteins was an absolute necessity for general homeostasis. As so many studies have shown, hormones and neuropeptides exert a tight control upon immune and inflammatory responses. They bind to specific cognate receptors that are expressed by different types of immune cells and thereby modulate their activity. If tolerance to these neuroendocrine signals and receptors were not firmly installed, the risk for developing autoimmune reactions toward these molecules would be very high and would compromise species survival (Geenen and Chrousos 2004).

1.5   Thymus and the Failure of Tolerance: Autoimmunity

The demonstration of the essential role of the thymus in self-tolerance raised the intuitive question about the potential role of a thymus dysfunction in the pathogenesis of autoimmune diseases. To answer this question, the transcription of insulin-related genes was investigated in the thymus of BioBreeding (BB) rats, which are with non-obese diabetic (NOD) mice a classical animal model of human type 1 diabetes (T1D). While insulin and IGF-1 gene expression was detected in the thymus of all BB rats examined, IGF-2 gene transcription was deficient in the majority of thymuses of diabetes-prone BB rats (BB-DP). This thymus deficiency in BB-DP rats might explain the characteristic lymphopenia of these animals, lymphopenia that also affects the population of Treg cells (Kecha-Kamoun et al. 2001).

Identification of the AutoImmune REgulator gene (*AIRE/Aire*) by a German-Finnish consortium in 1997 played a definitive role in the acceptation by the scientific community of the concept that a defect in thymus-dependent self-tolerance is a crucial event in the pathophysiology of many autoimmune diseases. *AIRE* mutations are responsible for a very rare congenital poly-autoimmune syndrome that associates hypoparathyroidism (with severe hypocalcemia), adrenal insufficiency (Addison's disease), as well as recurrent mucosal infections by *Candida albicans* (APECED or APS-1 syndrome) (The Finnish-German Consortium 1997). Further studies conducted in the laboratory of Diane Mathis and Christophe Benoist at Harvard University showed in 1992 that *Aire* expression is maximal in medullary TECs, and that $Aire^{-/-}$ mice develop several autoimmune processes in peripheral tissues (Anderson et al. 2002). Expression of many genes was diminished in the thymus of these transgenic mice, including genes coding for oxytocin, neuropeptide Y, insulin and IGF-2. Thus, the AIRE factor controls intrathymic transcription of many tissue-specific genes; *AIRE* mutations and *Aire* ablation lead to a marked decrease in the intrathymic transcription of these genes, which is associated with the development of autoimmune responses. So, it became evident that AIRE constitutes a major rampart against autoimmunity and thymus-dependant central self-tolerance became the principal research topic in laboratories interested in the pathophysiology of autoimmunity. More recently, it has been shown that the gene encoding fasciculation and elongation protein zeta family zinc finger 2 (*FEZF2/Fezf2*) protein also controls intrathymic transcription of self-peptides, the majority of which is not regulated by AIRE. Interestingly, *Fezf2* is also a transcription factor implicated in some developmental processes of the central nervous system (Takaba et al. 2015).

Although they do not develop autoimmune diabetes, $Igf2^{-/-}$ mice exhibit a tolerance to insulin that is markedly decreased in comparison with WT mice (Hansenne et al. 2006). IGF-2 expression seems thus necessary for establishment of a full tolerance to insulin, confirming that the close homology between IGF-2 and insulin mediates cross-tolerance between two members of the same family.

Congenital absence or acquired breakdown of central tolerance to the insulin family is condition necessary but not sufficient to the development of T1D. In homozygote twins, T1D concordance is only 30–40%, which confirms the importance of genetic factors in T1D susceptibility but particularly the influence of environmental factors such as infections with Coxsackie enteroviruses. Several data argue that a Coxsackie virus B4 (CV-B4) infection could induce a thymus dysfunction and disturb the programming of tolerance to the insulin family. CV-B4 was shown to be able to infect the epithelial and lymphoid compartments of human and murine thymus (Brilot et al. 2002). In human fetal thymic organ cultures, this infection induced an important decrease in thymocyte number, as well an increase in MHC expression by TECs and double-positive CD4+CD8+ thymocytes (Brilot et al. 2004). Also, CV-B4 infection of a medullary TEC line provoked a marked and specific decrease in IGF-2 expression (Jaïdane et al. 2012).

With regard to autoimmune thyroiditis, which is the most frequent autoimmune disease, the major thyroid antigens (thyroglobulin, thyroperoxydase, thyrotropin receptor/TSHR) are expressed in TECs in normal conditions. As observed first by Jan-August Hammar, thymus hyperplasia was repeatedly observed in Graves' hyperthyroidism and this hyperplasia progressively vanished along the treatment of this disease (Paschke and Geenen 1995; Murakami et al. 1996). A defective central tolerance linked to the level of intrathymic expression of TSHR has also been incriminated in the pathogenesis of Graves' disease (Colobran et al. 2011).

Moreover, the advent of tolerance as a cardinal property of the immune system contributed to a fundamental discrimination between what is antigenic, *i.e.* able to activate an immune response (immunogenic antigen), and what is tolerogenic (self-peptide/antigen) (Fig. 1.3).

1.6   Towards a Novel Type of Vaccination Against T1D?

IGF-2 is the dominant member of the insulin family expressed in the thymus and IGF-2 possesses tolerogenic properties (Yang et al. 2014; Geng et al. 2014). A thymus dysfunction drives the development of the diabetogenic autoimmune response and CV-B4 infection could be implicated as an environmental factor in T1D pathogenesis. Insulin is the primary antigen targeted in T1D and its high immunogenicity could be linked to the very low level of its transcription in medullary TEC subsets. In the same perspective, GAD65 is also well known as an important T1D antigen, but it is very

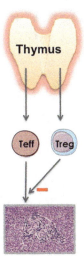

Thymus physiology

• AIRE- and FEZF2-regulated transcription of neuroendocrine self-peptides in thymus epithelium.

• Deletion of T cells with high affinity for MHC/neuroendocrine self-peptide complexes.

• Selection of CD4+ CD25+ Foxp3+ tTreg specific of neuroendocrine self-peptides.

Thymus physiopathology

• Absence or decrease in expression/presentation of neuroendocrine self-peptides in the thymus (APECED/APS-1, Graves' disease, Down syndrome, BB rat, etc.)

• Enrichment of T-cell repertoire with 'forbidden' self-reactive effector T cells (Teff).

• Decrease in selection of tTreg with specificity to neuroendocrine self-peptides.

Bridge between self-reactive Teff and target auto-antigens

Role of environmental factors (viruses, anti-cancer immunotherapy, gonadal steroids, gut microbiota, diet, endocrine disruptors, vitamin D deficiency, stress...)

Fig. 1.3 The central role of the thymus in the programming of self-tolerance en in the development of autoimmunity. In normal conditions, under the control of *Aire* and *Fezf2*, TECs express numerous genes related to neuroendocrine families or encoding tissue-restricted antigens. MHC presentation of these self-antigens induces both negative selection of self-reactive T cells and generation of Treg cells with the same specificity. This dual mechanism is responsible for the programming of central self-tolerance to neuroendocrine functions. In some pathological conditions, the decrease in intrathymic expression and presentation of the self-antigens leads to continuous generation in blood of self-reactive 'forbidden' T cells (Teff), as well as to a decrease in differentiation of thymic self-reactive Treg cells. This is a condition, necessary but not sufficient, for the development of an autoimmune response against target tissue antigens. For the clinical development of an autoimmune disease, environmental factors are also requested

weakly expressed in the thymus contrary to the isoform GAD67. On the basis of the fundamental tolerogenic properties of the thymus, IGF-2 and GAD67 in particular, a novel type of vaccination, '*negative self-vaccination*' is under current development (Geenen et al. 2010). In opposition to the classical 'positive' immunogenic vaccination, the final objective of negative self-vaccination would be to reprogram self-tolerance through elimination of anti-insulin and anti-GAD65 forbidden clones and recruitment of related Treg cells. This vaccination would be used both for preventing T1D and curing this disease with transplantation of novel insulin-secreting β cells and suppression of the autoimmune diabetogenic memory (Fig. 1.4).

1.7   Conclusion: The Prominent Role of the Thymus in Evolution

The role of the thymus may be also evaluated from a global evolutionary point of view (Geenen et al. 2013). The thymus appeared as a single organ in the same time or shortly after the immunological 'bing-bang', *i.e.* the emergence of the adaptive

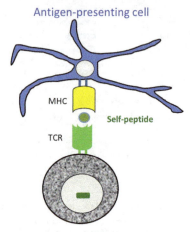

Classical « positive » vaccination

« Negative » self-vaccine

Antigen-presenting cell

Antigen-presenting cell

MHC

TCR   Antigen X

MHC

TCR   Self-peptide

+

−

Immunogenic response

Naïve T cell activation
Induction of memory T cells

Type 1 diabetes (TD1)

Antigens X = Insulin, GAD65,…

Tolerogenic response

Deletion of self-reactive T cells
Generation of self-specific Tregs

T1D-related self-peptides

IGF-2, GAD67

Fig. 1.4 Classical 'positive' vaccination as opposed to 'negative' self-vaccination. Classical vaccination essentially relies on the immunogenic response (naïve T cell activation and induction of memory immune cells) elicited by administration of antigen(s) representative of pathogens. Type 1 diabetes pathogenesis also includes an immunogenic response targeting T1D antigens such as insulin and GAD65. The novel type of 'negative' self-vaccination proposes to use thymus self-peptides for promoting a tolerogenic response (deletion of self-reactive T cells and generation of self-reactive Treg cells). For T1D prevention and cure, corresponding thymus self-peptides are IGF-2 and GAD67, respectively

immune response depending on RAG1 and RAG2 genes in cartilaginous fishes (rays and sharks) some 450 million years ago. Thymoid formations dispersed in the branchial apparatus of lamprey have preceded the advent of a unique thymus. They already expressed *Foxn4* (forkhead box N4), the paralog of *Foxn1*, the transcription factor specific of TEC differentiation (Bajoghli et al. 2011; Boehm 2011). The same study also provided evidence for a functional analogy between variable lymphocyte receptor (VLR) in thymoids and TCR recombination in the thymus, thus opening the question of the occurrence of autoimmune-like phenomena in jawless vertebrates. Thymus emergence allowed the orchestration of central self-tolerance, which was a rampart absolutely needed to counter the high risk of autotoxicity inherent to the tremendous lottery of the new adaptive system. A defect in thymus dependant self-tolerance, either genetic (*AIRE* and *FEZF2* mutations) or acquired (*i.e.* after CV-B4 infection) is a condition, necessary but not sufficient, for the development of many organ-specific autoimmune diseases (Fig. 1.5).

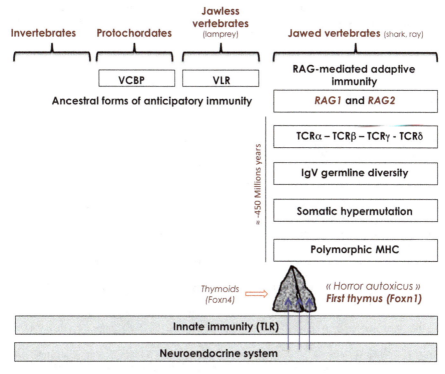

Fig. 1.5 Integrated evolution of the immune and neuroendocrine systems. Neuroendocrine precursors and principles are evolutionarily ancient and did not evolve extensively except by gene duplication and differential RNA splicing. Throughout evolution, the neuroendocrine and innate immune system have evolved in parallel, and still coexist in all living species without any aggression of the innate immune system toward neuroendocrine glands. Indeed, toll-like receptors (TLR), which are the main mediators of innate response, do not react against normal or undamaged self. A high risk of inherent autoimmunity toward neuroendocrine tissues resulted from the appearance of recombination-activating genes RAG1 and RAG2, and RAD-dependent adaptive immunity in jawed cartilaginous fishes some 450 millions years ago (the 'big-bang' of immunity). Preceded by ancestor *Foxn4+* thymoids in gill baskets of lamprey larvae, the first unique thymus (with *Foxn1+* TECs) also emerged in jawed vertebrates. The intrathymic presentation of neuroendocrine self-peptides (arrows) may be viewed *a posteriori* as a very efficient and economical way to instruct the adaptive T cell system in recognizing and tolerizing neuroendocrine functions already during thymus-dependent cell differentiation in foetal life. *VCBP* variable region-containing chitin-binding protein, *VLR* variable lymphocyte receptor

Acknowledgements This chapter is dedicated to the memory of Bruno Kyewski and Harald von Boehmer who passed away in 2018.

The studies summarized here have been supported by the Fonds Léon Fredericq (University Hospital of Liege), by the University of Liège, Wallonia, the F.S.R.-NFSR of Belgium, the Wallonia-Brussels Federation, the Fonds Alphonse Rahier for research in Diabetology (Belgium), the European Commission (FP6 Integrated Project Euro-Thymaide 2004–2008), the European Association for the Study of Diabetes (EASD, Germany), and the Juvenile Diabetes Research Foundation (JDRF, New York). From the Brazilian side, the work has been supported by the

National Institute of Science and Technology on Neuroimmunomodulation, Conselho Nacional de Desenvolvimento Científico e Tecnológico (CNPq), Coordenação de Aperfeiçoamento de Pessoal de Nível Superior (Capes) through Financial code 001, Faperj, Fiocruz and Mercosur (FOCEM).

References

Ader R (1983) Developmental psychoneuroimmunology. Dev Psychobiol 16:251–267. https://doi.org/10.1002/dev.420160402

Anderson MS, Venanzi ES, Klein L et al (2002) Projection of an immunological self shadow within the thymus by the aire protein. Science 298:1395–1401. https://doi.org/10.1126/science.1075958

Bajoghli B, Guo P, Aghaallaei N et al (2011) A thymus candidate in lampreys. Nature 470:90–94. https://doi.org/10.1038/nature09655

Besedovsky HO, Sorkin E (1974) Thymus involvement in female sexual maturation. Nature 249:356–358

Besedovsky HO, del Rey AE, Sorkin E (1985) Immune-neuroendocrine interactions. J Immunol 135(suppl 2):750s–754s

Boehm T (2011) Design principles of adaptive immune systems. Nat Rev Immunol 11:307–317. https://doi.org/10.1038/nri2944

Brilot F, Chehadeh W, Charlet-Renard C et al (2002) Persistent infection of human thymic epithelial cells by coxsackievirus B4. J Virol 76:5260–5265. https://doi.org/10.1128/JVI.76.10.5260-5265.2002

Brilot F, Geenen V, Hober D, Stoddart C (2004) Coxsackievirus B4 infection of human fetal thymus cells. J Virol 78:9854–9861. https://doi.org/10.1128/JVI.78.18.9854-9861.2004

Burnet FM, Mackay IR (1962) Lymphoepithelial structures and autoimmune disease. Lancet II:1030–1033

Colobran R, del Pilar Armengol M, Faner R et al (2011) Association of an SNP with intrathymic transcription of TSHR and Graves' disease: a role for defective tolerance. Hum Mol Genet 20:3415–3423. https://doi.org/10.1093/hmg/ddr247

Davis MM, Chien YH, Gascoigne MR, Hedrick SM (1984) A murine T cell receptor gene complex: isolation, structure and rearrangement. Immunol Rev 81:235–258

Derbinski J, Schulte A, Kyewski B, Klein L (2001) Promiscuous gene expression in medullary thymic epithelial cells mirrors the peripheral self. Nat Immunol 2:1032–1039. https://doi.org/10.1038/ni723

Ehrlich P (1900) The Croonian Lecture: on immunity. Proc Soc Lond Biol 66:424

Ericsson A, Geenen V, Vrindts-Gevaert Y et al (1990) Expression of preprotachykinin A and neuropeptide-Y messenger RNA in the thymus. Mol Endocrinol 4:1211–1218. https://doi.org/10.1210/mend-4-8-1211

Funder JW (1990) Paracrine, cryptocrine, acrocrine. Mol Cell Endocrinol 70:C21–C24

Geenen V (2012) Presentation of neuroendocrine self in the thymus: a necessity for integrated evolution of the immune and neuroendocrine systems. Ann N Y Acad Sci 1261:42–48. https://doi.org/10.1111/j.1749-6632.2012.06624.x

Geenen V, Chrousos GP (2004) Immunoendocrinology in health and disease. Marcel Dekker, New York, NY

Geenen V, Lefebvre PJ (1998) The intrathymic expression of insulin-related genes: implications for pathophysiology and treatment of type 1 diabetes. Diabetes Metab Rev 14:95–103

Geenen V, Legros JJ, Franchimont P et al (1986) The neuroendocrine thymus: coexistence of oxytocin and neurophysin in the human thymus. Science 232:508–511. https://doi.org/10.1126/science.3961493

Geenen V, Legros JJ, Franchimont P et al (1987) The thymus as a neuroendocrine organ. Synthesis of vasopressin and oxytocin in human thymic epithelium. Ann N Y Acad Sci 496:56–66

Geenen V, Defresne MP, Robert F et al (1988) The neurohormonal thymic microenviron-ment: immunocytochemical evidence that thymic nurse cells are neuroendocrine cells. Neuroendocrinology 47:365–368

Geenen V, Vandersmissen E, Cormann-Goffin N et al (1993a) Membrane translocation and rela-tionship with MHC class I of a human thymic neurophysin-like protein. Thymus 22:55–66

Geenen V, Achour I, Robert F et al (1993b) Evidence that insulin-like growth factor 2 (IGF2) is the dominant thymic peptide of the insulin superfamily. Thymus 21:115–127

Geenen V, Mottet M, Dardenne O et al (2010) Thymic self-antigens for the design of a negative/tolerogenic self-vaccination against type 1 diabetes. Curr Opin Pharmacol 10:461–472. https://doi.org/10.1016/j.coph.2010.04.005

Geenen V, Bodart G, Henry S et al (2013) Programming of neuroendocrine self in the thymus and its defect in the development of neuroendocrine autoimmunity. Front Neurosci 7:e17. https://doi.org/10.3389/fnins.2013.00187

Geng XR, Yang G, Li M et al (2014) Insulin-like growth factor 2 enhances functions of antigen-specific regulatory B cells. J Biol Chem 289:17941–17950. https://doi.org/10.1074/jbc.M113.51262

Gershon RK, Cohen P, Hencin R, Liebhaler SA (1972) Suppressor T cells. J Immunol 108:586–590

Guillemin R, Cohn M, Melnechuk T (eds) (1985) Neural modulation of immunity. Raven Press, New York, NY

Hammar JA (1921) The new views at the morphology of the thymus gland and their bearing on the problem of the function of the thymus. Endocrinology 5:543–573. https://doi.org/10.1210/endo-5-5-543

Hansenne I, Rasier G, Charlet-Renard C et al (2004) Neurohypophysial receptor gene expression by thymic T cell subsets and thymic T cell lymphoma cell lines. Clin Dev Immunol 11:45–51

Hansenne I, Renard-Charlet C, Greimers R, Geenen V (2006) Dendritic cell differentiation and immune tolerance to insulin-related peptides in Igf2-deficient mice. J Immunol 176:4651–4657

Jaïdane H, Caloone D, Lobert PE et al (2012) Persistent infection of thymic epithelial cells with coxsackievirus B4 results in decreased expression of type 2 insulin-like growth factor. J Virol 86:11151–11162. https://doi.org/10.1128/JVI.00726-12

Kappler JW, Roehm N, Marrack P (1987) T cell tolerance by clonal elimination in the thymus. Cell 49:273–280

Kecha-Kamoun O, Achour I, Martens H et al (2001) Thymic expression of insulin-related genes in an animal model of autoimmune type 1 diabetes. Diabetes Metab Res Rev 17:146–152. https://doi.org/10.1002/dmmr.182

Kisielow P, Blüthmann H, Staerz UD et al (1988) Tolerance in T-cell receptor transgenic mice involves deletion of nonmature CD4+8+thymocytes. Nature 333:742–746. https://doi.org/10.1038/333742a0

Kyewski B, Klein L (2006) A central role for central tolerance. Annu Rev Immunol 24:571–606. https://doi.org/10.1146/annurev.immunol.23.021704.115601

MacDonald HR, Schneider R, Lees RK et al (1988) T-cell receptor Vβ use predict reactivity and tolerance to Mlsa-encoded antigens. Nature 332:40–45

Malissen M, Minard K, Mjolsness S et al (1984) Mouse T cell antigen receptor: structure and orga-nization of constant and joining segments encoding the beta polypeptide. Cell 37:1101–1110

Martens H, Malgrange B, Robert F et al (1996a) Cytokine production by human thymic epithelial cells: control by the immune recognition of the neurohypophysial self-antigen. Regul Pept 67:39–45. https://doi.org/10.1016/S0167-0115(96)00105-X

Martens H, Goxe B, Geenen V (1996b) The thymic repertoire of neuroendocrine self-antigens: physiological implications in T-cell life and death. Immunol Today 17:312–317. https://doi.org/10.1016/0167-5699(96)10023-2

Martens H, Kecha O, Charlet-Renard C et al (1998) Neurohypophysial peptides activate phosphor-ylation of focal adhesion kinases in immature thymocytes. Neuroendocrinology 67:282–289. https://doi.org/10.1159/000054324

Medawar PB (1963) Discussion after Miller JFAP and Osoba D. Role of the thymus in the origin of immunological competence. In: Wolstenholme GEW, Knight J (eds) The immunologically competent cell: its nature and origin, vol 16. Ciba Foundation Study Group, London

Mendez-da-Cruz DA, Smaniotto S, Keller AC, Dardenne M, Savino W (2008) Multivectorial abnormal cell migration in the NOD mouse thymus. J Immunol 180:4639–4647

Miller JF (1961) Immunological function of the thymus. Lancet II:748–749. https://doi.org/10.1016/S0140-6736(61)90693-6

Miller JF (1964) The thymus and the development of immunologic responsiveness. Science 144:1544–1551

Murakami M, Hosoi Y, Negishi T et al (1996) Thymic hyperplasia in patients with Graves' disease. Identification of thyrotropin receptors in human thymus. J Clin Invest 98:2228–2234. https://doi.org/10.1172/JCI119032

Ohki H, Martin C, Corber C et al (1987) Tolerance induced by thymic epithelial grafts in birds. Science 237:1032–1035. https://doi.org/10.1126/science.3616623

Osoba D, Miller JF (1963) Evidence for a humoral factor responsible for the maturation of immunological faculty. Nature 199:653–654

Ott I, Scott JC (1909) The action of glandular extracts upon the contractions of the uterus. J Exp Med 11:326–330

Paschke R, Geenen V (1995) Messenger RNA expression for a TSH receptor variant in the thymus of a two-year child. J Mol Med 73:577–580. https://doi.org/10.1007/BF00195143

Rammensee HG, Falk K, Rötzschke O (1993) Peptides naturally presented by MHC class I molecules. Annu Rev Immunol 11:213–244. https://doi.org/10.1146/annurev.iy.11.040.193.001241

Sakagushi S, Sakagushi N, Asano M et al (1995) Immunologic self-tolerance maintained by activated T cells expressing IL-2 receptor alpha-chain (CD25). Breakdown of a single mechanism of self-tolerance causes various autoimmune diseases. J Immunol 155:1151–1164

Salkind J (1915) Contributions histologiques à la biologie comparée du thymus. Arch Zool Exp Gén 55:81–322

Savino W, Villa-Verde DM, Lannes-Vieira J (1993) Extracellular matrix proteins in intrathymic cell migration and differentiation? Immunol Today 14:158–161

Savino W (ed) (1999) Neuroendocrine Control of the Thymus. Karger AG, Basel

Savino W, Dardenne M (eds) (2011) Neuroendocrine Immunology of the thymus. Neuroimmunomodul 18:261–358

Savino W, Mendez-da-Cruz DA, Silva JS, Dardenne M, Cotta-de-Almeida V (2002) Intrathymic T-cell migration: a combinatorial interplay of extracellular matrix and chemokines? Trends Immunol 26:305–313

Selye H (1946) The general adaptation syndrome and the diseases of adaptation. J Clin Endocrinol Metab 6:117–230

Takaba H, Morishita Y, Tomofuji Y et al (2015) Fezf2 orchestrates a thymic program of self-antigen expression for immune tolerance. Cell 163:975–987. https://doi.org/10.1016/j.cell.2015.10.013

The Finnish-German Consortium (1997) An autoimmune disease, APECED, caused by mutations in a novel gene featuring two PHD-type zinc-finger domains. Nat Genet 17:799–403. https://doi.org/10.1038/ng1297-399

Tonegawa S (1976) Reiteration frequency of immunoglobulin light chain genes: further evidence for somatic generation of antibody diversity. Proc Natl Acad Sci U S A 73:203–207

Toyonaga B, Yanagi Y, Suciu-Foca N et al (1984) Rearrangements of T-cell receptor gene YT35 in human DNA from thymic leukaemia T-cell lines and functional T-cell clones. Nature 311:385–387

Vanneste Y, Ntodou-Thome A, Vandersmissen E et al (1997) Identification of neurotensin-related peptides in human thymic epithelial cell membranes and relationship with major histocompatibility complex class I molecules. J Neuroimmunol 76:161–166. https://doi.org/10.1016/S0165-5728(97)00052-0

du Vigneaud V, Ressler C, Trippett S (1953) The sequence of amino acids in ocytocin with a proposal for the structure of oxytocin. J Biol Chem 205:949–957

Wiemann M, Ehret G (1993) Subcellular characterization of immunoreactive oxytocin within thymic epithelial cells of the male mouse. Cell Tissue Res 273:573–575. https://doi.org/10.1007/BF00304614

Yang G, Geng XR, Song JP et al (2014) Insulin-like growth factor 2 enhances regulatory T-cell functions and suppresses food allergy in an experimental model. J Allergy Clin Immunol 133:1702–1708. https://doi.org/10.1016/j.jaci.2014.02.014

Chapter 2
Thymus Ontogeny and Development

J. J. Muñoz and A. G. Zapata

Abstract The thymus is a primary lymphoid organ constituted by a 3D epithelial network that provides a specialized microenvironment in which seeding lymphoid progenitors undergo phenotypical and functional maturation. During the earlier steps of thymic organogenesis, the specification of the pharyngeal endoderm to thymus fate takes place independently of the expression of the transcription factor Foxn1 that, however, governs the later organogenesis of thymus together with the colonizing lymphoid cells. In the present chapter, we will review evidence describing early development of thymus and its resemblance with the development of endoderm-derived epithelial organs based on tubulogenesis and branching morphogenesis as well as the molecules known to be involved in these processes.

2.1 Introduction

The thymus is a primary lymphoid organ in which seeding lymphoid progenitors undergo phenotypical and functional maturation in a 3D epithelial network that governs the process and allows the generation of immunocompetent T lymphocytes. These close thymocyte (T)/thymic epithelial cell (TEC) interactions initiate with the lymphoid seeding of thymic anlage during ontogeny. Previously, during the named early thymic organogenesis, the specification of the pharyngeal endoderm to thymus fate occurs, in a process independent of the expression of the transcription factor Foxn1 that, however, governs the later organogenesis of thymus together with the colonizing lymphoid cells.

The early specification of pharyngeal epithelium to thymic tissue and initial primordium formation is remarkably similar to that initiating the development of other epithelial organs, (i.e., salivary glands or pancreas) which, as the thymus, derive

J. J. Muñoz
Center for Cytometry and Fluorescence Microscopy, Complutense University, Madrid, Spain

A. G. Zapata (✉)
Center for Cytometry and Fluorescence Microscopy, Complutense University, Madrid, Spain

Department of Cell Biology, Complutense University, Madrid, Spain
e-mail: zapata@bio.ucm.es

© Springer Nature Switzerland AG 2019
G. A. Passos (ed.), *Thymus Transcriptome and Cell Biology*,
https://doi.org/10.1007/978-3-030-12040-5_2

from the gut endoderm and need an extensive surface for performing their major functions (Hauser and Hoffman 2015; Larsen and Grapin-Botton 2017). In the case of the thymus, the establishment of an extensive 3D epithelial network is mandatory for thymocyte maturation. In this respect, a large thymic epithelial surface is necessary for housing positive and negative thymocyte selection that results in huge numbers of apoptotic thymocytes that, before being engulfed by thymic macrophages, occupy a large extension.

In the present chapter, we will review evidence describing early development of thymus and its resemblance with the development of endoderm-derived epithelial organs based on tubulogenesis and branching morphogenesis as well as the molecules known to be involved in these processes.

2.2   Appearance and Earliest Stages of Thymic Development

The earliest stages of thymic development have been extensively studied (Gordon and Manley 2011; Manley et al. 2011; Takahama et al. 2017). They include: the specification of thymus tissue in the third pharyngeal pouch; the formation of a unique primordium common to parathyroid and thymus, the separation of both organs and the migration of lateral thymic lobes into the mediastinum.

The appearance of a thymic rudiment is related to the segmentation of the posterior pharynx that results in the specification of endodermal tissue into TECs (Takahama et al. 2017). For this purpose, an inner layer of endodermal cells and an outer one of ectodermal cells of the third branchial cleft fuse (Graham 2001). This fact and pioneer morphological results suggested that the thymic epithelium derived from both endoderm and ectoderm (Cordier and Heremans 1975; Cordier and Haumont 1980). More recent studies determined conclusively that all TECs derived exclusively from the endodermal embryonic layer (Le Douarin and Jotereau 1975; Gordon et al. 2004) and suggested that ectodermal cells could be just inductors of thymic tissue rather than contribute to its formation (Gordon et al. 2004). On the other hand, the endodermal origin of all thymic epithelium was confirmed by the identification of a clonal bipotent thymic epithelial progenitor capable of giving rise to both cortical and medullary TECs (Rossi et al. 2006).

As early as E9.5, the murine pharyngeal endoderm evaginates generating a common primordium that expresses Gcm2 (Glial cells missing homolog 2) antero-dorsally, an early parathyroid marker. 12 h later, the ventral domain of the common rudiment begins to express Foxn1, the distinctive transcription factor of thymic tissue (Gordon et al. 2001). At these early stages, the rudiment consists of a simple epithelium surrounded by neural crest (NC)-derived mesenchyme.

From E11.5, this common rudiment gradually detaches from the lateral surface of the pharynx in a process which involves apoptosis, probably mediated by neural crest (NC)-derived mesenchyme because Splotch mutants deficient in NC cells exhibit delayed or no detachment (Griffith et al. 2009; Chen et al. 2010). Other involved molecules are Shh, Pax and Frs2a, whose lack maintains the common rudiment connected to the pharynx (Peters et al. 1998; Moore-Scott and Manley 2005; Kameda et al. 2009).

At E12, the common primordium is totally isolated from the pharynx, and the parathyroid and thymic tissue begins to be individualised. In the case of the thymus, the lateral lobes descend caudally and medially to the midline. These processes are again governed by NC-derived mesenchyme and various molecules, such as BMP4 (Griffith et al. 2009); Ephrin B2 (Foster et al. 2010) and Hoxa 3 (Chen et al. 2010). Remarkably, these results support an important role for NC-derived mesenchyme in the earliest stages of thymus development, except in the initial specification of endoderm. Later, when the thymus is an independent organ, NC-derived mesenchyme also contributes to the organization of connective tissue of both capsule and septae (Le Lievre and Le Douarin 1975), but in the adult thymus the mesoderm-derived mesenchyme becomes more important (Jiang et al. 2000).

This detachment of the thymic lobes from the pharynx and their migration implies a difference with respect to other endoderm-derived organs but, in general, the formation of the thymus/parathyroid primordium is very similar to that of other branching organs, such as the salivary gland or the pancreas. At E9.5, thymus/parathyroid outpocket consists of a single epithelial cell layer surrounding a central lumen that is continuous with the pharynx. At E11–12, the epithelium proliferates (Gordon et al. 2004) to form a multilayered or pseudo-stratified epithelium polarized with respect to a ramified central lumen and lined by K5+ Cld 3/4+ cells (Gordon and Manley 2011; Munoz et al. 2015). This central lumen is the residual cavity of the original pouch cavity.

In the following days, the thymus grows and K5+ Cld 3/4 + cell cords increase their total length and branching degree. At the same time, external clefts determine an incipient lobulation that become clearly evident by day E14.5. Beyond E12.5, the initial lumen is almost totally closed although a central lumen is still visible at E12.5 and to some extent at E13.5 (Gordon et al. 2010; Munoz et al. 2015) and secondary forming lumens can be observed in the K5+Cld 3/4+ branching cell cords up to E13.5, similar to those found in a process of lumen formation (Munoz et al. 2015). Thus, the thymic primordium at these initial stages is very similar to the initial primordium of other branching developing organs constituted by a three-dimensional mass or bud formed by unpolarized cells in which lumen formation and elongation take place within the proliferating bud (Tucker 2007; Walker et al. 2008; Villasenor et al. 2010; Larsen and Grapin-Botton 2017). However, a definitive lumen is not developed in the thymus (Fig. 2.1). Essentially, the central Cld+K5+ cells will differentiate into the thymic medulla and the surrounding Cld 3/4− cells will generate the cortex (Hamazaki et al. 2007; Munoz et al. 2015), although lineage relationships between these two TEC populations are complex, unclear and change along the embryonic and postnatal life (reviewed elsewhere in this book).

The presence of lymphoid progenitors in the thymic primordium introduces another important difference in the thymus development with respect to other epithelial organs. Thymocyte precursors enter the thymus at around E11.5 through the surrounding mesenchyme (Masuda et al. 2005) inducing the specific three-dimensional network formed by dendritic shaped TECs necessary for thymocyte functional maturation. This 3D arrangement of thymic epithelium is, to some extent, precluded in the absence of thymocytes (Vroegindeweij et al. 2010) or FoxN1 (Cordier 1974; Dooley et al. 2005; Guo et al. 2011).

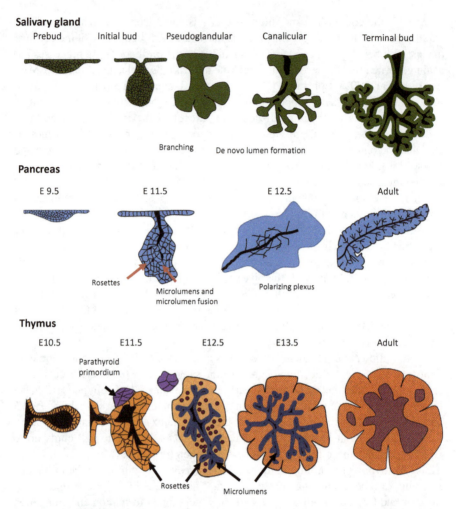

Fig. 2.1 Thymus follows a morphogenetic development similar to those of salivary gland (branching of an unpolarized primordium and later de novo lumen formation) and pancreas (polarization and remodeling of an unpolarized mass resulting in more synchronous branching, lumen formation and differentiation) modified by the expression of foxn1 and the presence of lymphoid precursors. A definitive lumen is not formed. Modified from Gordon and Manley (2011)

2.3 A Complex Network of Transcription Factors and Morphogens Appears to Govern the Early Development of Thymic Primordium

There are three important problems for a conclusive determination of the acting sequence of various molecular families involved in the control of early thymus development and their functioning mechanisms:

a) Any defect in the molecules involved in the formation and/or organization of pharyngeal pouches and/or arches affects the thymus development importantly, although such molecules are not directly involved.

b) Several molecules described as important agents for thymus development function at various stages of development, even showing opposite effects throughout development.

c) Most information on this issue has been obtained studying KO mice in which one or more molecules are compromised in many specific cell types, which does not allow establish the truc significance of the molecular deficits. When, in a few cases, the deletion of molecules was induced specifically in a concrete cell type, the results obtained were more realistic and, sometimes, different that those pioneer ones.

2.3.1 Transcription Factors Involved in Early Thymus Development

Two main groups of transcription factors govern the first stages of thymic primordium development: the complex including Hoxa3 and Pax1/9 together with Eya1 and Six 1/4 (Gordon and Manley 2011) and Tbx1, a transcription factor related to alterations in the human chromosome 22q11.2 that result in different syndromes, such as DiGeorge syndrome (DGS), velocardiofacial syndrome (VCFS) and conotruncal anomaly face syndrome (Baldini 2004). On the other hand, morphogens of the families FGF, BMP/TGFβ, Shh and Wnt are targets of the action of those transcription factors and in turn regulate their transcriptional activity. Furthermore, the retinoid acid, a vitamin A metabolite that regulates the postnatal development of TECs (Wendland et al. 2018), plays a major role in the early development of thymus, as reported in other epithelial organs. In the case of thymus, retinoid acid would diffuse from the NC-derived mesenchyme for specifying the pharyngeal pouch. Thus, retinoid acid antagonists or mutant mice deficient in the molecule or its signalling pathway show thymic agenesis (Wendling et al. 2000; Begemann et al. 2001).

Tbx1 is expressed in both the third pharyngeal pouch endoderm and the surrounding mesenchyme (Lindsay et al. 2001). Pioneer results related the absence of Tbx1 to thymus aplasia, although its lack indeed affected the pharyngeal pouches avoiding the thymus formation (Lindsay et al. 2001). Nevertheless, more recent results have demonstrated that the direct role of endodermal expression of Tbx1 on thymic epithelium development is not promoting but repressing TEC development. Thus, ectopic Tbx1 expression specifically in the ventral domain of the third pharyngeal pouch, where the thymic primordium will appear, blocks Foxn1 expression and inhibits both TEC proliferation and differentiation but does not reverse thymus fate (Reeh et al. 2014) indicating that, actually, disappearance of Tbx1 expression from the pharyngeal pouch endoderm is necessary for proper thymic organization. Furthermore, Shh is also involved in this process because its ectopic activation in the third pharyngeal pouch endoderm induces Tbx1 expression that blocks Foxn1 expression (Reeh et al. 2014).

Apart from Shh, other molecules regulating Tbx1 expression include the transcription factor Pbx1 (Manley et al. 2004) and morphogens of the BMP and FGF families. In turn, this last family is target of the Tbx1 action. Mice deficient in Chordin, a BMP antagonist, show reduced Tbx1 and FGF8 expression in the pharyngeal endoderm (Vitelli et al. 2002; Bachiller et al. 2003). These results support the existence of a network in which Tbx1 acts downstream of BMPs but upstream of FGF8. Accordingly, the selective deletion of FGF8 in Tbx1-expressing cells phenocopies the DG and VCF syndromes (Brown et al. 2004).

The second molecular complex known to be involved in the regulation of early thymus development consists of the paired box (Pax) proteins, intimately related to the transcription factors, Hoxa 3, Eya 1 and Tbx 1 (Manley and Capecchi 1995; Su et al. 2001). To date Pax 1, Pax 3 and Pax 9 seem to be necessary for thymus development (Wallin et al. 1996; Peters et al. 1998). As mentioned above, Splotch mutants, deficient in Pax 3, exhibit an important reduction of NC cells. Accordingly, those previously mentioned events that are governed by NC-derived mesenchyme are severely affected in these mutants. Thus, although they organize an early common primordium, it does not detach from the pharynx and the limits of parathyroid and thymus domains change resulting in an enlarged thymus and a small parathyroid (Griffith et al. 2009). Pax 3 appears therefore to be necessary to determine TEC fate in the third pharyngeal endoderm (Blackburn and Manley 2004).

Pax 1 and Pax 9, which belong to the same group of paired box proteins, exhibit similar patterns of expression firstly in the foregut endoderm and later in the endoderm of the third pharyngeal pouch (Wallin et al. 1996), although Pax 9 occurs also in the surrounding mesenchyme (Hetzer-Egger et al. 2002). Moreover, defects in Pax 1 appear to affect mainly the thymic epithelium whereas those observed in Pax 9-deficient mice are attributed principally to mesenchyme defects. Thus, Pax 1 mutant thymi contain large epithelial cysts (Wallin et al. 1996) whereas Pax 9$^{-/-}$ embryos do not fold away from foregut and the thymus rudiment remains in the pharynx. Nevertheless, the control of early thymus development exerted by both molecules seems to be closely associated to Hoxa 3 signaling pathway.

As Pax 9, Hoxa 3 is expressed in both third pharyngeal pouch endoderm and NC-derived mesenchyme and its lack decreases the endodermic expression of Pax 1 and Pax 9 (Manley and Capecchi 1995) and results in blockade of the parathyroid/thymus rudiment, increased apoptotic endodermal cells and reduced mesenchyme cell proliferation (Chisaka and Kameda 2005). Together, these results suggest that Pax 1 and Pax 9 function downstream of Hoxa 3 but the three molecules show synergic effects on early thymus maturation (Neubuser et al. 1995; Peters et al. 1998; Su and Manley 2000). More recent studies have compared the phenotype of Hoxa 3 knockout mice and those of either endoderm or NC cells conditioned mutant mice showed that Hoxa 3 expression in endoderm controls the timing of initial thymic primordium formation but is not responsible for inducing thymic fate (Chojnowski et al. 2014). In null mutants, thymus and parathyroid are absent but organ-specific domains in the pharyngeal pouch are present and express, although with certain delay, the regional thymic markers Foxn1, Foxg1 or IL7. The expression of the parathyroid marker Gcm2 is weak and quickly downregulated. In these mutants, both parathyroid and thymus fail to develop while Hoxa 3 endoderm or NC cells conditioned mutations result in small ectopic organs. Hoxa 3 deletion conditioned to endoderm causes delayed expression of Foxn1, which in turn results in a small thymus, Foxg1 and BMP4, as well as in delayed detachment from the pharynx inducing altered migration and location. By contrast, NC cells conditioned mutation has little effects on thymus development except for resulting in small size after E17.5. In this case, separation from the pharynx is completely precluded producing ectopic location of the organ. On the other hand, Foxn1Cre driven deletion of Hoxa 3 has no effect on thymus development indicating that its function is restricted to the initial primordium formation before E11.5 (Chojnowski et al. 2014). Likewise, Eya 1, Six and Pax co-localize in NC cells and pharyngeal endoderm, but the relationships between these molecules are presumably different. Six expression in endoderm is dependent on Eya 1 and in its absence Six is lost (Xu et al. 2002) and the thymic primordium is not formed, but the expression of Hoxa 3, Pax 1 and Pax 3 remain in the E10.5 pouch endoderm (Zou et al. 2006). Accordingly, Six 1 acts downstream of Eya 1 whereas Hoxa 3, Pax 1 and Pax 3 do it upstream or follow independent pathways of Eya 1 (Xu et al. 2002),

2.3.2 Morphogens and Early Thymus Development

Morphogens participate in numerous aspects of early thymic development, they frequently exhibit synergy but sometimes show opposite effects. FGF is a family of molecular mediators of the epithelial tissues. They affect the epithelial component in the thymus at different stages of development. FGF 8 and FGF 10 are involved in the maturation of the endoderm; the first one is expressed in the pouch epithelium whereas FGF 10 is produced by the surrounding NC-derived mesenchyme (Frank et al. 2002). In the fetal thymus, FGF7 and FGF10 are presumably produced by the mesenchyme of capsule and septae, as removal of surrounding mesenchyme from E12 foetal thymic lobes inhibits the effects on TECs (Jenkinson et al. 2003, 2007) although FGF 7 is also released from thymic blood vessels (Erickson et al. 2002). Both molecules FGF 7 and FGF 10 affect TEC proliferation, but not differentiation, acting through their FGFR2iiib receptor. Accordingly, deficient mice either in the receptor or FGF 10 show severe thymic hypoplasia and reduced TEC proliferation (Revest et al. 2001).

Shh is clearly a promoter of parathyroid development, therefore negatively regulating the development of thymus domain in the common rudiment. In this respect, Shh and BMP4 act through Tbx1 (Bain et al. 2016), as above indicated, as antagonists for governing thymus development (Moore-Scott and Manley 2005).

BMPs and Wnt, the other families of morphogens involved in thymus development are particularly interesting as regulators of Foxn1, the transcription factor that governs the maturation of thymus tissue from its appearance in the thymic area of common primordium (Balciunaite et al. 2002). BMP4 and Noggin, two antagonists of BMP family, govern the first stages of common primordium development before Foxn1 is expressed, induced in the NC-derived mesenchyme by FGF 8 (Neves et al. 2012). BMP4 is expressed in the ventral domain of the pouch and Noggin colocalizes with Gcm2 in the dorsal one (Patel et al. 2006). Moreover, BMP4 contributes to the definitive separation of both domains and, therefore, to the individualisation of parathyroid and thymus, as BMP4 elimination delays the process (Gordon and Manley 2011). The loss of BMP4 from pharyngeal endoderm previous to Foxn1 expression does not affect patterning/separation from the pharynx or even the initial organ formation (Gordon et al. 2010), but inhibition of BMP signalling provokes reduced Foxn1 expression in zebrafish thymus primordium (Chen et al. 2010) whereas BMP signalling activation promotes it (Soza-Ried et al. 2008). Together, these results suggest the following sequence of functioning of BMP4 in early thymus development. FGF 8-mediated mesenchyme BMP signalling targets the expression of both Foxn1 and BMP4 in the endodermal cells (Soza-Ried et al. 2008), endodermal BMP4 establishes a regulatory feedback loop (Metz et al. 1998) to ensure expression of both BMP4 and Foxn1 in the future thymic epithelium (Swann et al. 2017b). Thus, BMP signalling must be maintained for promoting Foxn1 positive TECs (Swann et al. 2017b).

Distinct members of Wnt family are importantly expressed in both TECs and fibroblasts of embryonic and adult thymus (Heinonen et al. 2011), whereas only the first ones express their specific receptors (Balciunaite et al. 2002). Non-canonical molecules, such as Wnt4 and Wnt5b and the canonical Wnt10b co-express with Foxn1 in the third pharyngeal pouch and in both embryonic and adult thymus (Balciunaite et al. 2002). However, the role played by these molecules in thymus development is controversial. Whereas the overexpression of Dkk1, a Wnt4 inhibitor, in TECs induces thymic atrophy and a reduction of both epithelial progenitor cells and TEC proliferation (Osada et al. 2010), it has been claimed that the absence of β-catenin dependent Wnt signalling is necessary for a proper thymus development (Swann et al. 2017a). Furthermore, β-catenin deficient mice express Foxn1 suggesting that the molecule is dispensable for the expression of Foxn1; on the contrary, stabilized β-catenin overexpression, which is modulated by other morphogens such as BMPs, results in decreased numbers of Foxn1 transcripts (Zuklys et al. 2009; Swann et al. 2017a). However, sustained Wnt signalling remarkably promotes production of Wnt antagonists (Niida et al. 2004) that block thymocyte maturation in FTOCs (Mulroy et al. 2002).

A summary of early thymic development and the main molecular families involved in it is shown in Fig. 2.2.

2.4 Later Thymic Organogenesis: The Role of Foxn1

As repeatedly claimed, both Foxn1 expression and interactions between thymocytes and TECs after lymphoid progenitor seeding of E11.5–E12 thymic primordium regulate the so-named late organogenesis of thymus (Munoz et al. 2018). The role

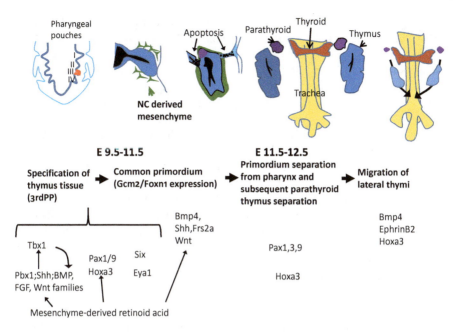

Fig. 2.2 Summary of thymus organogenesis and involved molecules

played by Foxn1 in the TEC development is intimately associated with first studies on the thymus of spontaneously mutated nude mice (see recent summary by Anderson and McCarthy 2015). Boehm and colleagues (Nehls et al. 1996) identified whn, later named Foxn1 gene, as the gene responsible of the nude mouse phenotype. Nude mice lack a proper thymus and, instead, a polarized tubular structure unable to be colonized by lymphoid precursors and to sustain thymocyte development is formed in these mutant mice (Nehls et al. 1996; Itoi et al. 2001; Dooley et al. 2005). Therefore, it is considered as an essential factor for triggering and promoting thymus development, although there is certain controversy on possible differences in its role in embryonic, postnatal and adult thymus, as well as on its relevance for cortical and medullary thymic epithelium.

Foxn1 expression is developmentally regulated by different factors. As above described, BMP4 is key for the onset of Foxn1 expression in the thymic primordium (Bleul and Boehm 2005; Patel et al. 2006; Soza-Ried et al. 2008). It has been claimed that BMP4 directly up-regulates Foxn1 (Tsai et al. 2003). In addition, Wnt ligands, including Wnt4 and Wnt5b are implicated in the control of Foxn1 (Balciunaite et al. 2002; Ma et al. 2013). Other molecules remarked as Foxn1 regulators are the members of transcription factors of E2F family, E2F3 and E2F4, and, as indicated, the transcription factor Tbx1 that represses Foxn1 expression in TECs (Reeh et al. 2014).

Foxn1 may be considered as essential to trigger TEC differentiation, although evidence indicates that it is not necessary to specify thymic fate. Different experimental approaches have determined the relevance of Foxn1 for inducing expression of genes concerned with TEC differentiation, including Trp 63 (transformation related protein 63), Pax1, FGFR2IIIb, IL7, chemokines, Dll4, Aire, CD40, CD80, cyclin D and several genes of Wnt signalling pathway (Nowell et al. 2011; Bredenkamp et al. 2014a; Zuklys et al. 2016). Zuklys et al. (2016) demonstrated the occurrence of 450 high-confidence Foxn1 target genes in postnatal thymus, confirming those previously reported by using other methodologies. These target genes include threonine peptidases, including Tasp1, components of thymoproteosome of the Psmb family, including Psmb11 that encodes β5t, and CD83 a surface molecule expressed in cTEC required for the development of CD4 thymocytes (Fujimoto et al. 2002). However, still few studies have established a direct transcriptional regulation of these genes by Foxn1. Recently, Uddin and colleagues (Uddin et al. 2017) identified a highly conserved Foxn1 binding sequence that is proximally located to the β5t coding sequence, confirming that in vitro Foxn1 protein binds the sequence promoting β5t gene transcription and in vivo this cis-regulatory element is key for β5t expression in cTEC. In accordance with this role of Foxn1 in promoting the expression of thymic function genes, Foxn1 ectopic expression in fetal fibroblasts induces thymic differentiation (Bredenkamp et al. 2014b).

The relevance of Foxn1 in promoting TEC differentiation seems to begin after the initial determination of thymic epithelium. Actually, foetal nude mice develop a thymic primordium that is very similar to the WT one (Nowell et al. 2011) and ectopically transplanted E9.0 third pharyngeal pouches (3PP) which do not yet express Foxn1, generate a functional thymus, indicating that 3PPs are committed to TEC lineage independently of FoxN1 expression (Gordon et al. 2004). Besides, Foxn1 reactivation in single cells of Foxn1 null thymic rudiments results in the generation of miniature thymuses, each containing well defined cortical and medullary areas (Bleul et al. 2006). These results would indicate that Foxn1 deficiency would arrest the maturation of TECs at a bipotent thymic precursor cell stage (Vaidya et al. 2016). However, E15 nude (Foxn1-deficient) fetal thymus contains K5^{hi} Cld4^{hi} areas suggesting that mTEC specification could be a Foxn1 independent process (Nowell et al. 2011).

As described before, thymic primordium of nude mice is similar, as the WT one, to the other branching organs. It express claudin 3/4 (Nowell et al. 2011) and develops into a branched tubular structure in which both ductal and acinar components appear (Dooley et al. 2005). This fact and the similarities of thymus morphogenesis with other branching organs would indicate that thymus development and initial cortex-medulla regional specification is driven by a primordial pattern of branching morphogenesis present in the nude thymus and later modified by both Foxn1 expression and the arrival of lymphoid precursors (Munoz et al. 2015). Actually, in Ikaros DNA-binding domain mutant thymus, that expresses Foxn1 but lacks lymphoid precursors, the tubular nature of nude thymus is not present (Georgopoulos et al. 1994). Therefore, a clear definition on the mechanisms of initial specification of both thymus primordium and cortex/medulla compartments is lacking and the role played by Foxn1 in them remains inconclusively determined.

By contrast, the role of Foxn1 in TEC differentiation and the maintenance of thymic structure and function is well stated (Vaidya et al. 2016). Foxn1 expression seems to be differentially modulated in distinct TEC subsets (Itoi et al. 2007; Chen et al. 2009). Adult cTECs express higher levels of Foxn1 than mTECs. Besides, the expression in both cell types is higher in CIII^{hi} cells than in CIII^{lo} cells (Nowell et al. 2011; Ki et al. 2014; Rode et al. 2015; O'Neill et al. 2016). Foxn1 is expressed in practically all embryonic TEC and apparently disappears gradually in postnatal thymus (Rode et al. 2015) but it also remains important for adult thymus (Chen et al. 2009; Cheng et al. 2010; Corbeaux et al. 2010). Thus, it is assumed that its continuous expression is necessary for TEC maintenance (Barsanti et al. 2017). The analysis of hypomorphic alleles of FoxN1 (Su et al. 2003) or methods allowing the temporal control of such expression (Chen et al. 2009) indicate that Foxn1 effects are dependent on its dose and are necessary for different stages of thymus development (Vaidya et al. 2016). Mice homozygous for a hypomorphic Foxn1 allele whose transcripts lack the amino-terminal domain of Foxn1 protein exhibit a hypoplastic cystic thymus without identifiable cortex and medulla compartment. In addition, they show lymphoid colonization and the thymocyte maturation is relatively normal in foetal but not in adult thymi in which DN3 thymocytes lack and TcR β expression is reduced (Su et al. 2003). A second mouse expressing other hypomorphic Foxn1 allele (Nowell et al. 2011) which produces 15% of WT levels of Foxn1 transcripts, partially supports T-cell development showing decreased proportions of thymocytes. These results suggest that Foxn1 allows initiate TEC differentiation but the process is later blocked in intermediate stages of development (Su et al. 2003; Nowell et al. 2011). Other recent studies report correlations between decreased Foxn1 levels and fewer CIII+ CD80+ CD86+ mTECs, severe reduction of thymic epithelial progenitor cells and increased clonal deletion, as evidenced by diminished negatively selected thymocytes and differentiation of T reg cells (Zuklys et al. 2016). Foxn1 is also necessary to maintain thymic structure. Thus, K14 conditioned removal of Foxn1 results in the progressive polarization of medullary cells, Cld3/4 expression and lumen formation (Guo et al. 2011).

Acknowledgments This work was supported by Grants BFU2013-41112-R from the Spanish Ministry of Economy and Competitiveness, Cell Therapy Network (RD12/0019/0007) from the Spanish Ministry of Health and Consume, and Avancell-CM (S2017/BMD-3692) from Community of Madrid

References

Anderson G, McCarthy NI (2015) Laying bare the nude mouse gene. J Immunol 194(3):847–848. https://doi.org/10.4049/jimmunol.1403061
Bachiller D, Klingensmith J, Shneyder N, Tran U, Anderson R, Rossant J, De Robertis EM (2003) The role of chordin/Bmp signals in mammalian pharyngeal development and DiGeorge syndrome. Development 130(15):3567–3578

Bain VE, Gordon J, O'Neil JD, Ramos I, Richie ER, Manley NR (2016) Tissue-specific roles for sonic hedgehog signaling in establishing thymus and parathyroid organ fate. Development 143(21):4027–4037. https://doi.org/10.1242/dev.141903

Balciunaite G, Keller MP, Balciunaite E, Piali L, Zuklys S, Mathieu YD, Gill J, Boyd R, Sussman DJ et al (2002) Wnt glycoproteins regulate the expression of FoxN1, the gene defective in nude mice. Nat Immunol 3(11):1102–1108. https://doi.org/10.1038/ni850

Baldini A (2004) DiGeorge syndrome: an update. Curr Opin Cardiol 19(3):201–204

Barsanti M, Lim JM, Hun ML, Lister N, Wong K, Hammett MV, Lepletier A, Boyd RL, Giudice A et al (2017) A novel Foxn1(eGFP/+) mouse model identifies Bmp4-induced maintenance of Foxn1 expression and thymic epithelial progenitor populations. Eur J Immunol 47(2):291–304. https://doi.org/10.1002/eji.201646553

Begemann G, Schilling TF, Rauch GJ, Geisler R, Ingham PW (2001) The zebrafish neckless mutation reveals a requirement for raldh2 in mesodermal signals that pattern the hindbrain. Development 128(16):3081–3094

Blackburn CC, Manley NR (2004) Developing a new paradigm for thymus organogenesis. Nat Rev Immunol 4(4):278–289. https://doi.org/10.1038/nri1331

Bleul CC, Boehm T (2005) BMP signaling is required for normal thymus development. J Immunol 175(8):5213–5221

Bleul CC, Corbeaux T, Reuter A, Fisch P, Monting JS, Boehm T (2006) Formation of a functional thymus initiated by a postnatal epithelial progenitor cell. Nature 441(7096):992–996. https://doi.org/10.1038/nature04850

Bredenkamp N, Nowell CS, Blackburn CC (2014a) Regeneration of the aged thymus by a single transcription factor. Development 141(8):1627–1637. https://doi.org/10.1242/dev.103614

Bredenkamp N, Ulyanchenko S, O'Neill KE, Manley NR, Vaidya HJ, Blackburn CC (2014b) An organized and functional thymus generated from FOXN1-reprogrammed fibroblasts. Nat Cell Biol 16(9):902–908. https://doi.org/10.1038/ncb3023

Brown CB, Wenning JM, Lu MM, Epstein DJ, Meyers EN, Epstein JA (2004) Cre-mediated excision of Fgf8 in the Tbx1 expression domain reveals a critical role for Fgf8 in cardiovascular development in the mouse. Dev Biol 267(1):190–202. https://doi.org/10.1016/j.ydbio.2003.10.024

Chen L, Xiao S, Manley NR (2009) Foxn1 is required to maintain the postnatal thymic microenvironment in a dosage-sensitive manner. Blood 113(3):567–574. https://doi.org/10.1182/blood-2008-05-156265

Chen L, Zhao P, Wells L, Amemiya CT, Condie BG, Manley NR (2010) Mouse and zebrafish Hoxa3 orthologues have nonequivalent in vivo protein function. Proc Natl Acad Sci U S A 107(23):10555–10560. https://doi.org/10.1073/pnas.1005129107

Cheng L, Guo J, Sun L, Fu J, Barnes PF, Metzger D, Chambon P, Oshima RG, Amagai T et al (2010) Postnatal tissue-specific disruption of transcription factor FoxN1 triggers acute thymic atrophy. J Biol Chem 285(8):5836–5847. https://doi.org/10.1074/jbc.M109.072124

Chisaka O, Kameda Y (2005) Hoxa3 regulates the proliferation and differentiation of the third pharyngeal arch mesenchyme in mice. Cell Tissue Res 320(1):77–89. https://doi.org/10.1007/s00441-004-1042-z

Chojnowski JL, Masuda K, Trau HA, Thomas K, Capecchi M, Manley NR (2014) Multiple roles for HOXA3 in regulating thymus and parathyroid differentiation and morphogenesis in mouse. Development 141(19):3697–3708. https://doi.org/10.1242/dev.110833

Corbeaux T, Hess I, Swann JB, Kanzler B, Haas-Assenbaum A, Boehm T (2010) Thymopoiesis in mice depends on a Foxn1-positive thymic epithelial cell lineage. Proc Natl Acad Sci U S A 107(38):16613–16618. https://doi.org/10.1073/pnas.1004623107

Cordier AC (1974) Ultrastructure of the thymus in "Nude" mice. J Ultrastruct Res 47(20):26–40

Cordier AC, Haumont SM (1980) Development of thymus, parathyroids, and ultimo-branchial bodies in NMRI and nude mice. Am J Anat 157(3):227–263. https://doi.org/10.1002/aja.1001570303

Cordier AC, Heremans JF (1975) Nude mouse embryo: ectodermal nature of the primordial thymic defect. Scand J Immunol 4(2):193–196

Dooley J, Erickson M, Roelink H, Farr AG (2005) Nude thymic rudiment lacking functional foxn1 resembles respiratory epithelium. Dev Dyn 233(4):1605–1612. https://doi.org/10.1002/dvdy.20495

Erickson M, Morkowski S, Lehar S, Gillard G, Beers C, Dooley J, Rubin JS, Rudensky A, Farr AG (2002) Regulation of thymic epithelium by keratinocyte growth factor. Blood 100(9):3269–3278. https://doi.org/10.1182/blood-2002-04-1036

Foster KE, Gordon J, Cardenas K, Veiga-Fernandes H, Makinen T, Grigorieva E, Wilkinson DG, Blackburn CC, Richie E et al (2010) EphB-ephrin-B2 interactions are required for thymus migration during organogenesis. Proc Natl Acad Sci U S A 107(30):13414–13419. https://doi.org/10.1073/pnas.1003747107

Frank DU, Fotheringham LK, Brewer JA, Muglia LJ, Tristani-Firouzi M, Capecchi MR, Moon AM (2002) An Fgf8 mouse mutant phenocopies human 22q11 deletion syndrome. Development 129(19):4591–4603

Fujimoto Y, Tu L, Miller AS, Bock C, Fujimoto M, Doyle C, Steeber DA, Tedder TF (2002) CD83 expression influences CD4+ T cell development in the thymus. Cell 108(6):755–767

Georgopoulos K, Bigby M, Wang JH, Molnar A, Wu P, Winandy S, Sharpe A (1994) The Ikaros gene is required for the development of all lymphoid lineages. Cell 79(1):143–156

Gordon J, Manley NR (2011) Mechanisms of thymus organogenesis and morphogenesis. Development 138(18):3865–3878. https://doi.org/10.1242/dev.059998

Gordon J, Bennett AR, Blackburn CC, Manley NR (2001) Gcm2 and Foxn1 mark early para-thyroid- and thymus-specific domains in the developing third pharyngeal pouch. Mech Dev 103(1-2):141–143

Gordon J, Wilson VA, Blair NF, Sheridan J, Farley A, Wilson L, Manley NR, Blackburn CC (2004) Functional evidence for a single endodermal origin for the thymic epithelium. Nat Immunol 5(5):546–553. https://doi.org/10.1038/ni1064

Gordon J, Patel SR, Mishina Y, Manley NR (2010) Evidence for an early role for BMP4 signaling in thymus and parathyroid morphogenesis. Dev Biol 339(1):141–154. https://doi.org/10.1016/j.ydbio.2009.12.026

Graham A (2001) The development and evolution of the pharyngeal arches. J Anat 199(Pt 1-2):133–141

Griffith AV, Cardenas K, Carter C, Gordon J, Iberg A, Engleka K, Epstein JA, Manley NR, Richie ER (2009) Increased thymus- and decreased parathyroid-fated organ domains in Splotch mutant embryos. Dev Biol 327(1):216–227. https://doi.org/10.1016/j.ydbio.2008.12.019

Guo J, Rahman M, Cheng L, Zhang S, Tvinnereim A, Su DM (2011) Morphogenesis and main-tenance of the 3D thymic medulla and prevention of nude skin phenotype require FoxN1 in pre- and post-natal K14 epithelium. J Mol Med (Berl) 89(3):263–277. https://doi.org/10.1007/s00109-010-0700-8

Hamazaki Y, Fujita H, Kobayashi T, Choi Y, Scott HS, Matsumoto M, Minato N (2007) Medullary thymic epithelial cells expressing aire represent a unique lineage derived from cells expressing claudin. Nat Immunol 8(3):304–311. https://doi.org/10.1038/ni1438

Hauser BR, Hoffman MP (2015) Regulatory mechanisms driving salivary gland organogenesis. Curr Top Dev Biol 115:111–130. https://doi.org/10.1016/bs.ctdb.2015.07.029

Heinonen KM, Vanegas JR, Brochu S, Shan J, Vainio SJ, Perreault C (2011) Wnt4 regulates thymic cellularity through the expansion of thymic epithelial cells and early thymic progenitors. Blood 118(19):5163–5173. https://doi.org/10.1182/blood-2011-04-350553

Hetzer-Egger C, Schorpp M, Haas-Assenbaum A, Balling R, Peters H, Boehm T (2002) Thymopoiesis requires Pax9 function in thymic epithelial cells. Eur J Immunol 32(4):1175–1181. https://doi.org/10.1002/1521-4141(200204)32:4<1175::Aid-immu1175>3.0.Co;2-u

Itoi M, Kawamoto H, Katsura Y, Amagai T (2001) Two distinct steps of immigration of hematopoi-etic progenitors into the early thymus anlage. Int Immunol 13(9):1203–1211

Itoi M, Tsukamoto N, Amagai T (2007) Expression of Dll4 and CCL25 in Foxn1-negative epi-thelial cells in the post-natal thymus. Int Immunol 19(2):127–132. https://doi.org/10.1093/intimm/dxl129

Jenkinson WE, Jenkinson EJ, Anderson G (2003) Differential requirement for mesenchyme in the proliferation and maturation of thymic epithelial progenitors. J Exp Med 198(2):325–332. https://doi.org/10.1084/jem.20022135

Jenkinson WE, Rossi SW, Parnell SM, Jenkinson EJ, Anderson G (2007) PDGFRalpha-expressing mesenchyme regulates thymus growth and the availability of intrathymic niches. Blood 109(3):954–960. https://doi.org/10.1182/blood-2006-05-023143

Jiang X, Rowitch DH, Soriano P, McMahon AP, Sucov HM (2000) Fate of the mammalian cardiac neural crest. Development 127(8):1607–1616

Kameda Y, Ito M, Nishimaki T, Gotoh N (2009) FRS2alpha is required for the separation, migration, and survival of pharyngeal-endoderm derived organs including thyroid, ultimobranchial body, parathyroid, and thymus. Dev Dyn 238(3):503–513. https://doi.org/10.1002/dvdy.21867

Ki S, Park D, Selden HJ, Seita J, Chung H, Kim J, Iyer VR, Ehrlich LIR (2014) Global transcriptional profiling reveals distinct functions of thymic stromal subsets and age-related changes during thymic involution. Cell Rep 9(1):402–415. https://doi.org/10.1016/j.celrep.2014.08.070

Larsen HL, Grapin-Botton A (2017) The molecular and morphogenetic basis of pancreas organogenesis. Semin Cell Dev Biol 66:51–68. https://doi.org/10.1016/j.semcdb.2017.01.005

Le Douarin NM, Jotereau FV (1975) Tracing of cells of the avian thymus through embryonic life in interspecific chimeras. J Exp Med 142(1):17–40

Le Lievre CS, Le Douarin NM (1975) Mesenchymal derivatives of the neural crest: analysis of chimaeric quail and chick embryos. J Embryol Exp Morphol 34(1):125–154

Lindsay EA, Vitelli F, Su H, Morishima M, Huynh T, Pramparo T, Jurecic V, Ogunrinu G, Sutherland HF et al (2001) Tbx1 haploinsufficieny in the DiGeorge syndrome region causes aortic arch defects in mice. Nature 410(6824):97–101. https://doi.org/10.1038/35065105

Ma D, Wei Y, Liu F (2013) Regulatory mechanisms of thymus and T cell development. Dev Comp Immunol 39(1-2):91–102. https://doi.org/10.1016/j.dci.2011.12.013

Manley NR, Capecchi MR (1995) The role of Hoxa-3 in mouse thymus and thyroid development. Development 121(7):1989–2003

Manley NR, Selleri L, Brendolan A, Gordon J, Cleary ML (2004) Abnormalities of caudal pharyngeal pouch development in Pbx1 knockout mice mimic loss of Hox3 paralogs. Dev Biol 276(2):301–312. https://doi.org/10.1016/j.ydbio.2004.08.030

Manley NR, Richie ER, Blackburn CC, Condie BG, Sage J (2011) Structure and function of the thymic microenvironment. Front Biosci (Landmark Ed) 16:2461–2477

Masuda K, Itoi M, Amagai T, Minato N, Katsura Y, Kawamoto H (2005) Thymic anlage is colonized by progenitors restricted to T, NK, and dendritic cell lineages. J Immunol 174(5):2525–2532

Metz A, Knochel S, Buchler P, Koster M, Knochel W (1998) Structural and functional analysis of the BMP-4 promoter in early embryos of Xenopus laevis. Mech Dev 74(1-2):29–39

Moore-Scott BA, Manley NR (2005) Differential expression of Sonic hedgehog along the anterior-posterior axis regulates patterning of pharyngeal pouch endoderm and pharyngeal endoderm-derived organs. Dev Biol 278(2):323–335. https://doi.org/10.1016/j.ydbio.2004.10.027

Mulroy T, McMahon JA, Burakoff SJ, McMahon AP, Sen J (2002) Wnt-1 and Wnt-4 regulate thymic cellularity. Eur J Immunol 32(4):967–971. https://doi.org/10.1002/1521-4141(200204)32:4<967::Aid-immu967>3.0.Co;2-6

Munoz JJ, Cejalvo T, Tobajas E, Fanlo L, Cortes A, Zapata AG (2015) 3D immunofluorescence analysis of early thymic morphogenesis and medulla development. Histol Histopathol 30(5):589–599. https://doi.org/10.14670/HH-30.589

Munoz JJ, Garcia-Ceca J, Montero-Herradon S, Sanchez Del Collado B, Alfaro D, Zapata A (2018) Can a proper T-cell development occur in an altered thymic epithelium? Lessons From EphB-Deficient Thymi. Front Endocrinol (Lausanne) 9:135. https://doi.org/10.3389/fendo.2018.00135

Nehls M, Kyewski B, Messerle M, Waldschutz R, Schuddekopf K, Smith AJ, Boehm T (1996) Two genetically separable steps in the differentiation of thymic epithelium. Science (New York, NY) 272(5263):886–889

Neubuser A, Koseki H, Balling R (1995) Characterization and developmental expression of Pax9, a paired-box-containing gene related to Pax1. Dev Biol 170(2):701–716. https://doi.org/10.1006/dbio.1995.1248

Neves H, Dupin E, Parreira L, Le Douarin NM (2012) Modulation of Bmp4 signalling in the epithelial-mesenchymal interactions that take place in early thymus and parathyroid development in avian embryos. Dev Biol 361(2):208–219. https://doi.org/10.1016/j.ydbio.2011.10.022

Niida A, Hiroko T, Kasai M, Furukawa Y, Nakamura Y, Suzuki Y, Sugano S, Akiyama T (2004) DKK1, a negative regulator of Wnt signaling, is a target of the beta-catenin/TCF pathway. Oncogene 23(52):8520–8526. https://doi.org/10.1038/sj.onc.1207892

Nowell CS, Bredenkamp N, Tetelin S, Jin X, Tischner C, Vaidya H, Sheridan JM, Stenhouse FH, Heussen R et al (2011) Foxn1 regulates lineage progression in cortical and medullary thymic epithelial cells but is dispensable for medullary sublineage divergence. PLoS Genet 7(11):e1002348. https://doi.org/10.1371/journal.pgen.1002348

O'Neill KE, Bredenkamp N, Tischner C, Vaidya HJ, Stenhouse FH, Peddie CD, Nowell CS, Gaskell T, Blackburn CC (2016) Foxn1 is dynamically regulated in thymic epithelial cells during embryogenesis and at the onset of thymic involution. PLoS One 11(3):e0151666. https://doi.org/10.1371/journal.pone.0151666

Osada M, Jardine L, Misir R, Andl T, Millar SE, Pezzano M (2010) DKK1 mediated inhibition of Wnt signaling in postnatal mice leads to loss of TEC progenitors and thymic degeneration. PLoS One 5(2):e9062. https://doi.org/10.1371/journal.pone.0009062

Patel SR, Gordon J, Mahbub F, Blackburn CC, Manley NR (2006) Bmp4 and Noggin expression during early thymus and parathyroid organogenesis. Gene Expr Patterns 6(8):794–799. https://doi.org/10.1016/j.modgep.2006.01.011

Peters H, Neubuser A, Kratochwil K, Balling R (1998) Pax9-deficient mice lack pharyngeal pouch derivatives and teeth and exhibit craniofacial and limb abnormalities. Genes Dev 12(17):2735–2747

Reeh KA, Cardenas KT, Bain VE, Liu Z, Laurent M, Manley NR, Richie ER (2014) Ectopic TBX1 suppresses thymic epithelial cell differentiation and proliferation during thymus organogenesis. Development 141(15):2950–2958. https://doi.org/10.1242/dev.111641

Revest JM, Suniara RK, Kerr K, Owen JJ, Dickson C (2001) Development of the thymus requires signaling through the fibroblast growth factor receptor R2-IIIb. J Immunol 167(4):1954–1961

Rode I, Martins VC, Kublbeck G, Maltry N, Tessmer C, Rodewald HR (2015) Foxn1 protein expression in the developing, aging, and regenerating thymus. J Immunol 195(12):5678–5687. https://doi.org/10.4049/jimmunol.1502010

Rossi SW, Jenkinson WE, Anderson G, Jenkinson EJ (2006) Clonal analysis reveals a common progenitor for thymic cortical and medullary epithelium. Nature 441(7096):988–991. https://doi.org/10.1038/nature04813

Soza-Ried C, Bleul CC, Schorpp M, Boehm T (2008) Maintenance of thymic epithelial phenotype requires extrinsic signals in mouse and zebrafish. J Immunol 181(8):5272–5277

Su DM, Manley NR (2000) Hoxa3 and pax1 transcription factors regulate the ability of fetal thymic epithelial cells to promote thymocyte development. J Immunol 164(11):5753–5760

Su D, Ellis S, Napier A, Lee K, Manley NR (2001) Hoxa3 and pax1 regulate epithelial cell death and proliferation during thymus and parathyroid organogenesis. Dev Biol 236(2):316–329. https://doi.org/10.1006/dbio.2001.0342

Su DM, Navarre S, Oh WJ, Condie BG, Manley NR (2003) A domain of Foxn1 required for crosstalk-dependent thymic epithelial cell differentiation. Nat Immunol 4(11):1128–1135. https://doi.org/10.1038/ni983

Swann JB, Happe C, Boehm T (2017a) Elevated levels of Wnt signaling disrupt thymus morphogenesis and function. Sci Rep 7(1):785. https://doi.org/10.1038/s41598-017-00842-0

Swann JB, Krauth B, Happe C, Boehm T (2017b) Cooperative interaction of BMP signalling and Foxn1 gene dosage determines the size of the functionally active thymic epithelial compartment. Sci Rep 7(1):8492. https://doi.org/10.1038/s41598-017-09213-1

Takahama Y, Ohigashi I, Baik S, Anderson G (2017) Generation of diversity in thymic epithelial cells. Nat Rev Immunol 17(5):295–305. https://doi.org/10.1038/nri.2017.12

Tsai PT, Lee RA, Wu H (2003) BMP4 acts upstream of FGF in modulating thymic stroma and regulating thymopoiesis. Blood 102(12):3947–3953. https://doi.org/10.1182/blood-2003-05-1657

Tucker AS (2007) Salivary gland development. Semin Cell Dev Biol 18(2):237–244. https://doi.org/10.1016/j.semcdb.2007.01.006

Uddin MM, Ohigashi I, Motosugi R, Nakayama T, Sakata M, Hamazaki J, Nishito Y, Rode I, Tanaka K et al (2017) Foxn1-beta5t transcriptional axis controls CD8(+) T-cell production in the thymus. Nat Commun 8:14419. https://doi.org/10.1038/ncomms14419

Vaidya HJ, Briones Leon A, Blackburn CC (2016) FOXN1 in thymus organogenesis and development. Eur J Immunol 46(8):1826–1837. https://doi.org/10.1002/eji.201545814

Villasenor A, Chong DC, Henkemeyer M, Cleaver O (2010) Epithelial dynamics of pancreatic branching morphogenesis. Development 137(24):4295–4305. https://doi.org/10.1242/dev.052993

Vitelli F, Taddei I, Morishima M, Meyers EN, Lindsay EA, Baldini A (2002) A genetic link between Tbx1 and fibroblast growth factor signaling. Development 129(19):4605–4611

Vroegindeweij E, Crobach S, Itoi M, Satoh R, Zuklys S, Happe C, Germeraad WT, Cornelissen JJ, Cupedo T et al (2010) Thymic cysts originate from Foxn1 positive thymic medullary epithelium. Mol Immunol 47(5):1106–1113. https://doi.org/10.1016/j.molimm.2009.10.034

Walker JL, Menko AS, Khalil S, Rebustini I, Hoffman MP, Kreidberg JA, Kukuruzinska MA (2008) Diverse roles of E-cadherin in the morphogenesis of the submandibular gland: insights into the formation of acinar and ductal structures. Dev Dyn 237(11):3128–3141. https://doi.org/10.1002/dvdy.21717

Wallin J, Eibel H, Neubuser A, Wilting J, Koseki H, Balling R (1996) Pax1 is expressed during development of the thymus epithelium and is required for normal T-cell maturation. Development 122(1):23–30

Wendland K, Niss K, Kotarsky K, Wu NYH, White AJ, Jendholm J, Rivollier A, Izarzugaza JMG, Brunak S et al (2018) Retinoic acid signaling in thymic epithelial cells regulates thymopoiesis. J Immunol 201(2):524–532. https://doi.org/10.4049/jimmunol.1800418

Wendling O, Dennefeld C, Chambon P, Mark M (2000) Retinoid signaling is essential for patterning the endoderm of the third and fourth pharyngeal arches. Development 127(8):1553–1562

Xu PX, Zheng W, Laclef C, Maire P, Maas RL, Peters H, Xu X (2002) Eya1 is required for the morphogenesis of mammalian thymus, parathyroid and thyroid. Development 129(13):3033–3044

Zou D, Silvius D, Davenport J, Grifone R, Maire P, Xu PX (2006) Patterning of the third pharyngeal pouch into thymus/parathyroid by Six and Eya1. Dev Biol 293(2):499–512. https://doi.org/10.1016/j.ydbio.2005.12.015

Zuklys S, Gill J, Keller MP, Hauri-Hohl M, Zhanybekova S, Balciunaite G, Na KJ, Jeker LT, Hafen K et al (2009) Stabilized beta-catenin in thymic epithelial cells blocks thymus development and function. J Immunol 182(5):2997–3007. https://doi.org/10.4049/jimmunol.0713723

Zuklys S, Handel A, Zhanybekova S, Govani F, Keller M, Maio S, Mayer CE, Teh HY, Hafen K et al (2016) Foxn1 regulates key target genes essential for T cell development in postnatal thymic epithelial cells. Nat Immunol 17(10):1206–1215. https://doi.org/10.1038/ni.3537

Chapter 3
The Ins and Outs of Thymic Epithelial Cell Differentiation and Function

Minoru Matsumoto, Pedro M. Rodrigues, Laura Sousa, Koichi Tsuneyama, Mitsuru Matsumoto, and Nuno L. Alves

Given the importance of the thymus for T cell development and general immune function, it is hard to conceive of the time when the origin of lymphocytes – where they developed and their complexity – was completely unknown. I was fortunate to have found myself situated at the right time and place to evaluate the function of the thymus, which at the time was considered superfluous.

Jacques F. A. P. Miller (2006).

Minoru Matsumoto and Pedro M. Rodrigues contributed equally to this work.
Mitsuru Matsumoto and Nuno L. Alves supervised equally to this work.

M. Matsumoto
Division of Molecular Immunology, Institute for Enzyme Research, Tokushima University, Tokushima, Japan

Department of Molecular and Environmental Pathology, Institute of Biomedical Sciences, The University of Tokushima Graduate School, Tokushima, Japan

P. M. Rodrigues · N. L. Alves (✉)
Instituto de Investigação e Inovação em Saúde, Universidade do Porto, Porto, Portugal

Thymus Development and Function Laboratory, Instituto de Biologia Molecular e Celular, Porto, Portugal
e-mail: nalves@ibmc.up.pt

L. Sousa
Instituto de Investigação e Inovação em Saúde, Universidade do Porto, Porto, Portugal

K. Tsuneyama
Department of Molecular and Environmental Pathology, Institute of Biomedical Sciences, The University of Tokushima Graduate School, Tokushima, Japan

M. Matsumoto (✉)
Division of Molecular Immunology, Institute for Enzyme Research, Tokushima University, Tokushima, Japan
e-mail: mitsuru@tokushima-u.ac.jp

© Springer Nature Switzerland AG 2019
G. A. Passos (ed.), *Thymus Transcriptome and Cell Biology*,
https://doi.org/10.1007/978-3-030-12040-5_3

35

Abstract The thymus is the organ dedicated to the generation of T cells, which are key effector cells in immune clearance of pathogens and tumours. However, T cells can also react to our own tissues or turn into cancer cells, such as in the case of autoimmunity and leukaemia, respectively. Therefore, the development and selection of T cells is a tightly regulated process that proceeds within inductive thymic microenvironments formed by cortical (c) and medullary (m) thymic epithelial cells (TECs). Herein, we critically summarize our current knowledge on the molecular principles underlying the development and diversification of TEC compartments and highlight their specialized roles in specific stages of T cell development. Knowledge in this area is of fundamental and clinical relevance to understand how the immune system reaches the equilibrium between immunity and tolerance induction.

3.1   Introduction to the Thymus

As the site for the production of functionally diverse and self-tolerant T cells, the thymus represents an irreplaceable constituent of the immune system (Miller 1961). Anatomically divided into two main regions, the outer cortex and the inner medulla, the thymus is continually colonized by bone marrow-derived hematopoietic precursors that differentiate into T cells while migrating through dedicated cortical and medullary regions (Fig. 3.1a) (Takahama 2006).

The differentiation of T lymphocytes is a temporally coordinated process that generally selects T cells bearing T cell receptors (TCRs) restricted to self-major histocompatibility complex (MHC) and tolerant to self-antigens (self-Ags), over T cells that express non-functional or auto-reactive TCRs. Upon entry into the thymus, thymic seeding precursors (TSP) proceed through four classical double negative (DN) stages (DN1–DN4), as they traffic towards the outer regions of the thymic cortex. Under physiological conditions, thymocytes at DN2 stage commit into the T cell lineage and initiate the rearrangement of the TCR β, γ and δ genes (DN2–DN3). Following β-selection, DN4 thymocytes initiate the rearrangement of the TCRα locus and transit to the double positive (DP) stage, which is characterized by the upregulation of CD4 and CD8 co-receptors and the expression of a fully assembled

---

→

Fig. 3.1 (continued) results in their commitment into the single positive (SP) CD4 or CD8 lineage. Positively selected SP thymocytes continue their migration into in the medulla, wherein cells expressing TCRs that are simultaneously restricted to self-MHC and tolerant to self-antigens complete their maturation and egress into the peripheral lymphoid organs. Solid arrows represent transition through the distinct stages of thymocyte development. (**b**) Phenotypic and functional proprieties of thymic epithelial cells (TECs). TECs are the most predominant cell type of the thymic stroma and are mainly divided into cortical (c) and medullary (m) TECs. Both TEC lineages can be identified through the usage of common or specific panel of markers. While cTECs regulate early stages of T cell development, including lineage specification and positive selection, mTECs have an essential role in the establishment of central tolerance, both via negative selection of autoreactive thymocytes or their diversion towards the regulatory T cell (Treg) lineage

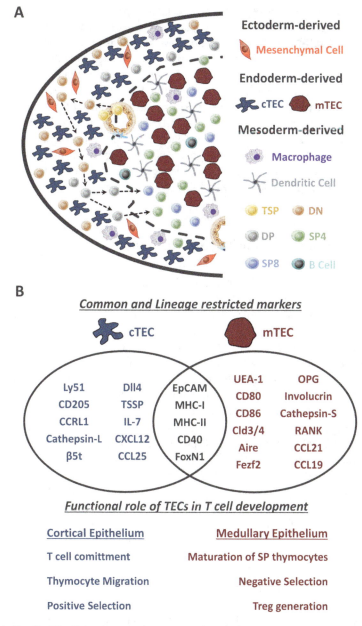

Fig. 3.1 Detailed T cell developmental stages and phenotypic characterization of distinct cellular microenvironment of the thymus. (**a**) Schematic illustration of the cellular composition and organization of the adult thymus. Representation of distinct stromal cells that compose the specialized thymic NICHE. Once within the thymus, thymic seeding progenitors (TSPs) undergo a complex differentiation pathway towards the outer regions of the thymic cortex as CD4⁻CD8⁻ double negative (DN) thymocytes. Following T cell receptor (TCR) rearrangement and β-selection, DN cells mature into CD4⁺CD8⁺ double positive (DP) thymocytes. Positive selection of DP thymocytes

αβ TCR. Still in the cortex, the newly generated repertoire of TCRs expressed by DP thymocytes is screened for their capacity to recognize endogenous peptides in the context of self-MHC molecules, a process known as positive selection. At this stage, cells that express TCRs with the capacity to bind self-peptide/MHC with the right affinity survive and continue their development into the single positive (SP) CD4 or CD8 stage. Following positive selection, SP thymocytes revert their migration into the medulla, wherein their TCRs are further scanned for the presence of TCR that binds to self-Ags with high-affinity. Herein, auto-reactive SP cells are negatively selected or redirected into regulatory T cells (Tregs), while SP thymocytes expressing TCRs that are simultaneously restricted to self-MHC and tolerant to self-Ags complete their maturation and egress into the peripheral lymphoid organs (Fig. 3.1a) (Petrie and Zuniga-Pflucker 2007).

3.2 Constituents of the Thymic Stroma

The development of T cells is not a cell-autonomous process and relies on an organized three-dimensional network of thymic stromal cells that provides the essential instructive cues for thymopoiesis (Ciofani and Zuniga-Pflucker 2007). Important elements of the thymic microenvironment include endothelial cells, mesenchymal cells and thymic epithelial cells (TECs), as well as cells of hematopoietic origin, such as dendritic cells (DCs), macrophages and B cells (Boyd et al. 1993) (Fig. 3.1a). The thymic stroma directly or indirectly influences T cell development, through the cooperation with different stromal cells for the establishment of dedicated functional thymopoietic niches. For example, mesenchymal cells may impact on T cell development (Anderson et al. 1993; Anderson et al. 1997), possibly by producing extracellular matrix (ECM) proteins that concentrate and present essential soluble growth factors (e.g. cytokines), to developing thymocytes (Banwell et al. 2000). Additionally, mesenchymal cells affect the proliferation of TECs through the provision of fibroblast growth factors (FGF)-7 and -10 (Revest et al. 2001), as well as retinoic acid (RA) and insulin-like growth factor (IGF)-1 and -2 (Jenkinson et al. 2007; Sitnik et al. 2012). The endothelium plays a critical role in the homing to the thymus through the expression of the adhesion molecules VCAM-1, ICAM-1 and P-selectin whose ligands are expressed by TSPs (Rossi et al. 2005; Scimone et al. 2006; Gossens et al. 2009). On the other hand, macrophages are important for the clearance of apoptotic thymocytes, and thymic DCs (further discussed below) and B cells have been shown to contribute to the induction of central tolerance through the expression and presentation of self-Ags (Boyd et al. 1993; Esashi et al. 2003; Proietto et al. 2008; Yamano et al. 2015). In this chapter, we will focus on TECs, which are the major constituent of the thymic stroma in the pre-involuted thymus and play a non-redundant role in T cell development and selection.

TECs are classically subdivided into cortical (cTECs) and medullary (mTECs) subtypes based on their spatial location, functional and phenotypic proprieties

(Abramson and Anderson 2017) (Fig. 3.1b). While cTECs regulate early stages of T cell development, including lineage specification and positive selection, mTECs have an essential role in the establishment of central tolerance, both via negative selection of auto-reactive thymocytes or their diversion towards the Treg lineage (Klein et al. 2014) (Fig. 3.1b). Phenotypically, cTECs are defined by the expression of Cytokeratin 8 (K8) and K18, Ly51 (also known as BP1/CD249), CD205 (also known as DEC-205) and the atypical chemokine receptor 4 (ACKR4, also known as CCRL1). Furthermore, the expression of Delta-like ligand 4 (Dll4), high levels of Interleukin 7 (IL-7) and the thymoproteasome subunit β5t may also aid in the identification of discrete cTEC subsets (Ohigashi et al. 2016). On the other hand, mTECs are identified by the expression of K5 and K14 as well, by the selective binding to the lectin *Ulex Europeus Agglutinin 1* (UEA-1) and the binding of ER-TR5 and mouse thymic stroma 10 (MTS10) antibodies to unknown mTEC-expressing molecules (Van Vliet et al. 1984; Farr and Anderson 1985; Godfrey et al. 1990). A further degree of maturation heterogeneity within mTEC subpopulations can be resolved by the expression of CD40, CD80, MHC class II (MHC-II), CCL21, Aire, Fezf2 and Involucrin (Alves et al. 2009a; Abramson and Anderson 2017).

3.3   The Origins of Thymic Epithelial Cells

3.3.1   Lineage Specification of TECs During Embryonic Development

The establishment of functionally competent TEC microenvironment is a critical prerequisite for the development of a diverse and self-tolerant peripheral T cell pool (Abramson and Anderson 2017). This process commences during early thymic organogenesis and involves the participation of cells from all three embryonic germ layers (endoderm-derived epithelium, ectoderm-derived neural crest (NC) mesenchyme and mesoderm-derived hematopoietic cells) (Gordon and Manley 2011). The murine TEC differentiation process has been temporally divided into early thymocyte-independent and later thymocyte-depend stages (Manley 2000). During early organogenesis (from embryonic day (E) 9.5–12.5), a series of morphogenetic events induce the detachment of the endoderm-derived epithelium from the third pharyngeal pouch, which becomes surrounded by NC-derived cells (Gordon and Manley 2011). During the later phase of thymic development (from E13.5–birth), the differentiation of functionally mature cortical and medullary TEC primarily relies on signals derived from developing thymocytes (discussed later in this chapter) (van Ewijk et al. 1994; van Ewijk et al. 2000).

   The early ontogeny of the murine thymus is intimately connected with the formation of the parathyroid gland, with both organs emerging from a common primordium derived from the third pharyngeal pouches (3PP) of the foregut endodermal tube around E11 of gestation (Blackburn and Manley 2004). The patterning of specific parathyroid and thymic domains occurs prior to the formation of the organ rudiments and is dependent on the respective expression of the transcription factors Glial cells missing homolog 2 (Gcm2) and Forkhead box N1 (Foxn1). Gcm2 is a key regulator of parathyroid differentiation and is expressed in the anterior-dorsal region of the 3PP epithelium as early as E9.5 (Gordon et al. 2001; Liu et al. 2007). On the other hand, Foxn1 is a critical determinant of TEC identity and function and is first detected in the posterior-ventral side of the pharyngeal pouch at E11.25 (Blackburn et al. 1996; Gordon et al. 2001). Among the hierarchy of transcription factors involved in thymic organogenesis, Foxn1 remains to date as a prominent element with a nonredundant role in TEC differentiation (Romano et al. 2013). Foxn1 is a member of the forkhead box transcription factor with a winged-helix/forkhead DNA-binding domain and a transcriptional activation domain (Schuddekopf et al. 1996). Foxn1 regulates the expression of chief targets of thymopoiesis in TECs, including genes that contribute to T cell lineage commitment (Dll4), antigen processing and presentation (β5t) and lymphocyte migration (CCL25) (Zuklys et al. 2016). Consequently, inactivation of Foxn1 gene causes the athymic "nude" phenotype in humans and mice, due to a developmental arrest in TEC differentiation (Nehls et al. 1996; Bleul and Boehm 2000; Romano et al. 2013). Still, early stages of thymic organogenesis proceed normally in nude mice until E11 with the formation of the common thymus-parathyroid primordium, suggesting that Foxn1 is not involved in the initial thymic/parathyroid fate specification (Nehls et al. 1996). Interestingly, reduction in Foxn1 expression and in the number of Foxn1-expressing TECs has been coupled to thymic involution (Corbeaux et al. 2010; Rode et al. 2015; O'Neill et al. 2016). Conversely, tissue-specific activation of *Foxn1* expression overcomes the defective thymic development in nude mice (Bleul et al. 2006) and transgenic expression of Foxn1 in TECs augments thymopoiesis in aged mice, delaying age-associated thymic involution (Chen et al. 2009; Zook et al. 2011; Bredenkamp et al. 2014a). Moreover, introduction of Foxn1 reprograms mouse embryonic fibroblasts into functional TECs (Bredenkamp et al. 2014b). These findings indicate that modulation of Foxn1 levels has therapeutic potential. Although evidence indicates that Bone morphogenetic protein (Bmp), Wingless-related MMTV integration (Wnt) and retinoblastoma (Rb) family of proteins control Foxn1 expression (Balciunaite et al. 2002; Tsai et al. 2003; Soza-Ried et al. 2008; Garfin et al. 2013), the molecular mechanisms that regulate the induction of this chief thymopoietic factor remain elusive. Solving this question might provide means to functionalize TECs in therapies targeting thymic disorders.

3.3.2   Developmental Models of TECs

TEC differentiation is a dynamic process that is initiated during fetal development and proceeds throughout post-natal life (Abramson and Anderson 2017). Despite anatomically and functionally distinct, cTEC and mTEC develop from common bipotent TEC progenitors (TEPs) present within the embryonic and postnatal thymus (Bleul et al. 2006; Rossi et al. 2006). The existence of TEPs was first suggested by the identification of cells that reside at the cortico-medullary junction in the adult thymus and co-expressed cortical (K8) and medullary (K5) markers (Ropke et al. 1995; Klug et al. 1998). Years later, two complementary studies provided experimental evidence for the existence of bipotent TEPs. Firstly, lineage tracing experiments showed that microinjected single yellow fluorescent protein (YFP)-tagged E12 TEC give rise to both cTECs and mTECs upon integration into native thymus (Rossi et al. 2006). Secondly, using a gain-of-function model in which mice bearing a conditional mutant allele of Foxn1 can be reverted to the functional form in TECs, it was demonstrated that bipotent TEPs persist in the postnatal thymus and contribute to the generation of functional thymic niches (Bleul et al. 2006). These observations support the notion that TEPs arise during thymic organogenesis independently of Foxn1, but its expression is central to initiate the transcriptional program involved in the differentiation of TEPs into cTEC and mTEC lineages.

The identification of markers to define TEPs has been under intense scrutiny, with the disclosure of new developments that contribute to our understanding of the blueprint of TEC differentiation. Although initial studies suggested that E14 MTS24+ TECs have the capacity to generate a fully functional organized thymic epithelial microenvironment (Bennett et al. 2002; Gill et al. 2002), subsequent reassessment has revealed that progenitor activity was not restricted to the MTS24+ fraction (Rossi et al. 2007a), arguing against the usage this marker to exclusively identify bipotent TEPs. Despite the paucity of markers to recognize TEPs, the initial studies led to the concept that cTEC and mTEC developed independently from each other through distinct lineage-restricted precursors. This reductionist model has been recently questioned with a series of complementary studies that have addressed the timing and relationship between the establishment of cTEC and mTEC microenvironments (Baik et al. 2013; Ohigashi et al. 2013; Ribeiro et al. 2013). Particularly, it has been shown that embryonic TECs expressing cortical-restricted markers (e.g. CD205, β5t and high expression of IL-7) can give rise to both cTECs and mTECs (Baik et al. 2013; Ohigashi et al. 2013; Ribeiro et al. 2013). These reports paved the way for a refined model of TEC differentiation, wherein TEPs progress first through the cortical lineage prior to the commitment into mTEC differentiation (Fig. 3.2) (reviewed in (Alves et al. 2014)). Supporting the notion that the thymic cortex represents a reservoir of bipotent TEC progenitors, recent studies identified distinct cTEC-associated subsets in the postnatal and adult thymus that are enriched in purported TEPs (Wong et al. 2014; Ulyanchenko et al. 2016; Meireles et al. 2017).

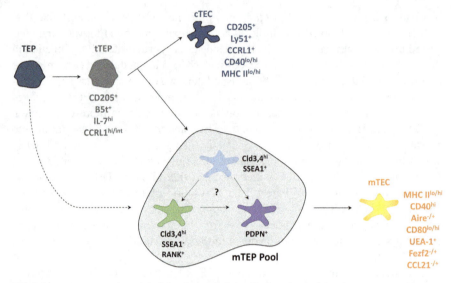

Fig. 3.2 Developmental model of thymic epithelial cells. Despite deriving from common bipotent TEC progenitors (TEPs), cortical (c) and medullary (m) thymic epithelial cells (TECs) specialize into functionally and phenotypically distinct cell types. TEPs progress through a "transitional progenitor (tTEP)" stage that exhibit cellular and molecular traits associated with cTEC before commitment to a cTEC or mTEC lineage. In recent years, three novel subsets of mTEC-restricted progenitors (mTEP) have been characterized. While some of these subsets are derived from tTEPs during the embryonic and post-natal period, one cannot formally exclude the direct contribution of TEPs for the mTEC pool (represented by the dashed arrow). Furthermore, the precursor-product relationship among the different mTEC-committed progenitors (represented by '?') and their spatiotemporal contribution of the maintenance of distinct mTEC lineages along life remains poorly defined

3.4   Specialization of the Thymic Microenvironments: Lineage Restricted TEC Progenitors

3.4.1   cTEC

Initial studies on the precursor-product relationship of the cTEC lineage revealed that TECs exhibiting a cortical phenotype (Ly51[+], CD205[+], K8[+]) are detected during early thymic organogenesis (Klug et al. 2002; Alves et al. 2010; Ribeiro et al. 2013). Interestingly, the emergence of cTECs is not compromised in mice with a profound arrest in early T cell development ($Rag2^{-/-}Il2rg^{-/-}$; CD3εTge26 and Ikaros$^{-/-}$) (Klug et al. 2002; Alves et al. 2010; Ribeiro et al. 2013; Ribeiro et al. 2014), supporting the notion that the initial stages of cTEC differentiation can proceed in a thymocyte-independent fashion. Still, the presence of DN3 thymocytes in $Rag^{-/-}$ mice promote the formation of a three-dimensional reticular cTEC network (van Ewijk et al. 2000). In this respect, signals from DN thymocytes drive the complete differentiation of cTECs, as an arrest in early T cell development results in a partial block in the expression of MHC-II, CD40 and CCRL1 in cTECs

(Shakib et al. 2009; Ribeiro et al. 2014). These findings indicate the existence of a DN stage-specific requirement for the maturation of the cortical epithelium. Still, the precise nature of signals provided by DN thymocytes to cTECs remains to be determined.

Despite the lack of appropriate markers to better define the developmental maturation of cTECs, the differential expression of the CCRL1, CD205, CD40 and MHC-II molecules provided clues on the identity of lineage-committed cTEC progenitors and established new sequential stages of cTEC development (Shakib et al. 2009; Ribeiro et al. 2014). Cells with a CD205$^+$CD40$^-$ phenotype are proposed to represent a population of cTEC progenitors that lies in between bipotent progenitors and mature cTECs. These purported intermediate cTEC progenitors are enriched in proliferating cells and express lower levels of cTEC-specific transcripts such as β5t and cathepsin L, when compared with more mature cortical cells (CD205$^+$CD40$^+$MHC-II^{high}) (Shakib et al. 2009). Still, a fraction of fetal cTEC-like cells presents mTEC progenitor activity (Baik et al. 2013; Ohigashi et al. 2013; Ribeiro et al. 2013; Ribeiro et al. 2014), raising the possibility for the existence of heterogeneous niches of bipotent and unipotent progenitors within the cortical microenvironment. Future studies must aim at better characterizing the developmental checkpoints in cTEC differentiation.

3.4.2 mTEC

In contrast to our knowledge of cTEC development, the characterization of the mTEC-committed progenitors (mTEP) as well as the molecular mechanisms underlying the differentiation of mTECs downstream of bipotent progenitors have been better defined in recent years (Hamazaki et al. 2016). The observation that single mTEC islets are clonal provided pioneering indications for the existence of mTEP (Rodewald et al. 2001). Subsequent studies identified mTEP based on the expression of the tight junction components Claudin-3 and -4 (Cld3,4). The Cld3,4^{high} subset contains cells with the capacity to generate mTECs and to sustain a functional medullary epithelium throughout life (Hamazaki et al. 2007; Sekai et al. 2014). The identification of stage-specific embryonic antigen (SSEA1)-expressing cells within the Cld3,4^{high} TECs subset further resolved the pool of mTEC precursors (Sekai et al. 2014). Interestingly, SSEA1$^+$Cld3,4^{high} TECs are rare in the adult thymus and have a reduced capacity to generate mature mTECs when compared to their embryonic counterparts (Sekai et al. 2014). These findings suggest that the abundance and functional properties of unipotent mTEPs decrease throughout life. More recently, a further degree of heterogeneity has been revealed with the identification of cells expressing the receptor activator of NF-κB (RANK) within the SSEA1$^-$Cld3,4^{high} subset of the embryonic thymus, which can preferentially give rise to mTECs (Baik et al. 2016). Other fate-mapping study reported that podoplanin (PDPN)-expressing TECs located at the cortical-medullary junction (jTECs) contribute to the generation of nearly half of the adult mTEC compartment (Onder et al. 2015). While

SSEA1⁺Cld3,4ʰⁱᵍʰ and PDPN⁺ jTECs derive from embryonic and postnatal cells with a past history of β5t expression (Ohigashi et al. 2015; Mayer et al. 2016), the exact precursors of the RANK⁺SSEA1⁻Cld3,4ʰⁱᵍʰ population remain unknown. Additionally, it is still undefined whether PDPN⁺, RANK⁺ and SSEA1⁺Cld3,4ʰⁱᵍʰ TECs share a direct lineage relationship (Fig. 3.2).

Despite the remarkable progress in the field of TEC biology, the prospective isolation of TEPs at the single cell level has been difficult to attain due to the paucity of markers. As such, distinct TEC precursors are still defined at the population level. Moreover, regarding TEC characterization, there is still a lack of consensus regarding their molecular and phenotypic characteristics in the embryonic and adult thymus. Therefore, further studies are required to better define the identity of distinct TEC progenitors and also their spatiotemporal contribution to the maintenance of thymic epithelial niches along life.

3.5   Thymic Epithelial Cell Homeostasis and Differentiation

The characterization of the regulatory network that governs the homeostasis of the thymic epithelium has received considerable attention over the last decade. Knowledge in this area will have important repercussions for the development of therapeutic strategies aimed at modulating thymic function under different pathological conditions. Although little is known about the mechanisms that balance cTEC homeostasis, the rules controlling the maintenance of mTECs have been better defined. In the next section, we summarize recent studies that contributed to unravel cellular and molecular mediators of mTEC physiology.

3.5.1   mTEC

The development of mTEC depends on bidirectional interactions with distinct hematopoietic cells. The emergence of embryonic mTECs requires signals from lymphoid tissue inducer cells and γδ T cells (Rossi et al. 2007b; Roberts et al. 2012), and involves signaling through TNFR superfamily receptor activator of NF-κB (RANK) and lymphotoxin *beta* receptor (LTβR) expressed on TEC precursors (Rossi et al. 2007b; Mouri et al. 2011). The mTEC microenvironment continues to expand during postnatal life, fostered by additional interactions between TECs and mature thymocytes, namely positively selected and CD4 single positive (SP4) thymocytes (Anderson and Takahama 2012). Here, the concerted activation of RANK-, LTβR-, and CD40-mediated signaling on mTECs and their precursors completes the formation of the adult medullary niche.

Additionally, it might be also important to notice that thymic crosstalk drives mTEC differentiation as for cTEC differentiation (van Ewijk et al. 1994; Anderson et al. 2007) as described in Sect. 3.3. This is most clearly demonstrated by Rag2-deficient mice where the development of mTECs is impaired, although no defect in particular mTEC subset (e.g., Aire⁺ mTECs) has been noticed (Nishikawa et al. 2014).

Of note, the crosstalk between thymocytes and mTECs is bidirectional. Furthermore, the development and function of thymic DCs are also influenced by mTECs. In *aly* mice (see Section 3.6.1), numbers of thymic DCs and their expression of costimulatory molecules were affected in a thymic stroma-dependent manner (Mouri et al. 2014). The results suggest a pivotal role of NIK in the thymic stroma in establishing self-tolerance by orchestrating crosstalk between mTECs and DCs. Given that thymic DCs can affect the function of thymocytes in turn (Hofmann et al. 2011), thymic crosstalk can be viewed as an integrated circuit involving three components: mTECs, thymocytes, and thymic DCs.

3.6   Signals and Factors That Affect mTEC Development and Function

3.6.1   TNFR Superfamily Members and NF-κB

NF-κB is a major effector molecule downstream of many TNFR superfamily members. The first molecular mechanism underlying the defective mTEC integrity involving TNFR superfamily members was demonstrated by knockout mice LTβR together with *aly* mice (Boehm et al. 2003; Kajiura et al. 2004): *aly* mice harbor a natural mutation in NF-κB-inducing kinase (NIK) gene (Shinkura et al. 1999) which is a downstream signaling molecule of LTβR (Matsumoto et al. 1999; Matsushima et al. 2001). LTβR has been originally demonstrated to be essential for the secondary lymphoid organogenesis including peripheral lymph nodes and Peyer's patches (De Togni et al. 1994) and for the formation of germinal centers in the spleen (Matsumoto et al. 1997). The phenotypes of the thymus from *aly* mice were more severe compared with those from LTβR-deficient mice (Kajiura et al. 2004; Mouri et al. 2011) because NIK acts as a signaling molecule not only downstream of LTβR but also other TNFR superfamily members including RANK, an essential molecule for the mTEC development (Rossi et al. 2007b), and CD40 (Akiyama et al. 2008). Later on, IκB kinase (IKK) α (Kinoshita et al. 2006), a downstream molecule of NIK in the NF-κB signaling cascade, and the final effector non-canonical NF-κB components including RelB (Riemann et al. 2017) and p52 (Zhu et al. 2006) have been demonstrated to be essential for the normal thymic organogenesis. Another downstream molecule of TNFR superfamily member TRAF6 (Akiyama et al. 2005) also turned out to be essential for the mTEC development. The TNFR superfamily members upstream of TRAF6 are RANK and CD40 (Akiyama et al. 2008).

3.6.2   Stat3

Initial study suggested that Stat3 plays a role in the homeostasis of both cTECs and mTECs (Sano et al. 2001). However, later studies pointed out that the role of STAT3 is confined to mTECs (e.g., cellularity and architectural organization) but not for cTECs (Lomada et al. 2016; Satoh et al. 2016). Activation of Stat3 is mainly considered to be positioned downstream of EGF-R (Satoh et al. 2016).

3.6.3 Apoptosis-Related Molecules

Among the major prosurvival molecules, Mcl-1 has been demonstrated to control the survival of both mTECs and cTECs (Jain et al. 2017). Rather unexpectedly, BCL-2 and BCL-XL were dispensable for this action.

3.6.4 p53 Family

Another new dimension of the controller of mTEC homeostasis is tumor suppressor gene p53. p53 exerts broad actions in many biological activities. However, its role for the TEC development and function has been investigated only recently (Rodrigues et al. 2017). Conditional deletion of p53 in TECs (p53cKO) showed the abnormal integrity of mTEC niche, and the defect also spread to the cTEC compartment in the adult. Consequently, deficiency of p53 in TECs altered multiple functional modules of mTEC transcriptome including tissue-restricted antigens (TRAs) for the establishment of self-tolerance. In addition, p53cKO mice showed premature defects in mTEC-dependent Treg differentiation and thymocyte maturation (Rodrigues et al. 2017). p63, a homolog of p53, is a key, lineage-specific determinant of the proliferative capacity in stem cells of stratified epithelia, and p63-deficient mice showed abnormal TEC development (Senoo et al. 2007). The function of p53 and p63 in TEC homeostasis is complex, and we still need to know better the exact interrelationship between these molecules in TEC biology.

3.6.5 Autophagy

Role of autophagy in TEC homeostasis has been somewhat controversial (Nedjic et al. 2008; Sukseree et al. 2012). More recent data suggested that C-type lectin domain family 16A (Clec16a) is involved in TEC autophagy (Schuster et al. 2015). *CLEC16A* encodes a large protein of 1053 amino acids that contains several putative functional domains, including a C-type lectin domain. Its role in autophagy was recently confirmed in mice, where pancreatic β cell-specific deletion of *Clec16a* impaired glucose-stimulated insulin release: loss of Clec16a led to an increase in the Parkin, a master regulator of mitophagy (Soleimanpour et al. 2014). Interestingly, *Clec16a* knockdown in NOD prevented autoimmunity by reducing the pathogenicity of T cells, which was secondary to the attenuated Clec16a activity in TECs

(Schuster et al. 2015). Of note, CLEC16A variation has been associated with multiple immune-mediated diseases, including type 1 diabetes, multiple sclerosis and systemic lupus erythematosus.

3.6.6   MicroRNA

Dicer-deficient TECs showed the progressive loss of epithelium from the thymus due to an enhanced apoptosis which was more prominent in mTECs than in cTECs, resulting in the reduced thymus cellularity (Papadopoulou et al. 2011). The competence of Dicer-deficient TECs to support T cell development was progressively compromised, and this resulted in the defect both in the generation of DP thymocytes and their maturation to SP stage (Zuklys et al. 2012).

3.6.7   mTOR

mTOR has been known as an essential molecule in many metabolic pathways. TEC-specific deletion of mTOR caused the severe reduction of mTECs, the blockade of thymocyte differentiation, the reduced production of Tregs and the impaired expression of TRAs (Liang et al. 2018). Those are the phenotypes we often observe when the function of mTECs is abnormal. Accordingly, conditional deletion of mTOR in TECs caused autoimmune diseases characterized by enhanced tissue immune cell infiltration and the production of autoantibodies (Liang et al. 2018).

3.6.8   Interferon Regulatory Factors (IRF)

The IRF group of proteins, which includes nine members in mammals, plays a critical function in the regulation of many aspects of innate and adaptive immune responses. Two recent reports highlighted the participation of IRF4 and IRF7 in controlling distinct aspects of mTEC physiology (Otero et al. 2013; Haljasorg et al. 2017). TEC-specific disruption of IRF4 perturbs the regular composition of immature and mature mTECs (increase in mature cells), without affecting the size of the medullary niche. Furthermore, IRF4cKO mice exhibit an impaired expression of chemokines and costimulatory molecules on mTECs, which is accompanied by a reduction in thymic Tregs numbers and peripheral manifestations of autoimmunity (Haljasorg et al. 2017). In contrast to IRF4, loss of IRF7 has a detrimental impact on the medullary niche, causing a reduction in the overall number of mTECs and diminished levels of *Aire* transcripts (Otero et al. 2013). Yet, it remains undetermined the physiological consequences of TEC-specific IRF7 loss in terms of T cell development and tolerance induction. Despite the differential impact of IRF4 and IRF7 on mTEC homeostasis, both reports suggest that RANK-mediated signaling regulates the expression of these molecules.

3.6.9   Other Factors

There are several growth/differentiation factors reported to regulate the development of cTECs, mTECs or both. For example, secreted Wnt glycoproteins, expressed by TECs and thymocytes, regulate Foxn1 expression in both autocrine and paracrine fashions, thereby regulating the TEC development (Balciunaite et al. 2002). Additionally, Keratinocyte growth factor (KGF) enhances postnatal T-cell development via enhancements in proliferation and function of TECs (Rossi et al. 2007c).

3.7   Functional Diversity of cTECs and mTECs

3.7.1   The Multifunctional Contributions of cTECs in Early Stages of T Cell Development

3.7.1.1   Chemokines, Growth Factors and Lineage Commitment of Thymocytes

cTECs play an important role in the recruitment of thymic seeding progenitors through the production of chemotactic ligands CCL25 and CXCL12 (Plotkin et al. 2003; Liu et al. 2006; Krueger et al. 2010; Zlotoff et al. 2010; Calderon and Boehm 2011). Furthermore, these chemokines also contribute to the proper intrathymic migration of thymocytes across distinct regions of the thymic cortex (Ciofani and Zuniga-Pflucker 2007; Love and Bhandoola 2011). cTECs are also essential in providing chief instructive signals that commit hematopoietic progenitors towards the T cell lineage, while repressing alternative ones. This is achieved by the expression of Notch ligand Dll4 in cTECs (Fiorini et al. 2008; Koch et al. 2008). Additionally, cTECs represent physiological sources of stem cell factor (SCF or kit ligand) and IL-7 (Alves et al. 2009b; Buono et al. 2016), which mediate the survival and proliferation of DN thymocytes (Petrie and Zuniga-Pflucker 2007). Moreover, IL-7 is also important for the differentiation of γδ T cells (Moore et al. 1996). As such, cTECs are fundamental in orchestrating early stages of thymopoiesis.

3.7.1.2   Positive Selection: Shaping the TCR Repertoire

Another critical event in the choreography of T cell differentiation is the positive selection of DP thymocytes expressing αβ TCRs that are restricted to self-peptide/MHC molecules presented on cTECs (Klein et al. 2009; Klein et al. 2014). At this stage, the fate of DP thymocytes is dictated by both the timing and strength of the interaction between the TCRs and self-peptide/MHC complexes. While DP thymocytes expressing TCRs that are able to interact with self-peptide/MHC-II molecules differentiate into CD4+ SP cells, DP thymocytes bearing TCRs restricted to self-peptide/MHC-I differentiate to CD8+ SP cells. On the other hand, thymocytes expressing TCRs that do not bind or interact with high affinity with self-peptide/ MHC complexes die by apoptosis. Apart from MHC, another critical element during positive selection is the nature of the self-peptide presented by cTECs. The distinctive proteolytic and antigen-processing capacities of cTECs ensures the presentation of a specialized array of selecting self-peptides that are distinct from those displayed by mTECs and thymic DCs in the medulla (Takada et al. 2017). The generation of a broad repertoire of selecting MHC-II-associated self-peptides depends in part on the thymus-specific serine protease (TSSP) and the lysosomal protease cathepsin L, as deficiency in any of these proteases impairs positive selection of CD4+ SP cells (Nakagawa et al. 1998; Gommeaux et al. 2009). Furthermore, positive selection of some MHC-II-restricted transgenic TCRs seems compromised upon ablation of genes involved in macroauthophagy in the thymic epithelium (Nedjic et al. 2008), suggesting a role for this pathway in shaping the composition of the MHC-II ligandome on cTECs. On the other hand, the presentation of selecting MHC-I-associated self-peptides is critically dependent on the presence of a distinct form of the proteasome in cTECs (thymoproteasome), which is assembled by the incorporation of a cTEC-restricted β5t subunit (encoded by the gene *Psmb11*). The β5t subunit confers unique proteolytic activity to thymoproteasome as compared to the immuno- and constitutive-proteosome (Murata et al. 2007). As such, mice lacking β5t exhibit a severe and specific impairment in positive selection of CD8+ SP cells (Murata et al. 2007). Moreover, the residual CD8+ T cells that develop in *Psmb11*−/− mice are functionally compromised and present an altered αβTCR repertoire (Nitta et al. 2010; Takada et al. 2015), indicating that thymoproteasome generates a unique peptidome for MHC-I presentation and CD8 T cell selection. Hence, through the expression of a unique set of intracellular proteolytic enzymes, cTECs are able to produce a distinct array of selecting self-peptides that are pivotal for the production of a functional and diverse T cell repertoire.

3.8 The Essential role of mTEC in the Establishment of Self-Tolerance

Thymic medulla is a primary site for the establishment of self-tolerance providing an unique microenvironment to negatively select autoreactive T cells and/or differentiate Tregs. Tregs develop through high affinity interactions with self-Ags expressed from mTECs in cooperation with bone marrow (BM)-derived antigen presenting cells (APCs) (Sakaguchi 2004). Unique actions of several factors within TECs for the establishment of self-tolerance have recently been highlighted by studies using gene-targeted mice together with a strain of mouse bearing a natural mutation in NIK gene (*aly* mice) (Matsumoto 2007). Thus, it is highly likely that a group of genes controls self-tolerance within TECs through unique and coordinated actions. The comprehensive understanding of this process would help to unravel the pathogenesis of autoimmune disease.

3.8.1 Aire

3.8.1.1 Promiscuous Gene Expression (PGE) Controlled by Aire

Important for tolerance induction is the capacity of mTECs to express a set of self-Ags encompassing all the self-Ags expressed by parenchymal organs (i.e., TRAs), a phenomenon termed promiscuous gene expression (PGE) (Hanahan 1998; Klein et al. 2014). This fascinating finding was originally discovered by the transgenic expression of SV40 T antigen (Tag) using a rat insulin promoter (Smith et al. 1997). The authors found that transgenic mice expressed low levels of Tag mRNA in the thymus. Subsequently, they found that, in addition to the insulin-promoter driven Tag expression, thymic medulla contained rare cells that express the endogenous insulin and somatostatin genes (Smith et al. 1997). They called these cells 'peripheral antigen-expressing (PAE) cells'. Interestingly, transplantation of thymus from RIP-Tag mice into athymic hosts was sufficient to confer the tolerance by CD4[+] and/or CD8[+] T cells. Conversely, RIP-Tag mouse line that did not show Tag expression in the thymus suffered autoimmune attack in the β-islets. Based on these findings, Hanahan has proposed that PAE cells in thymic medulla play a crucial role in establishing self-tolerance (Hanahan 1998). Then, Kyewski and his colleagues have clearly demonstrated that PAE cells are mTECs, and they expanded the concept of PGE (Derbinski et al. 2001; Kyewski and Klein 2006). At the right timing, phenotypic analysis of mice deficient for Aire, a gene responsible for the hereditary type of autoimmune disease (Consortium 1997; Nagamine et al. 1997), was reported. Aire is strongly expressed by mature mTECs, and mTECs from Aire-deficient mice showed dramatically reduced TRA gene expression (Anderson et al. 2002).

3.8.1.2 Autoimmune Pathogenesis by Aire Deficiency

Aire-deficient mice were immediately demonstrated to have the defect in the negative selection of autoreactive T cells with the use of TCR transgenic mice expressing model self-Ag under the control of a tissue-specific promoter (Liston et al. 2003; Anderson et al. 2005). Using the same experimental system, it was demonstrated that the model self-Ag expressed from mTECs, including Aire[+] mTECs, could positively select the self-Ag-specific Tregs (Aschenbrenner et al. 2007; Mouri et al. 2017). Conversely, Aire-deficient mice showed a defect in the Treg production (Malchow et al. 2013; Yang et al. 2015). One critical question using these transgenic animals expressing model self-Ag in mTECs is whether the defective tolerance induction is due to the altered levels of model self-Ag expression in an Aire-dependent fashion. This important issue has not been consistent among the reports: mTECs from Aire-deficient mTECs showed reduced (Hubert et al. 2011) or unaltered expression level of model self-Ag (Anderson et al. 2005; Mouri et al. 2017). This argument was not confined to the TCR transgenic system. Instead, non-transgenic Aire-deficient mice developed autoimmunity against unaltered levels of self-Ags expressed by mTECs (Kuroda et al. 2005; Niki et al. 2006). Thus, it still remains to be determined whether the quantitative changes in self-Ag levels influenced by Aire can fully account for the autoimmune pathogenesis caused by Aire deficiency.

Of note, two studies pointed out that Aire regulates the process of the transfer of self-Ag expressed from mTECs to the BM-derived APCs using mice expressing OVA under the control of rat insulin-promoter (Hubert et al. 2011; Mouri et al. 2017). Consistent with this idea, analysis of TCR repertoire in mice deficient for Batf3, which is important for the production of CD103[+] DCs (CD8α[+] DCs) specialized for Ag transfer (Edelson et al. 2010), showed defective production of Tregs specific for Aire-dependent self-Ags (Perry et al. 2014). How Aire regulates Ag transfer process in an mTEC-dependent manner awaits further study. In this regard, it would be important to bear in mind that Aire[+] mTECs are not a particular subset of mTECs. Instead, Aire expression is inherent to all mTECs but may occur at the particular stage(s) and/or cellular states during their differentiation (Metzger et al. 2013; Kawano et al. 2015).

3.8.1.3   Molecular Action of Aire

Given that Aire controls the expression of wide varieties of TRAs, beyond the argument on the Aire-mediated autoimmune pathogenesis discussed above, it is important to understand how Aire regulates this process. There are two possible models to explain this phenomenon (Matsumoto et al. 2013). The current prevailing view is that Aire is directly involved in the transcriptional control of many TRA genes in mTECs by employing various transcriptional pathways. In other words, TRAs are considered to be the direct target genes of Aire's transcriptional activity (transcription model) (Org et al. 2009; Abramson et al. 2010; Zumer et al. 2013). Interestingly, transgenic expression of Aire in cTECs utilizing cTEC-specific β5t promoter did not induce TRA expression from cTECs (Nishijima et al. 2015). One possible reason for this might be that the (epi)genetic context of cTECs does not match that Aire finds in mTECs. As an example, Sirt1, which plays an important role in the deacetylation of Aire, is not expressed in cTECs (Chuprin et al. 2015).

The other model suggests that decreased TRA expression in Aire-deficient mice is secondary to the Aire-mediated mTEC differentiation program (maturation model) (Fig. 3.3) (Matsumoto et al. 2013). Detailed studies of Aire-deficient thymi revealed several important aspects of the Aire-dependent differentiation programs of mTECs, such as increased numbers of mTECs with a globular cell shape (Gillard et al. 2007; Yano et al. 2008) and reduced numbers of terminally differentiated mTECs expressing involucrin, the latter being associated with reduced numbers of

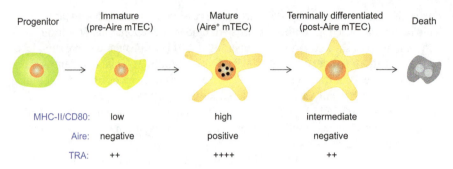

| Progenitor | Immature (pre-Aire mTEC) | | Mature (Aire+ mTEC) | | Terminally differentiated (post-Aire mTEC) | | Death |

MHC-II/CD80:   low   high   intermediate

Aire: negative   positive   negative

TRA: ++   ++++   ++

Fig. 3.3 Differentiation program of mTECs. mTECs develop from their progenitors, which are characterized by the expression of SSEA-1 and Cld3,4 (depicted in Fig. 3.2). Aire is not expressed at the immature stage (MHC-II/CD80^{low}), and it becomes detectable at the mature stage (MHC-II/CD80^{high}). Expression levels of TRAs are concomitantly highest at this stage. Before cell death, there is a stage in which Aire expression is terminated and their CD80 expression levels go down (Aire$^-$CD80^{intermediate}: post-Aire mTEC). Although expression levels of TRAs, in general, are low in post-Aire mTECs, keratinocyte-related genes such as *involucrin* become rather high at this stage. Thus, the spectrum of TRAs expressed differs depending on the maturation stage of mTECs

Hassall's corpuscles (Yano et al. 2008; Wang et al. 2012) (Fig. 3.3). An increased percentage of mTECs expressing high levels of CD80 (CD80^{high}) is another suggested aspect of the Aire-dependent mTEC differentiation program (Nishikawa et al. 2014). Given that acquisition of the properties of TRA gene expression depends on the maturation status of mTECs (Anderson et al. 2007), any defect in such an Aire-dependent maturation program could indirectly account for the defects of TRA gene expression in Aire-deficient mTECs. In this scenario, Aire-deficient mTECs would have defective TRA gene expression, because they are not fully differentiated to stage(s) where other undetermined transcriptional means for TRA gene expression beyond Aire become available and/or active. Although Aire is expressed by mature mTECs, the Aire$^+$ subset does not define a terminal differentiation stage, given the existence of post-Aire cells (Nishikawa et al. 2010) (Fig. 3.3).

The exact point, however, in the differentiation process at which Aire-deficient mTECs are prevented has not been clearly determined. When human AIRE protein was expressed under the MHC-II promoter (i.e., human AIRE expression at the timing of MHC-II expression), mTEC differentiation program was perturbed, and mTECs expressing human AIRE showed rather reduced TRA expression instead of augmented TRA expression (Nishijima et al. 2018). This observation suggests that human AIRE/Aire is not neutral for mTEC maturation program. Rather, Aire expression at an inappropriate timing and/or dose may perturb mTEC maturation. It goes without saying that precise elucidation of the target gene(s) relevant to the progression of mTEC differentiation controlled by Aire is an essential task to support the maturation model.

3.8.1.4 Alternative Functions of Aire That Goes Beyond PGE

Although many studies were conducted on the basis of positive regulation of TRA genes by Aire, one recent report suggested that Aire may function to negatively control the TRA expression. Aire was demonstrated to restrain the transcriptional duration and amplitude of TRAs by opposing the activity of the chromatin remodeler Brg1: accessibility at TRA gene loci was promoted by Brg1, but it was repressed by Aire, suggesting a dual role of Aire for TRA expression (Koh et al. 2018).

3.9 Thymic Involution

One of the hallmarks of the aged vertebrate immune system is the progressive deterioration of the thymic function, an evolutionary conserved physiological process termed "Thymic involution" (Shanley et al. 2009). Age-related thymic involution occurs earlier than other acknowledged features of aging, although with different onsets across species. For instance, whereas atrophy of the murine thymus was shown to commence around the time of sexual maturity (Sempowski et al. 2002; Gray et al. 2006), thymic involution in man begins from the first year of life (Steinmann et al. 1985). Nevertheless, in both mice and human, the major features of thymic atrophy include perturbations in TEC microenvironment, including a reduction in TEC cellularity, loss of cortex and medulla compartmentalization, appearance of areas devoid of epithelium and the expansion of adipose tissue and fibroblasts in perivascular space (Takeoka et al. 1996; Taub and Longo 2005; Gray et al. 2006; Gui et al. 2007; Aw et al. 2008; Aw et al. 2009; Yang et al. 2009). Albeit the aged thymus still presents a residual capacity to support T cell development (Hale et al. 2006), its regression is intimately linked to quantitative and qualitative abnormalities in the peripheral T cell compartment (Linton and Dorshkind 2004). Under normal circumstances, thymic involution is of minimal consequences to healthy individuals, but has been associated with an augmented incidence of diseases, such as autoimmunity, cancer or opportunistic infections, in the elderly or immune compromised individuals (Taub and Longo 2005; Chinn et al. 2012). Still, the precise mechanism(s) responsible for the atrophy of the thymus remains poorly understood. Over the years, several studies revealed multiple layers of complexity regarding the cellular and molecular mediators of thymic involution. Thymic atrophy is currently recognized to unfold in part due to intrinsic and extrinsic age-dependent alterations of the thymic stroma, in particular in TECs (Chinn et al. 2012). Aged TECs present a reduced expression of several genes that are crucial in controlling their function and maintenance (Ki et al. 2014). Specifically, the expression of Foxn1 was observed to diminish with age, concomitantly with the progressive expansion of Foxn1-negative cells within both cortical and medullary thymic epithelium (Corbeaux et al. 2010; Rode et al. 2015; O'Neill et al. 2016). Interestingly, evidence suggests that cortex-specific downregulation of Foxn1 may be the main driver of thymic involution as a result of an impaired expression of genes involved in cTEC-dependent T cell differentiation (O'Neill et al. 2016). Accordingly, attenuation of Foxn1 levels in the perinatal thymus leads to a premature atrophy (Chen et al. 2009). Hence, these observations suggest that both the onset and progression of thymic involution may be linked to a decline in Foxn1 expression, although the existence of other factors cannot be formally excluded.

The intrathymic microenvironment is characterized by a complex network of paracrine, autocrine and endocrine signals involving a number of soluble mediators (Hadden 1998), which may act as secondary agents in exacerbating the age-related decline of the thymus. Different studies revealed a pro-inflammatory signature in the microenvironment of aged thymus, which results in an increased expression of cytokines *Il1a*, *Il1b*, *Il6*, *Il12b*, *Il18*, and *Tnf* (Sempowski et al. 2000; Ki et al. 2014). Some of these cytokines were shown to have suppressive effect on the thymic function. In line with this, IL-1 and IL-6 administration to mice resulted in thymic involution (Morrissey et al. 1988; Sempowski et al. 2000). On the other hand, mice lacking the Nlrp3 inflammasome were protected from age-related thymic deterioration, due to the inability to produce active IL-1 (Youm et al. 2012). Among the different hormonal factors proposed to stimulate thymic involution are the sex steroids (Hince et al. 2008). The rapid decline in thymic function during puberty has been associated with a rise in sex hormone levels, which are known to exert profound inhibitory effects on immune functions (Grossman 1985; Velardi et al. 2015). This concept is consistent with observations that both thymocytes and TECs express sex steroid receptor and that the exogenous administration of sexual hormones caused a collapse in thymic function (Barr et al. 1982; Viselli et al. 1995; Rijhsinghani et al. 1996; Staples et al. 1999; Olsen et al. 2001; Lee et al. 2013). Interestingly, blockade of sex steroid production, either surgically or via luteinising hormone releasing hormone (LHRH) analogues, stimulates thymic rejuvenation in old mice (Greenstein et al. 1987; Heng et al. 2005; Sutherland et al. 2005). These alterations include the rebuilding of the TEC microenvironment, both at phenotypic, numerical and functional levels, and a boost in the thymopoietic capacity of the aged thymus to the levels observed in their young counterpart (Heng et al. 2005; Sutherland et al. 2005; Gray et al. 2006; Williams et al. 2008).

Besides aging, infection is an important factor that affects the process of thymic involution. Although increased endogenous corticosteroid activity during infection of microorganisms (i.e., bacteria, viruses, parasites, fungi) might account for the thymic involution, TEC intrinsic factor(s) also affects thymic involution. For example, Dicer-deficient TECs had higher expression of the IFN-γ receptor (Ifnar), which allowed suboptimal signals to trigger rapid loss of thymic cellularity during innate immune response (Papadopoulou et al. 2011). Among the miRNAs expressed in TECs, miR-29a showed highest expression predicted to have multiple interactions with Ifnar1 mRNA. Consistent with this finding, miR-29a-deficient mouse showed an inappropriate thymic involution due to the hypersensitivity against pathogen-associated molecular patterns (PAMP) (Papadopoulou et al. 2011).

3.10   Concluding Remarks

The molecular mechanisms underpinning TEC development and function have received considerable attention in the last decade. Apart of the fundamental importance, the therapeutic implications of several key findings, as highlighted above, are extremely relevant, because genetic defects that impair TEC function leads to immunodeficiency or autoimmunity. The challenges that lie ahead are to better elucidate the progenitor–product relationship that culminates with the development of fully functional cortical and medullary TECs remains. In addition, we need a more comprehensive understanding of how TECs impose immune tolerance to a myriad of peripheral-restricted antigens. Knowledge in this area will certainly open the door to a range of therapeutic interventions to regenerate and repair thymopoiesis under different pathological settings.

Acknowledgments   The European Research Council (ERC) under the EU's Horizon 2020 research and innovation program (grant agreement No 637843—TEC_Pro)—starting grant attributed to N.L.A—supports the studies from the laboratory of Nuno L. Alves. The studies from the laboratory of Mitsuru Matsumoto are supported by JSPS KAKENHI Grant Numbers JP16H06496 and JP16H05342, and by the Japan Agency for Medical Research and Development-Core Research for Evolutional Science and Technology. We apologize for not referring to all of the primary literature owing to the space limitations.

References

Abramson J, Anderson G (2017) Thymic epithelial cells. Annu Rev Immunol 35:85–118

Abramson J, Giraud M, Benoist C, Mathis D (2010) Aire's partners in the molecular control of immunological tolerance. Cell 140(1):123–135

Akiyama T, Maeda S, Yamane S, Ogino K, Kasai M, Kajiura F, Matsumoto M, Inoue J (2005) Dependence of self-tolerance on TRAF6-directed development of thymic stroma. Science 308(5719):248–251

Akiyama T, Shimo Y, Yanai H, Qin J, Ohshima D, Maruyama Y, Asaumi Y, Kitazawa J, Takayanagi H, Penninger JM, Matsumoto M, Nitta T, Takahama Y, Inoue J (2008) The tumor necrosis factor family receptors RANK and CD40 cooperatively establish the thymic medullary microenvironment and self-tolerance. Immunity 29(3):423–437

Alves NL, Huntington ND, Rodewald HR, Di Santo JP (2009a) Thymic epithelial cells: the multitasking framework of the T cell "cradle". Trends Immunol 30(10):468–474

Alves NL, Richard-Le Goff O, Huntington ND, Sousa AP, Ribeiro VS, Bordack A, Vives FL, Peduto L, Chidgey A, Cumano A, Boyd R, Eberl G, Di Santo JP (2009b) Characterization of the thymic IL-7 niche in vivo. Proc Natl Acad Sci U S A 106(5):1512–1517

Alves NL, Huntington ND, Mention JJ, Richard-Le Goff O, Di Santo JP (2010) Cutting edge: a thymocyte-thymic epithelial cell cross-talk dynamically regulates intrathymic IL-7 expression in vivo. J Immunol 184(11):5949–5953

Alves NL, Takahama Y, Ohigashi I, Ribeiro AR, Baik S, Anderson G, Jenkinson WE (2014) Serial progression of cortical and medullary thymic epithelial microenvironments. Eur J Immunol 44(1):16–22

Anderson G, Takahama Y (2012) Thymic epithelial cells: working class heroes for T cell development and repertoire selection. Trends Immunol 33(6):256–263

Anderson G, Jenkinson EJ, Moore NC, Owen JJ (1993) MHC class II-positive epithelium and mesenchyme cells are both required for T-cell development in the thymus. Nature 362(6415):70–73

Anderson G, Anderson KL, Tchilian EZ, Owen JJ, Jenkinson EJ (1997) Fibroblast dependency during early thymocyte development maps to the CD25+ CD44+ stage and involves interactions with fibroblast matrix molecules. Eur J Immunol 27(5):1200–1206

Anderson MS, Venanzi ES, Klein L, Chen Z, Berzins SP, Turley SJ, von Boehmer H, Bronson R, Dierich A, Benoist C, Mathis D (2002) Projection of an immunological self shadow within the thymus by the aire protein. Science 298(5597):1395–1401

Anderson MS, Venanzi ES, Chen Z, Berzins SP, Benoist C, Mathis D (2005) The cellular mechanism of aire control of T cell tolerance. Immunity 23(2):227–239

Anderson G, Lane PJ, Jenkinson EJ (2007) Generating intrathymic microenvironments to establish T-cell tolerance. Nat Rev Immunol 7(12):954–963

Aschenbrenner K, D'Cruz LM, Vollmann EH, Hinterberger M, Emmerich J, Swee LK, Rolink A, Klein L (2007) Selection of Foxp3+ regulatory T cells specific for self antigen expressed and presented by aire+ medullary thymic epithelial cells. Nat Immunol 8(4):351–358

Aw D, Silva AB, Maddick M, von Zglinicki T, Palmer DB (2008) Architectural changes in the thymus of aging mice. Aging Cell 7(2):158–167

Aw D, Taylor-Brown F, Cooper K, Palmer DB (2009) Phenotypical and morphological changes in the thymic microenvironment from ageing mice. Biogerontology 10(3):311–322

Baik S, Jenkinson EJ, Lane PJ, Anderson G, Jenkinson WE (2013) Generation of both cortical and aire(+) medullary thymic epithelial compartments from CD205(+) progenitors. Eur J Immunol 43(3):589–594

Baik S, Sekai M, Hamazaki Y, Jenkinson WE, Anderson G (2016) Relb acts downstream of medullary thymic epithelial stem cells and is essential for the emergence of RANK(+) medullary epithelial progenitors. Eur J Immunol 46(4):857–862

Balciunaite G, Keller MP, Balciunaite E, Piali L, Zuklys S, Mathieu YD, Gill J, Boyd R, Sussman DJ, Hollander GA (2002) Wnt glycoproteins regulate the expression of FoxN1, the gene defective in nude mice. Nat Immunol 3(11):1102–1108

Banwell CM, Partington KM, Jenkinson EJ, Anderson G (2000) Studies on the role of IL-7 presentation by mesenchymal fibroblasts during early thymocyte development. Eur J Immunol 30(8):2125–2129

Barr IG, Khalid BA, Pearce P, Toh BH, Bartlett PF, Scollay RG, Funder JW (1982) Dihydrotestosterone and estradiol deplete corticosensitive thymocytes lacking in receptors for these hormones. J Immunol 128(6):2825–2828

Bennett AR, Farley A, Blair NF, Gordon J, Sharp L, Blackburn CC (2002) Identification and characterization of thymic epithelial progenitor cells. Immunity 16(6):803–814

Blackburn CC, Manley NR (2004) Developing a new paradigm for thymus organogenesis. Nat Rev Immunol 4(4):278–289

Blackburn CC, Augustine CL, Li R, Harvey RP, Malin MA, Boyd RL, Miller JF, Morahan G (1996) The nu gene acts cell-autonomously and is required for differentiation of thymic epithelial progenitors. Proc Natl Acad Sci U S A 93(12):5742–5746

Bleul CC, Boehm T (2000) Chemokines define distinct microenvironments in the developing thymus. Eur J Immunol 30(12):3371–3379

Bleul CC, Corbeaux T, Reuter A, Fisch P, Monting JS, Boehm T (2006) Formation of a functional thymus initiated by a postnatal epithelial progenitor cell. Nature 441(7096):992–996

Boehm T, Scheu S, Pfeffer K, Bleul CC (2003) Thymic medullary epithelial cell differentiation, thymocyte emigration, and the control of autoimmunity require lympho-epithelial cross talk via LT{beta}R. J Exp Med 198(5):757–769

Boyd RL, Tucek CL, Godfrey DI, Izon DJ, Wilson TJ, Davidson NJ, Bean AG, Ladyman HM, Ritter MA, Hugo P (1993) The thymic microenvironment. Immunol Today 14(9):445–459

Bredenkamp N, Nowell CS, Blackburn CC (2014a) Regeneration of the aged thymus by a single transcription factor. Development 141(8):1627–1637

Bredenkamp N, Ulyanchenko S, O'Neill KE, Manley NR, Vaidya HJ, Blackburn CC (2014b) An organized and functional thymus generated from FOXN1-reprogrammed fibroblasts. Nat Cell Biol 16(9):902–908

Buono M, Facchini R, Matsuoka S, Thongjuea S, Waithe D, Luis TC, Giustacchini A, Besmer P, Mead AJ, Jacobsen SE, Nerlov C (2016) A dynamic niche provides Kit ligand in a stage-specific manner to the earliest thymocyte progenitors. Nat Cell Biol 18(2):157–167

Calderon L, Boehm T (2011) Three chemokine receptors cooperatively regulate homing of hematopoietic progenitors to the embryonic mouse thymus. Proc Natl Acad Sci U S A 108(18):7517–7522

Chen L, Xiao S, Manley NR (2009) Foxn1 is required to maintain the postnatal thymic microenvironment in a dosage-sensitive manner. Blood 113(3):567–574

Chinn IK, Blackburn CC, Manley NR, Sempowski GD (2012) Changes in primary lymphoid organs with aging. Semin Immunol 24(5):309–320

Chuprin A, Avin A, Goldfarb Y, Herzig Y, Levi B, Jacob A, Sela A, Katz S, Grossman M, Guyon C, Rathaus M, Cohen HY, Sagi I, Giraud M, McBurney MW, Husebye ES, Abramson J (2015) The deacetylase Sirt1 is an essential regulator of aire-mediated induction of central immunological tolerance. Nat Immunol 16(7):737–745

Ciofani M, Zuniga-Pflucker JC (2007) The thymus as an inductive site for T lymphopoiesis. Annu Rev Cell Dev Biol 23:463–493

Consortium TF-GA (1997) An autoimmune disease, APECED, caused by mutations in a novel gene featuring two PHD-type zinc-finger domains. Nat Genet 17(4):399–403

Corbeaux T, Hess I, Swann JB, Kanzler B, Haas-Assenbaum A, Boehm T (2010) Thymopoiesis in mice depends on a Foxn1-positive thymic epithelial cell lineage. Proc Natl Acad Sci U S A 107(38):16613–16618

De Togni P, Goellner J, Ruddle NH, Streeter PR, Fick A, Mariathasan S, Smith SC, Carlson R, Shornick LP, Strauss-Schoenberger J et al (1994) Abnormal development of peripheral lymphoid organs in mice deficient in lymphotoxin. Science 264(5159):703–707

Derbinski J, Schulte A, Kyewski B, Klein L (2001) Promiscuous gene expression in medullary thymic epithelial cells mirrors the peripheral self. Nat Immunol 2(11):1032–1039

Edelson BT, Kc W, Juang R, Kohyama M, Benoit LA, Klekotka PA, Moon C, Albring JC, Ise W, Michael DG, Bhattacharya D, Stappenbeck TS, Holtzman MJ, Sung SS, Murphy TL, Hildner K, Murphy KM (2010) Peripheral CD103+ dendritic cells form a unified subset developmentally related to CD8alpha+ conventional dendritic cells. J Exp Med 207(4):823–836

Esashi E, Sekiguchi T, Ito H, Koyasu S, Miyajima A (2003) Cutting edge: a possible role for CD4+ thymic macrophages as professional scavengers of apoptotic thymocytes. J Immunol 171(6):2773–2777

van Ewijk W, Shores EW, Singer A (1994) Crosstalk in the mouse thymus. Immunol Today 15(5):214–217

van Ewijk W, Hollander G, Terhorst C, Wang B (2000) Stepwise development of thymic microenvironments in vivo is regulated by thymocyte subsets. Development 127(8):1583–1591

Farr AG, Anderson SK (1985) Epithelial heterogeneity in the murine thymus: fucose-specific lectins bind medullary epithelial cells. J Immunol 134(5):2971–2977

Fiorini E, Ferrero I, Merck E, Favre S, Pierres M, Luther SA, MacDonald HR (2008) Cutting edge: thymic crosstalk regulates delta-like 4 expression on cortical epithelial cells. J Immunol 181(12):8199–8203

Garfin PM, Min D, Bryson JL, Serwold T, Edris B, Blackburn CC, Richie ER, Weinberg KI, Manley NR, Sage J, Viatour P (2013) Inactivation of the RB family prevents thymus involution and promotes thymic function by direct control of Foxn1 expression. J Exp Med 210(6):1087–1097

Gill J, Malin M, Hollander GA, Boyd R (2002) Generation of a complete thymic microenvironment by MTS24(+) thymic epithelial cells. Nat Immunol 3(7):635–642

Gillard GO, Dooley J, Erickson M, Peltonen L, Farr AG (2007) Aire-dependent alterations in medullary thymic epithelium indicate a role for aire in thymic epithelial differentiation. J Immunol 178(5):3007–3015

Godfrey DI, Izon DJ, Tucek CL, Wilson TJ, Boyd RL (1990) The phenotypic heterogeneity of mouse thymic stromal cells. Immunology 70(1):66–74

Gommeaux J, Gregoire C, Nguessan P, Richelme M, Malissen M, Guerder S, Malissen B, Carrier A (2009) Thymus-specific serine protease regulates positive selection of a subset of CD4+ thymocytes. Eur J Immunol 39(4):956–964

Gordon J, Manley NR (2011) Mechanisms of thymus organogenesis and morphogenesis. Development 138(18):3865–3878

Gordon J, Bennett AR, Blackburn CC, Manley NR (2001) Gcm2 and Foxn1 mark early para-thyroid- and thymus-specific domains in the developing third pharyngeal pouch. Mech Dev 103(1-2):141–143

Gossens K, Naus S, Corbel SY, Lin S, Rossi FM, Kast J, Ziltener HJ (2009) Thymic progeni-tor homing and lymphocyte homeostasis are linked via S1P-controlled expression of thymic P-selectin/CCL25. J Exp Med 206(4):761–778

Gray DH, Seach N, Ueno T, Milton MK, Liston A, Lew AM, Goodnow CC, Boyd RL (2006) Developmental kinetics, turnover, and stimulatory capacity of thymic epithelial cells. Blood 108(12):3777–3785

Greenstein BD, Fitzpatrick FT, Kendall MD, Wheeler MJ (1987) Regeneration of the thymus in old male rats treated with a stable analogue of LHRH. J Endocrinol 112(3):345–350

Grossman CJ (1985) Interactions between the gonadal steroids and the immune system. Science 227(4684):257–261

Gui J, Zhu X, Dohkan J, Cheng L, Barnes PF, Su DM (2007) The aged thymus shows normal recruitment of lymphohematopoietic progenitors but has defects in thymic epithelial cells. Int Immunol 19(10):1201–1211

Hadden JW (1998) Thymic endocrinology. Ann N Y Acad Sci 840:352–358

Hale JS, Boursalian TE, Turk GL, Fink PJ (2006) Thymic output in aged mice. Proc Natl Acad Sci U S A 103(22):8447–8452

Haljasorg U, Dooley J, Laan M, Kisand K, Bichele R, Liston A, Peterson P (2017) Irf4 expres-sion in thymic epithelium is critical for thymic regulatory T cell homeostasis. J Immunol 198(5):1952–1960

Hamazaki Y, Fujita H, Kobayashi T, Choi Y, Scott HS, Matsumoto M, Minato N (2007) Medullary thymic epithelial cells expressing aire represent a unique lineage derived from cells expressing claudin. Nat Immunol 8(3):304–311

Hamazaki Y, Sekai M, Minato N (2016) Medullary thymic epithelial stem cells: role in thymic epithelial cell maintenance and thymic involution. Immunol Rev 271(1):38–55

Hanahan D (1998) Peripheral-antigen-expressing cells in thymic medulla: factors in self-tolerance and autoimmunity. Curr Opin Immunol 10(6):656–662

Heng TS, Goldberg GL, Gray DH, Sutherland JS, Chidgey AP, Boyd RL (2005) Effects of cas-tration on thymocyte development in two different models of thymic involution. J Immunol 175(5):2982–2993

Hince M, Sakkal S, Vlahos K, Dudakov J, Boyd R, Chidgey A (2008) The role of sex steroids and gonadectomy in the control of thymic involution. Cell Immunol 252(1-2):122–138

Hofmann J, Mair F, Greter M, Schmidt-Supprian M, Becher B (2011) NIK signaling in dendritic cells but not in T cells is required for the development of effector T cells and cell-mediated immune responses. J Exp Med 208(9):1917–1929

Hubert FX, Kinkel SA, Davey GM, Phipson B, Mueller SN, Liston A, Proietto AI, Cannon PZ, Forehan S, Smyth GK, Wu L, Goodnow CC, Carbone FR, Scott HS, Heath WR (2011) Aire regulates the transfer of antigen from mTECs to dendritic cells for induction of thymic toler-ance. Blood 118(9):2462–2472

Jain R, Sheridan JM, Policheni A, Heinlein M, Gandolfo LC, Dewson G, Smyth GK, Sansom SN, Fu NY, Visvader JE, Hollander GA, Strasser A, Gray DHD (2017) A critical epithelial survival axis regulated by MCL-1 maintains thymic function in mice. Blood 130(23):2504–2515

Jenkinson WE, Rossi SW, Parnell SM, Jenkinson EJ, Anderson G (2007) PDGFRalpha-expressing mesenchyme regulates thymus growth and the availability of intrathymic niches. Blood 109(3):954–960

Kajiura F, Sun S, Nomura T, Izumi K, Ueno T, Bando Y, Kuroda N, Han H, Li Y, Matsushima A, Takahama Y, Sakaguchi S, Mitani T, Matsumoto M (2004) NF-kappaB-inducing kinase establishes self-tolerance in a thymic stroma-dependent manner. J Immunol 172(4):2067–2075

Kawano H, Nishijima H, Morimoto J, Hirota F, Morita R, Mouri Y, Nishioka Y, Matsumoto M (2015) Aire expression is inherent to most medullary thymic epithelial cells during their differentiation program. J Immunol 195(11):5149–5158

Ki S, Park D, Selden HJ, Seita J, Chung H, Kim J, Iyer VR, Ehrlich LIR (2014) Global transcriptional profiling reveals distinct functions of thymic stromal subsets and age-related changes during thymic involution. Cell Rep 9(1):402–415

Kinoshita D, Hirota F, Kaisho T, Kasai M, Izumi K, Bando Y, Mouri Y, Matsushima A, Niki S, Han H, Oshikawa K, Kuroda N, Maegawa M, Irahara M, Takeda K, Akira S, Matsumoto M (2006) Essential role of IkappaB kinase alpha in thymic organogenesis required for the establishment of self-tolerance. J Immunol 176(7):3995–4002

Klein L, Hinterberger M, Wirnsberger G, Kyewski B (2009) Antigen presentation in the thymus for positive selection and central tolerance induction. Nat Rev Immunol 9(12):833–844

Klein L, Kyewski B, Allen PM, Hogquist KA (2014) Positive and negative selection of the T cell repertoire: what thymocytes see (and don't see). Nat Rev Immunol 14(6):377–391

Klug DB, Carter C, Crouch E, Roop D, Conti CJ, Richie ER (1998) Interdependence of cortical thymic epithelial cell differentiation and T-lineage commitment. Proc Natl Acad Sci U S A 95(20):11822–11827

Klug DB, Carter C, Gimenez-Conti IB, Richie ER (2002) Cutting edge: thymocyte-independent and thymocyte-dependent phases of epithelial patterning in the fetal thymus. J Immunol 169(6):2842–2845

Koch U, Fiorini E, Benedito R, Besseyrias V, Schuster-Gossler K, Pierres M, Manley NR, Duarte A, Macdonald HR, Radtke F (2008) Delta-like 4 is the essential, nonredundant ligand for Notch1 during thymic T cell lineage commitment. J Exp Med 205(11):2515–2523

Koh AS, Miller EL, Buenrostro JD, Moskowitz DM, Wang J, Greenleaf WJ, Chang HY, Crabtree GR (2018) Rapid chromatin repression by aire provides precise control of immune tolerance. Nat Immunol 19(2):162–172

Krueger A, Willenzon S, Lyszkiewicz M, Kremmer E, Forster R (2010) CC chemokine receptor 7 and 9 double-deficient hematopoietic progenitors are severely impaired in seeding the adult thymus. Blood 115(10):1906–1912

Kuroda N, Mitani T, Takeda N, Ishimaru N, Arakaki R, Hayashi Y, Bando Y, Izumi K, Takahashi T, Nomura T, Sakaguchi S, Ueno T, Takahama Y, Uchida D, Sun S, Kajiura F, Mouri Y, Han H, Matsushima A, Yamada G, Matsumoto M (2005) Development of autoimmunity against transcriptionally unrepressed target antigen in the thymus of aire-deficient mice. J Immunol 174(4):1862–1870

Kyewski B, Klein L (2006) A central role for central tolerance. Annu Rev Immunol 24:571–606

Lee H, Kim H, Chung Y, Kim J, Yang H (2013) Thymocyte differentiation is regulated by a change in estradiol levels during the estrous cycle in mouse. Dev Reprod 17(4):441–449

Liang Z, Zhang L, Su H, Luan R, Na N, Sun L, Zhao Y, Zhang X, Zhang Q, Li J, Zhang L, Zhao Y (2018) MTOR signaling is essential for the development of thymic epithelial cells and the induction of central immune tolerance. Autophagy 14(3):505–517

Linton PJ, Dorshkind K (2004) Age-related changes in lymphocyte development and function. Nat Immunol 5(2):133–139

Liston A, Lesage S, Wilson J, Peltonen L, Goodnow CC (2003) Aire regulates negative selection of organ-specific T cells. Nat Immunol 4(4):350–354

Liu C, Saito F, Liu Z, Lei Y, Uehara S, Love P, Lipp M, Kondo S, Manley N, Takahama Y (2006) Coordination between CCR7- and CCR9-mediated chemokine signals in prevascular fetal thymus colonization. Blood 108(8):2531–2539

Liu Z, Yu S, Manley NR (2007) Gcm2 is required for the differentiation and survival of parathyroid precursor cells in the parathyroid/thymus primordia. Dev Biol 305(1):333–346

Lomada D, Jain M, Bolner M, Reeh KA, Kang R, Reddy MC, DiGiovanni J, Richie ER (2016) Stat3 signaling promotes survival and maintenance of medullary thymic epithelial cells. PLoS Genet 12(1):e1005777

Love PE, Bhandoola A (2011) Signal integration and crosstalk during thymocyte migration and emigration. Nat Rev Immunol 11(7):469–477

Malchow S, Leventhal DS, Nishi S, Fischer BI, Shen L, Paner GP, Amit AS, Kang C, Geddes JE, Allison JP, Socci ND, Savage PA (2013) Aire-dependent thymic development of tumor-associated regulatory T cells. Science 339(6124):1219–1224

Manley NR (2000) Thymus organogenesis and molecular mechanisms of thymic epithelial cell differentiation. Semin Immunol 12(5):421–428

Matsumoto M (2007) Transcriptional regulation in thymic epithelial cells for the establishment of self tolerance. Arch Immunol Ther Exp (Warsz) 55:27–34

Matsumoto M, Fu YX, Molina H, Chaplin DD (1997) Lymphotoxin-alpha-deficient and TNF receptor-I-deficient mice define developmental and functional characteristics of germinal centers. Immunol Rev 156:137–144

Matsumoto M, Iwamasa K, Rennert PD, Yamada T, Suzuki R, Matsushima A, Okabe M, Fujita S, Yokoyama M (1999) Involvement of distinct cellular compartments in the abnormal lymphoid organogenesis in lymphotoxin-alpha-deficient mice and alymphoplasia (aly) mice defined by the chimeric analysis. J Immunol 163(3):1584–1591

Matsumoto M, Nishikawa Y, Nishijima H, Morimoto J, Matsumoto M, Mouri Y (2013) Which model better fits the role of aire in the establishment of self-tolerance: the transcription model or the maturation model? Front Immunol 4:210

Matsushima A, Kaisho T, Rennert PD, Nakano H, Kurosawa K, Uchida D, Takeda K, Akira S, Matsumoto M (2001) Essential role of nuclear factor (NF)-kappaB-inducing kinase and inhibitor of kappaB (IkappaB) kinase alpha in NF-kappaB activation through lymphotoxin beta receptor, but not through tumor necrosis factor receptor I. J Exp Med 193(5):631–636

Mayer CE, Zuklys S, Zhanybekova S, Ohigashi I, Teh HY, Sansom SN, Shikama-Dorn N, Hafen K, Macaulay IC, Deadman ME, Ponting CP, Takahama Y, Hollander GA (2016) Dynamic spatio-temporal contribution of single beta5t+ cortical epithelial precursors to the thymus medulla. Eur J Immunol 46(4):846–856

Meireles C, Ribeiro AR, Pinto RD, Leitao C, Rodrigues PM, Alves NL (2017) Thymic crosstalk restrains the pool of cortical thymic epithelial cells with progenitor properties. Eur J Immunol 47(6):958–969

Metzger TC, Khan IS, Gardner JM, Mouchess ML, Johannes KP, Krawisz AK, Skrzypczynska KM, Anderson MS (2013) Lineage tracing and cell ablation identify a post-aire-expressing thymic epithelial cell population. Cell Rep 5(1):166–179

Miller JFAP (1961) Imunological function of the thymus. Lancet 2(7205):748–749

Miller JF (2006) Vestigial no more. Nat Immunol 7(1):3–5

Moore TA, von Freeden-Jeffry U, Murray R, Zlotnik A (1996) Inhibition of gamma delta T cell development and early thymocyte maturation in IL-7 -/- mice. J Immunol 157(6):2366–2373

Morrissey PJ, Charrier K, Alpert A, Bressler L (1988) In vivo administration of IL-1 induces thymic hypoplasia and increased levels of serum corticosterone. J Immunol 141(5):1456–1463

Mouri Y, Yano M, Shinzawa M, Shimo Y, Hirota F, Nishikawa Y, Nii T, Kiyonari H, Abe T, Uehara H, Izumi K, Tamada K, Chen L, Penninger JM, Inoue J, Akiyama T, Matsumoto M (2011) Lymphotoxin signal promotes thymic organogenesis by eliciting RANK expression in the embryonic thymic stroma. J Immunol 186(9):5047–5057

Mouri Y, Nishijima H, Kawano H, Hirota F, Sakaguchi N, Morimoto J, Matsumoto M (2014) NF-kappaB-inducing kinase in thymic stroma establishes central tolerance by orchestrating cross-talk with not only thymocytes but also dendritic cells. J Immunol 193(9):4356–4367

Mouri Y, Ueda Y, Yamano T, Matsumoto M, Tsuneyama K, Kinashi T, Matsumoto M (2017) Mode of tolerance induction and requirement for aire are governed by the cell types that express self-antigen and those that present antigen. J Immunol 199(12):3959–3971

Murata S, Sasaki K, Kishimoto T, Niwa S, Hayashi H, Takahama Y, Tanaka K (2007) Regulation of CD8+ T cell development by thymus-specific proteasomes. Science 316(5829):1349–1353

Nagamine K, Peterson P, Scott HS, Kudoh J, Minoshima S, Heino M, Krohn KJ, Lalioti MD, Mullis PE, Antonarakis SE, Kawasaki K, Asakawa S, Ito F, Shimizu N (1997) Positional cloning of the APECED gene. Nat Genet 17(4):393–398

Nakagawa T, Roth W, Wong P, Nelson A, Farr A, Deussing J, Villadangos JA, Ploegh H, Peters C, Rudensky AY (1998) Cathepsin L: critical role in Ii degradation and CD4 T cell selection in the thymus. Science 280(5362):450–453

Nedjic J, Aichinger M, Emmerich J, Mizushima N, Klein L (2008) Autophagy in thymic epithelium shapes the T-cell repertoire and is essential for tolerance. Nature 455(7211):396–400

Nehls M, Kyewski B, Messerle M, Waldschutz R, Schuddekopf K, Smith AJ, Boehm T (1996) Two genetically separable steps in the differentiation of thymic epithelium. Science 272(5263):886–889

Niki S, Oshikawa K, Mouri Y, Hirota F, Matsushima A, Yano M, Han H, Bando Y, Izumi K, Matsumoto M, Nakayama KI, Kuroda N, Matsumoto M (2006) Alteration of intra-pancreatic target-organ specificity by abrogation of aire in NOD mice. J Clin Invest 116(5):1292–1301

Nishijima H, Kitano S, Miyachi H, Morimoto J, Kawano H, Hirota F, Morita R, Mouri Y, Masuda K, Imoto I, Ikuta K, Matsumoto M (2015) Ectopic aire expression in the thymic cortex reveals inherent properties of aire as a tolerogenic factor within the medulla. J Immunol 195(10):4641–4649

Nishijima H, Kajimoto T, Matsuoka Y, Mouri Y, Morimoto J, Matsumoto M, Kawano H, Nishioka Y, Uehara H, Izumi K, Tsuneyama K, Okazaki IM, Okazaki T, Hosomichi K, Shiraki A, Shibutani M, Mitsumori K, Matsumoto M (2018) Paradoxical development of polymyositis-like autoimmunity through augmented expression of autoimmune regulator (AIRE). J Autoimmun 86:75–92

Nishikawa Y, Hirota F, Yano M, Kitajima H, Miyazaki J, Kawamoto H, Mouri Y, Matsumoto M (2010) Biphasic aire expression in early embryos and in medullary thymic epithelial cells before end-stage terminal differentiation. J Exp Med 207(5):963–971

Nishikawa Y, Nishijima H, Matsumoto M, Morimoto J, Hirota F, Takahashi S, Luche H, Fehling HJ, Mouri Y, Matsumoto M (2014) Temporal lineage tracing of aire-expressing cells reveals a requirement for aire in their maturation program. J Immunol 192(6):2585–2592

Nitta T, Murata S, Sasaki K, Fujii H, Ripen AM, Ishimaru N, Koyasu S, Tanaka K, Takahama Y (2010) Thymoproteasome shapes immunocompetent repertoire of CD8+ T cells. Immunity 32(1):29–40

O'Neill KE, Bredenkamp N, Tischner C, Vaidya HJ, Stenhouse FH, Peddie CD, Nowell CS, Gaskell T, Blackburn CC (2016) Foxn1 is dynamically regulated in thymic epithelial cells during embryogenesis and at the onset of thymic involution. PLoS One 11(3):e0151666

Ohigashi I, Zuklys S, Sakata M, Mayer CE, Zhanybekova S, Murata S, Tanaka K, Hollander GA, Takahama Y (2013) Aire-expressing thymic medullary epithelial cells originate from beta5t-expressing progenitor cells. Proc Natl Acad Sci U S A 110(24):9885–9890

Ohigashi I, Zuklys S, Sakata M, Mayer CE, Hamazaki Y, Minato N, Hollander GA, Takahama Y (2015) Adult thymic medullary epithelium is maintained and regenerated by lineage-restricted cells rather than bipotent progenitors. Cell Rep 13(7):1432–1443

Ohigashi I, Kozai M, Takahama Y (2016) Development and developmental potential of cortical thymic epithelial cells. Immunol Rev 271(1):10–22

Olsen NJ, Olson G, Viselli SM, Gu X, Kovacs WJ (2001) Androgen receptors in thymic epithelium modulate thymus size and thymocyte development. Endocrinology 142(3):1278–1283

Onder L, Nindl V, Scandella E, Chai Q, Cheng HW, Caviezel-Firner S, Novkovic M, Bomze D, Maier R, Mair F, Ledermann B, Becher B, Waisman A, Ludewig B (2015) Alternative NF-kappaB signaling regulates mTEC differentiation from podoplanin-expressing precursors in the cortico-medullary junction. Eur J Immunol 45(8):2218–2231

Org T, Rebane A, Kisand K, Laan M, Haljasorg U, Andreson R, Peterson P (2009) AIRE activated tissue specific genes have histone modifications associated with inactive chromatin. Hum Mol Genet 18(24):4699–4710

Otero DC, Baker DP, David M (2013) IRF7-dependent IFN-beta production in response to RANKL promotes medullary thymic epithelial cell development. J Immunol 190(7):3289–3298

Papadopoulou AS, Dooley J, Linterman MA, Pierson W, Ucar O, Kyewski B, Zuklys S, Hollander GA, Matthys P, Gray DH, De Strooper B, Liston A (2011) The thymic epithelial microRNA network elevates the threshold for infection-associated thymic involution via miR-29a mediated suppression of the IFN-alpha receptor. Nat Immunol 13(2):181–187

Perry JS, Lio CW, Kau AL, Nutsch K, Yang Z, Gordon JI, Murphy KM, Hsieh CS (2014) Distinct contributions of aire and antigen-presenting-cell subsets to the generation of self-tolerance in the thymus. Immunity 41(3):414–426

Petrie HT, Zuniga-Pflucker JC (2007) Zoned out: functional mapping of stromal signaling microenvironments in the thymus. Annu Rev Immunol 25:649–679

Plotkin J, Prockop SE, Lepique A, Petrie HT (2003) Critical role for CXCR4 signaling in progenitor localization and T cell differentiation in the postnatal thymus. J Immunol 171(9):4521–4527

Proietto AI, van Dommelen S, Zhou P, Rizzitelli A, D'Amico A, Steptoe RJ, Naik SH, Lahoud MH, Liu Y, Zheng P, Shortman K, Wu L (2008) Dendritic cells in the thymus contribute to T-regulatory cell induction. Proc Natl Acad Sci U S A 105(50):19869–19874

Revest JM, Suniara RK, Kerr K, Owen JJ, Dickson C (2001) Development of the thymus requires signaling through the fibroblast growth factor receptor R2-IIIb. J Immunol 167(4):1954–1961

Ribeiro AR, Rodrigues PM, Meireles C, Di Santo JP, Alves NL (2013) Thymocyte selection regulates the homeostasis of IL-7-expressing thymic cortical epithelial cells in vivo. J Immunol 191(3):1200–1209

Ribeiro AR, Meireles C, Rodrigues PM, Alves NL (2014) Intermediate expression of CCRL1 reveals novel subpopulations of medullary thymic epithelial cells that emerge in the postnatal thymus. Eur J Immunol 44(10):2918–2924

Riemann M, Andreas N, Fedoseeva M, Meier E, Weih D, Freytag H, Schmidt-Ullrich R, Klein U, Wang ZQ, Weih F (2017) Central immune tolerance depends on crosstalk between the classical and alternative NF-kappaB pathways in medullary thymic epithelial cells. J Autoimmun 81:56–67

Rijhsinghani AG, Thompson K, Bhatia SK, Waldschmidt TJ (1996) Estrogen blocks early T cell development in the thymus. Am J Reprod Immunol 36(5):269–277

Roberts NA, White AJ, Jenkinson WE, Turchinovich G, Nakamura K, Withers DR, McConnell FM, Desanti GE, Benezech C, Parnell SM, Cunningham AF, Paolino M, Penninger JM, Simon AK, Nitta T, Ohigashi I, Takahama Y, Caamano JH, Hayday AC, Lane PJ, Jenkinson EJ, Anderson G (2012) Rank signaling links the development of invariant gammadelta T cell progenitors and aire(+) medullary epithelium. Immunity 36(3):427–437

Rode I, Martins VC, Kublbeck G, Maltry N, Tessmer C, Rodewald HR (2015) Foxn1 protein expression in the developing, aging, and regenerating thymus. J Immunol 195(12):5678–5687

Rodewald HR, Paul S, Haller C, Bluethmann H, Blum C (2001) Thymus medulla consisting of epithelial islets each derived from a single progenitor. Nature 414(6865):763–768

Rodrigues PM, Ribeiro AR, Perrod C, Landry JJM, Araujo L, Pereira-Castro I, Benes V, Moreira A, Xavier-Ferreira H, Meireles C, Alves NL (2017) Thymic epithelial cells require p53 to support their long-term function in thymopoiesis in mice. Blood 130(4):478–488

Romano R, Palamaro L, Fusco A, Giardino G, Gallo V, Del Vecchio L, Pignata C (2013) FOXN1: a master regulator gene of thymic epithelial development program. Front Immunol 4:187

Ropke C, Van Soest P, Platenburg PP, Van Ewijk W (1995) A common stem cell for murine cortical and medullary thymic epithelial cells? Dev Immunol 4(2):149–156

Rossi FM, Corbel SY, Merzaban JS, Carlow DA, Gossens K, Duenas J, So L, Yi L, Ziltener HJ (2005) Recruitment of adult thymic progenitors is regulated by P-selectin and its ligand PSGL-1. Nat Immunol 6(6):626–634

Rossi SW, Jenkinson WE, Anderson G, Jenkinson EJ (2006) Clonal analysis reveals a common progenitor for thymic cortical and medullary epithelium. Nature 441(7096):988–991

Rossi SW, Chidgey AP, Parnell SM, Jenkinson WE, Scott HS, Boyd RL, Jenkinson EJ, Anderson G (2007a) Redefining epithelial progenitor potential in the developing thymus. Eur J Immunol 37(9):2411–2418

Rossi SW, Kim MY, Leibbrandt A, Parnell SM, Jenkinson WE, Glanville SH, McConnell FM, Scott HS, Penninger JM, Jenkinson EJ, Lane PJ, Anderson G (2007b) RANK signals from

CD4(+)3(-) inducer cells regulate development of aire-expressing epithelial cells in the thymic medulla. J Exp Med 204(6):1267–1272

Rossi SW, Jeker LT, Ueno T, Kuse S, Keller MP, Zuklys S, Gudkov AV, Takahama Y, Krenger W, Blazar BR, Hollander GA (2007c) Keratinocyte growth factor (KGF) enhances postnatal T-cell development via enhancements in proliferation and function of thymic epithelial cells. Blood 109(9):3803–3811

Sakaguchi S (2004) Naturally arising CD4+ regulatory t cells for immunologic self-tolerance and negative control of immune responses. Annu Rev Immunol 22:531–562

Sano S, Takahama Y, Sugawara T, Kosaka H, Itami S, Yoshikawa K, Miyazaki J, van Ewijk W, Takeda J (2001) Stat3 in thymic epithelial cells is essential for postnatal maintenance of thymic architecture and thymocyte survival. Immunity 15(2):261–273

Satoh R, Kakugawa K, Yasuda T, Yoshida H, Sibilia M, Katsura Y, Levi B, Abramson J, Koseki Y, Koseki H, van Ewijk W, Hollander GA, Kawamoto H (2016) Requirement of Stat3 signaling in the postnatal development of thymic medullary epithelial cells. PLoS Genet 12(1):e1005776

Schuddekopf K, Schorpp M, Boehm T (1996) The whn transcription factor encoded by the nude locus contains an evolutionarily conserved and functionally indispensable activation domain. Proc Natl Acad Sci U S A 93(18):9661–9664

Schuster C, Gerold KD, Schober K, Probst L, Boerner K, Kim MJ, Ruckdeschel A, Serwold T, Kissler S (2015) The autoimmunity-associated gene CLEC16A modulates thymic epithelial cell autophagy and alters T cell selection. Immunity 42(5):942–952

Scimone ML, Aifantis I, Apostolou I, von Boehmer H, von Andrian UH (2006) A multistep adhesion cascade for lymphoid progenitor cell homing to the thymus. Proc Natl Acad Sci U S A 103(18):7006–7011

Sekai M, Hamazaki Y, Minato N (2014) Medullary thymic epithelial stem cells maintain a functional thymus to ensure lifelong central T cell tolerance. Immunity 41(5):753–761

Sempowski GD, Hale LP, Sundy JS, Massey JM, Koup RA, Douek DC, Patel DD, Haynes BF (2000) Leukemia inhibitory factor, oncostatin M, IL-6, and stem cell factor mRNA expression in human thymus increases with age and is associated with thymic atrophy. J Immunol 164(4):2180–2187

Sempowski GD, Gooding ME, Liao HX, Le PT, Haynes BF (2002) T cell receptor excision circle assessment of thymopoiesis in aging mice. Mol Immunol 38(11):841–848

Senoo M, Pinto F, Crum CP, McKeon F (2007) p63 Is essential for the proliferative potential of stem cells in stratified epithelia. Cell 129(3):523–536

Shakib S, Desanti GE, Jenkinson WE, Parnell SM, Jenkinson EJ, Anderson G (2009) Checkpoints in the development of thymic cortical epithelial cells. J Immunol 182(1):130–137

Shanley DP, Aw D, Manley NR, Palmer DB (2009) An evolutionary perspective on the mechanisms of immunosenescence. Trends Immunol 30(7):374–381

Shinkura R, Kitada K, Matsuda F, Tashiro K, Ikuta K, Suzuki M, Kogishi K, Serikawa T, Honjo T (1999) Alymphoplasia is caused by a point mutation in the mouse gene encoding Nf-kappa b-inducing kinase. Nat Genet 22(1):74–77

Sitnik KM, Kotarsky K, White AJ, Jenkinson WE, Anderson G, Agace WW (2012) Mesenchymal cells regulate retinoic acid receptor-dependent cortical thymic epithelial cell homeostasis. J Immunol 188(10):4801–4809

Smith KM, Olson DC, Hirose R, Hanahan D (1997) Pancreatic gene expression in rare cells of thymic medulla: evidence for functional contribution to T cell tolerance. Int Immunol 9(9):1355–1365

Soleimanpour SA, Gupta A, Bakay M, Ferrari AM, Groff DN, Fadista J, Spruce LA, Kushner JA, Groop L, Seeholzer SH, Kaufman BA, Hakonarson H, Stoffers DA (2014) The diabetes susceptibility gene Clec16a regulates mitophagy. Cell 157(7):1577–1590

Soza-Ried C, Bleul CC, Schorpp M, Boehm T (2008) Maintenance of thymic epithelial phenotype requires extrinsic signals in mouse and zebrafish. J Immunol 181(8):5272–5277

Staples JE, Gasiewicz TA, Fiore NC, Lubahn DB, Korach KS, Silverstone AE (1999) Estrogen receptor alpha is necessary in thymic development and estradiol-induced thymic alterations. J Immunol 163(8):4168–4174

Steinmann GG, Klaus B, Muller-Hermelink HK (1985) The involution of the ageing human thymic epithelium is independent of puberty. A morphometric study. Scand J Immunol 22(5):563–575

Sukseree S, Mildner M, Rossiter H, Pammer J, Zhang CF, Watanapokasin R, Tschachler E, Eckhart L (2012) Autophagy in the thymic epithelium is dispensable for the development of self-tolerance in a novel mouse model. PLoS One 7(6):e38933

Sutherland JS, Goldberg GL, Hammett MV, Uldrich AP, Berzins SP, Heng TS, Blazar BR, Millar JL, Malin MA, Chidgey AP, Boyd RL (2005) Activation of thymic regeneration in mice and humans following androgen blockade. J Immunol 175(4):2741–2753

Takada K, Van Laethem F, Xing Y, Akane K, Suzuki H, Murata S, Tanaka K, Jameson SC, Singer A, Takahama Y (2015) TCR affinity for thymoproteasome-dependent positively selecting peptides conditions antigen responsiveness in CD8(+) T cells. Nat Immunol 16(10):1069–1076

Takada K, Kondo K, Takahama Y (2017) Generation of peptides that promote positive selection in the thymus. J Immunol 198(6):2215–2222

Takahama Y (2006) Journey through the thymus: stromal guides for T-cell development and selection. Nat Rev Immunol 6(2):127–135

Takeoka Y, Chen SY, Yago H, Boyd R, Suehiro S, Shultz LD, Ansari AA, Gershwin ME (1996) The murine thymic microenvironment: changes with age. Int Arch Allergy Immunol 111(1):5–12

Taub DD, Longo DL (2005) Insights into thymic aging and regeneration. Immunol Rev 205:72–93

Tsai PT, Lee RA, Wu H (2003) BMP4 acts upstream of FGF in modulating thymic stroma and regulating thymopoiesis. Blood 102(12):3947–3953

Ulyanchenko S, O'Neill KE, Medley T, Farley AM, Vaidya HJ, Cook AM, Blair NF, Blackburn CC (2016) Identification of a bipotent epithelial progenitor population in the adult thymus. Cell Rep 14(12):2819–2832

Van Vliet E, Melis M, Van Ewijk W (1984) Monoclonal antibodies to stromal cell types of the mouse thymus. Eur J Immunol 14(6):524–529

Velardi E, Dudakov JA, van den Brink MR (2015) Sex steroid ablation: an immunoregenerative strategy for immunocompromised patients. Bone Marrow Transplant 50(Suppl 2):S77–S81

Viselli SM, Olsen NJ, Shults K, Steizer G, Kovacs WJ (1995) Immunochemical and flow cytometric analysis of androgen receptor expression in thymocytes. Mol Cell Endocrinol 109(1):19–26

Wang X, Laan M, Bichele R, Kisand K, Scott HS, Peterson P (2012) Post-aire maturation of thymic medullary epithelial cells involves selective expression of keratinocyte-specific autoantigens. Front Immunol 3(March):19

Williams KM, Lucas PJ, Bare CV, Wang J, Chu YW, Tayler E, Kapoor V, Gress RE (2008) CCL25 increases thymopoiesis after androgen withdrawal. Blood 112(8):3255–3263

Wong K, Lister NL, Barsanti M, Lim JM, Hammett MV, Khong DM, Siatskas C, Gray DH, Boyd RL, Chidgey AP (2014) Multilineage potential and self-renewal define an epithelial progenitor cell population in the adult thymus. Cell Rep 8(4):1198–1209

Yamano T, Nedjic J, Hinterberger M, Steinert M, Koser S, Pinto S, Gerdes N, Lutgens E, Ishimaru N, Busslinger M, Brors B, Kyewski B, Klein L (2015) Thymic B cells are licensed to present self antigens for central T cell tolerance induction. Immunity 42(6):1048–1061

Yang H, Youm YH, Sun Y, Rim JS, Galban CJ, Vandanmagsar B, Dixit VD (2009) Axin expression in thymic stromal cells contributes to an age-related increase in thymic adiposity and is associated with reduced thymopoiesis independently of ghrelin signaling. J Leukoc Biol 85(6):928–938

Yang S, Fujikado N, Kolodin D, Benoist C, Mathis D (2015) Immune tolerance. Regulatory T cells generated early in life play a distinct role in maintaining self-tolerance. Science 348(6234):589–594

Yano M, Kuroda N, Han H, Meguro-Horike M, Nishikawa Y, Kiyonari H, Maemura K, Yanagawa Y, Obata K, Takahashi S, Ikawa T, Satoh R, Kawamoto H, Mouri Y, Matsumoto M (2008) Aire controls the differentiation program of thymic epithelial cells in the medulla for the establishment of self-tolerance. J Exp Med 205(12):2827–2838

Youm YH, Kanneganti TD, Vandanmagsar B, Zhu X, Ravussin A, Adijiang A, Owen JS, Thomas MJ, Francis J, Parks JS, Dixit VD (2012) The Nlrp3 inflammasome promotes age-related thymic demise and immunosenescence. Cell Rep 1(1):56–68

Zhu M, Chin RK, Christiansen PA, Lo JC, Liu X, Ware C, Siebenlist U, Fu YX (2006) NF-kappaB2 is required for the establishment of central tolerance through an aire-dependent pathway. J Clin Invest 116(11):2964–2971

Zlotoff DA, Sambandam A, Logan TD, Bell JJ, Schwarz BA, Bhandoola A (2010) CCR7 and CCR9 together recruit hematopoietic progenitors to the adult thymus. Blood 115(10):1897–1905

Zook EC, Krishack PA, Zhang S, Zeleznik-Le NJ, Firulli AB, Witte PL, Le PT (2011) Overexpression of Foxn1 attenuates age-associated thymic involution and prevents the expansion of peripheral CD4 memory T cells. Blood 118(22):5723–5731

Zuklys S, Mayer CE, Zhanybekova S, Stefanski HE, Nusspaumer G, Gill J, Barthlott T, Chappaz S, Nitta T, Dooley J, Nogales-Cadenas R, Takahama Y, Finke D, Liston A, Blazar BR, Pascual-Montano A, Hollander GA (2012) MicroRNAs control the maintenance of thymic epithelia and their competence for T lineage commitment and thymocyte selection. J Immunol 189(8):3894–3904

Zuklys S, Handel A, Zhanybekova S, Govani F, Keller M, Maio S, Mayer CE, Teh HY, Hafen K, Gallone G, Barthlott T, Ponting CP, Hollander GA (2016) Foxn1 regulates key target genes essential for T cell development in postnatal thymic epithelial cells. Nat Immunol 17(10):1206–1215

Zumer K, Saksela K, Peterlin BM (2013) The mechanism of tissue-restricted antigen gene expression by AIRE. J Immunol 190(6):2479–2482

Chapter 4
T-Cell Development: From T-Lineage Specification to Intrathymic Maturation

Kogulan Yoganathan, Edward L. Y. Chen, Jastaranpreet Singh, and Juan Carlos Zúñiga-Pflücker

Abstract T-cell development occurs in the thymus in both mice and humans. Upon entry into the thymus, bone marrow-derived blood-borne progenitors receive instructive signals, including Notch signaling, to extinguish their potential to develop into alternative immune lineages while committing to the T-cell fate. Upon T-lineage commitment, developing T-cells receive further instructional cues to generate different T-cell sublineages, which together possess diverse immunological functions to provide host immunity. Over the years, numerous studies have contributed to a greater understanding of key thymic signals that govern T-cell differentiation and subset generation. Here, we review these critical signaling factors that govern the different stages of both mouse and human T-cell development, while also focusing on the transcriptional changes that mediate T-cell identity and diversity.

4.1 Overview of Mouse T-Cell Development

The development of mouse T-cells can be characterized into distinct stages by the expression of key cell surface markers, as outlined in Fig. 4.1 (upper panels). Bone marrow (BM)-derived blood-borne thymus-seeding progenitors (TSPs) enter the thymus at the cortical medullary junction (CMJ) and lack surface expression of the CD4 and CD8 coreceptors and are termed CD4$^-$CD8$^-$ double negative (DN) cells, which can be further subclassified. Early T-cell progenitors (ETPs), express the receptor tyrosine kinase cKit (SCFR or CD117) and the hyaluronic acid receptor (CD44), but lack surface expression of the IL-2Rα chain (CD25) (Rothenberg et al. 2016). The CD44$^+$CD25$^-$ population is termed DN1 cells. As the ETPs progress and migrate through the thymic cortex towards the subcapsular zone (SCZ) of the thymus, they

Kogulan Yoganathan and Edward L. Y. Chen contributed equally to this work.

K. Yoganathan · E. L. Y. Chen · J. Singh · J. C. Zúñiga-Pflücker (✉)
Department of Immunology, Sunnybrook Research Institute, University of Toronto, Toronto, ON, Canada
e-mail: jczp@sri.utoronto.ca

© Springer Nature Switzerland AG 2019
G. A. Passos (ed.), *Thymus Transcriptome and Cell Biology*,
https://doi.org/10.1007/978-3-030-12040-5_4

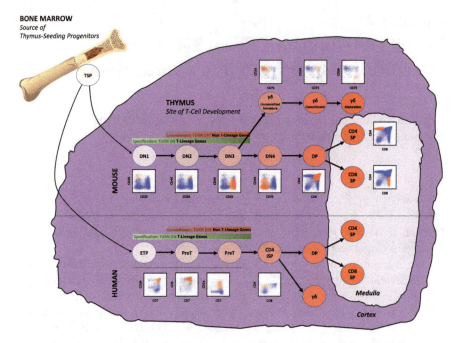

Fig. 4.1 Overview of mouse and human T-cell development. Thymus-seeding progenitors (TSPs) from the bone marrow migrate to the thymus and undergo differentiation and maturation through developmental stages that can be broadly characterized by cell surface markers, leading to the production of mature T-cells. In mice, TSPs enter the thymus and are characterized as double negative (DN)1 (CD44$^+$CD25$^-$) cells that primarily enter the T-lineage, although they still retain developmental potential for alternative immune cell fates. As DN1 cells turn on genes conducive for the T-cell lineage in a process called specification, they become DN2 (CD44$^+$CD25$^+$) cells. The loss of cell surface expression of CD44 marks DN3 (CD44$^-$CD25$^+$) cells. At this stage, access to alternative lineage genes is lost and cells become committed to the T-lineage. It is at the DN3 stage that developing T-cells bifurcate to adopt either the αβ- or the γδ-lineage. DN3 cells that successfully rearrange their TCRβ chains progress through αβ-lineage development to the DN4 (CD44$^-$CD25$^-$) stage and upregulate surface co-receptors CD4 and CD8 and are termed double positive (DP; CD4$^+$CD8$^+$) cells. DP cells rearrange their TCRα chains and undergo positive and negative selection in the thymic cortex and medulla, respectively, subsequently maturing into CD4$^+$ or CD8$^+$ single positive (SP) cells that exit the thymus and migrate to the periphery. DN3 cells that rearrange their TCRγ and TCRδ chains progress through γδ-lineage development as uncommitted, immature γδ T-cells (CD24$^+$CD73$^-$). Upon strong ligand engagement, they undergo commitment to the γδ-lineage (CD24$^+$CD73$^+$), and finally exit the thymus as mature γδ T-cells (CD24$^-$CD73$^+$). In humans, CD34$^+$ TSPs enter the thymus as ETPs and pass through distinct developmental stages. The early stages of T-cell development are defined by a loss of CD34 expression, and upregulation of CD7 and CD5 expression as progenitors progress through the progenitor T (proT)-cell stages. During these stages, progenitors are specified towards the human T-lineage. The subsequent expression of CD1a marks commitment to the T-lineage and cells are termed precursor T (preT)-cells. In contrast to mouse T-cell development, human T-lineage cells transition through a CD4$^+$ ISP stage, rather than a CD8$^+$ ISP stage, where they bifurcate to the αβ- or γδ-lineage. For αβ T-cell development, CD4 ISPs transition to the DP stage. Based on insights in the mouse, it is thought that positive and negative selection subsequently ensue for the generation of CD4$^+$ and CD8$^+$ SP cells

upregulate CD25 on their surface and become CD44$^+$CD25$^+$ DN2 cells. At this stage, they undergo proliferation to expand the developing T-cell progenitor pool (Petrie and Zuniga-Pflucker 2007; Rothenberg et al. 2016). These DN2 cells can be further divided into DN2a (CD117^{++}) and DN2b (CD117$^+$) cells based on a slight decrease in CD117 surface expression at the DN2b stage as cells become committed to the T-lineage and lose the ability to differentiate into the natural killer (NK) and myeloid fates (Rothenberg et al. 2008, 2016). As developing T-cells begin to express the recombination activating genes 1 or 2 (*Rag1* or *Rag2*) and rearrange their T-cell receptor (TCR) β locus (or their TCRγ and TCRδ loci to become γδTCR$^+$ T-cells through another process of differentiation), they slow in their proliferation and downregulate surface expression of CD44 and CD117 to become CD44$^-$CD25$^+$ DN3 cells (Rothenberg et al. 2008). *β-selection* is an important developmental checkpoint that takes place near the SCZ where DN3 cells that have successfully rearranged their TCRβ loci (CD27^{lo} CD44$^-$ CD25$^+$ DN3a cells) express the TCRβ chain and pair with the invariant pre-Tα chain to assemble at the cell surface along with CD3 chains (Michie and Zuniga-Pflucker 2002; Rothenberg et al. 2008, 2016). Those DN3a cells that undergo pre-TCR signaling have successfully passed β-selection and can now undergo a proliferative expansion as they progress towards the inner cortex, while differentiating sequentially into DN3b (CD27^{hi} CD44$^-$ CD25$^+$) and DN4 (CD44$^-$CD25$^-$) cells, and subsequently CD8$^+$ (CD3$^-$TCRβ$^-$) immature single positive (ISP) cells and CD4$^+$CD8$^+$ double positive (DP) cells (Rothenberg et al. 2008, 2016). DP thymocytes arrest in proliferation and re-express *Rag1* and *Rag2* genes to enable rearrangement of the TCRα locus. Successful rearrangement of the TCRα chain leads to cell surface expression of the αβTCR and subsequent *positive selection* of a productively rearranged αβTCR that can recognize self-antigen in the context of major histocompatibility complex (MHC) class I or II. αβTCR$^+$ T-cells that are positively selected on MHC-I or MHC-II express either a CD8 or a CD4 coreceptor, respectively. These positively selected thymocytes pass through the medulla as CD4$^+$ or CD8$^+$ single-positive (SP) cells where removal of potentially autoreactive T-cells occurs in a process called *negative selection*. During the stages of positive and negative selection, a series of alternate differentiation processes can yield invariant natural killer T-cells (iNKTs), mucosal-associated invariant T-cells (MAITs) and regulatory T-cell (Treg) subsets, respectively. Following selection, mature T-cells exit the thymus for circulation into the periphery.

4.1.1 Pre-Thymic T-Cell Development: Settling on Thymus-Seeding Progenitors

4.1.1.1 Identity of the Thymus-Seeding Progenitor

Prior to any intrathymic event, initiation of postnatal T-cell development requires constant supply of BM-derived TSPs (Foss et al. 2001), as the thymus itself does not typically contain its own self-renewing source of these cells (Goldschneider et al. 1986). In the adult thymus, the area spanning the CMJ provides key signals for the recruitment of TSPs, such as CCL25, CCL21, and P-Selectin, while in turn TSPs express CCR9, CCR7, and PSGL-1, respectively (Rossi et al. 2005; Uehara et al. 2002; Ueno et al. 2002). Upon entry to the thymus, TSPs become qualified as ETPs, and upon engagement of Notch ligands, proliferate and specify to the T-cell lineage (Petrie and Zuniga-Pflucker 2007). It is estimated that as few as ten TSPs enter the thymus per day, making the phenotypic characterization of these cells extremely difficult (Wallis et al. 1975). Over the course of trying to decipher the identity of the TSP, several potential candidates included cells within the Lineage$^-$Sca1^{hi}cKit^{hi} (LSK) compartment, which is comprised of hematopoietic stem cells (HSCs; Flt3$^-$), multipotent progenitors (MPPs; Flt3^{lo}), and lymphoid-primed multipotent progenitors (LMPPs; Flt3^{hi}) (Schwarz and Bhandoola 2004). In addition, common lymphoid progenitors (CLPs; Lineage$^-$Sca1^{lo}cKit^{lo}Flt 3^{hi}CD127$^+$) have also been described (Kondo et al. 1997). Therefore, numerous potential candidates exist for the mouse TSP, even as further experimentations over the years have narrowed the identity to more specific subpopulations.

Although all four progenitor subsets are capable of giving rise to T-cells, only Flt3-expressing progenitors (MPPs, LMPPs, and CLPs) have thymus-seeding capacity (Schwarz et al. 2007). Indeed, Flt3$^-$ HSCs are only able to give rise to a T-cell population in the thymus when injected intrathymically (Schwarz et al. 2007). In addition, only LMPPs and CLPs have expression of CCR7, and among the LMPPs and CLPs that are CCR7$^+$, a rare subset are also CCR9$^+$ (Krueger et al. 2010). As efficient thymic-homing is reliant on dual expression of these two chemokine receptors (Krueger et al. 2010), it is now currently understood that at steady-state conditions, TSPs lie within the LMPP/CLP compartment. However, the current debate is whether TSPs arise from LMPP/CLP exclusively or from both. ETPs possess strong T-lineage potential but still contain potential for alternative lineages including B, NK, dendritic cell (DC), and myeloid lineage fates (Petrie and Zuniga-Pflucker 2007). Thus, the general consensus that ETPs contain myeloid potential would argue against the notion of a lymphoid-exclusive progenitor being the source of TSPs. In addition, ETPs possess high CD117 expression, which is consistent with an LMPP, but not a CLP, phenotype (Porritt et al. 2004). However, a study by Serwold *et al.* showed that ETPs can be isolated from the thymus 70% of the time when mice were injected with CLPs, but only 10% of the time when mice were injected with LMPPs, indicating that CLPs are more efficient at thymus-seeding (Serwold et al. 2009).

Although still inconclusive, work by Ramond *et al.* may bring us closer to resolving this debate. Their study demonstrated that the mouse fetal thymus is colonized by two distinct progenitor types during two different temporal periods (Ramond et al. 2014). They argued that from embryonic day (E)12 to E15, the thymus is seeded by cells that resemble CLPs, but from E16 onwards, it is seeded by cells that resemble LMPPs. ETPs taken from the earlier period were not able to differentiate into myeloid cells when cultured *in vitro*, but those taken from the later period had the potential for myeloid differentiation. In addition, ETPs from the later period were found to have higher expression levels of Sca1 but lower expression levels of CD127 (IL-7Rα), a phenotype consistent with LMPPs. CLP-like cells based on cell surface marker expression can be detected in the fetal blood by E13. However, by E18 only LMPP-like cells were present while CLP-like cells were almost completely absent. Thus, based on these experimental results, it would seem that CLPs are the source of TSPs during early fetal T-cell development, and they are subsequently replaced by LMPPs during late fetal, postnatal, and adult time periods. This finding is corroborated by Schwarz *et al.* showing that in adult mice, LMPPs are detected in the blood while CLPs are not. Importantly, these circulating LMPPs had the capacity to efficiently give rise to T-cells in the thymus (Schwarz and Bhandoola 2004).

4.1.1.2   Notch Signaling and T-Lineage Specification of Thymus-Seeding Progenitors

T-cell specification and commitment occurs in the thymus and the requirement for Notch during these intrathymic processes has been well established (Hozumi et al. 2008; Petrie and Zuniga-Pflucker 2007; Radtke et al. 1999). However, just as the answer to the identity of the physiological TSP has become an elusive one, it has also been difficult to answer whether T-cell priming occurs in a pre-thymic setting, and if so, whether Notch signaling plays a role during this process. It is well understood that ETPs, in addition to having strong T-lineage potential, lack B-lineage potential (Porritt et al. 2004). Since B-lineage repression is sensitive to Notch signals (Pui et al. 1999; Wilson et al. 2001), it would appear that ETPs receive these Notch-instructive signals pre-thymically or immediately upon their entry into the thymus. Numerous studies have also implicated Notch in being important for the generation/maintenance of ETPs, as mouse models with perturbations in Notch signaling show a near diminished presence of ETPs in the thymus (Sambandam et al. 2005; Tan et al. 2005). However, it remains unclear whether this is due to a requirement for Notch to act pre-thymically for the generation and/or recruitment of TSPs, or whether Notch is required to promote their survival intrathymically. Taken together, there is a clear indication that Notch is responsible for enforcing the T- versus B-lineage outcomes of ETPs, as well as for maintaining their numbers in the thymus, but the anatomical location of this effect is an ongoing debate. Nonetheless, there is convincing evidence that Notch acts on fetal liver TSPs (Ramond et al. 2014), but there are conflicting reports on whether this is true for BM TSPs (De Obaldia et al. 2013; Lai and Kondo 2007; Perry et al. 2004; Sambandam et al. 2005; Weber et al. 2011; Yu et al. 2015).

In the same study by Ramond *et al.* described earlier, the authors also compared the expression levels of Notch target genes from TSPs isolated during the early versus late waves of fetal T-lymphopoiesis (Ramond et al. 2014). They noted that TSPs from the first wave had higher levels of *Hey1*, *Deltex1*, and *Nrarp* compared to TSPs from the second wave. In addition, TSPs from the first wave already showed up-regulated expression of T-lineage specific genes, such as those encoding for CD3γ and pre-Tα. Consistent with this observation, the first wave TSPs showed faster T-cell differentiation kinetics compared to TSPs from the second wave. Thus, it would seem that Notch signaling plays a pre-thymic role during early fetal T-lymphopoiesis, but not during late fetal T-cell development and onwards. Indeed, later studies confirmed that adult BM LMPPs and CLPs do not express the Notch-target genes *Hes1* and *Tcf7*, while thymic ETPs have very high levels of both genes (De Obaldia et al. 2013; Weber et al. 2011). Furthermore, work by Sambandam *et al.* described the near absence of ETPs in the thymus when Notch signaling was inhibited, but they did not observe perturbations in BM or blood LSKs (which contain LMPPs) (Sambandam et al. 2005). Thus, their study concluded that Notch is needed only after thymic entry of TSPs. However, what these studies may have failed to consider is that true LMPPs/CLPs that seed the thymus are rare and constitute a small proportion of the heterogeneous progenitor population studied. Thus, analysis of a more refined population within these BM subsets could have revealed additional insights.

A study by Lai *et al.* showed that VCAM-1⁻CCR9⁺ cells can define a population of BM LMPPs that are released into the blood, proposing a more defined TSP candidate (Lai and Kondo 2007). When analyzing this population specifically, the authors found that these cells expressed higher levels of *Notch1* compared to VCAM-1⁻CCR9⁻ or VCAM-1⁺ LMPPs, indicating that they have an increased capacity to undergo Notch signaling. Consistent with this notion, VCAM-1⁻CCR9⁺ LMPPs do indeed express higher levels of *Hes1*, *Deltex1*, *Gata3*, and *Il2ra*. Another study described CD62L as a useful marker to narrow down a specific subset of BM LMPPs that could represent a true TSP (Perry et al. 2004). CD62L⁺ LMPPs, as compared to CD62L⁻ cells, were faster at thymus engraftment and their colonization pattern was consistent with the expected TSP activity. The CD62L⁺ compartment was comprised of cells that showed a bias towards the T-cell fate, and reduced B-cell and myeloid potentials, all indicative of active Notch signaling in these cells. Work done by Yu *et al.* supports this notion, as conditional deletion of Delta-like 4 (Dll4) in BM osteoblasts showed a reduction in TSPs (specifically defined as Ly6D⁻ CLPs in their study), and consequently a reduction in thymic ETPs (Yu et al. 2015). In summary, TSPs may undergo Notch signaling in the BM, but it remains unclear whether Notch is solely required for T-cell priming of TSPs prior to thymic entry, or whether it is needed for their generation altogether.

4.1.2   Intrathymic T-Cell Development: Early Stages

4.1.2.1   Early T-Cell Progenitor Maintenance and Retention of T-Lineage Potential

Among the DN1 cell population in the thymus, those that are CD117^{hi}CD24$^{-/lo}$ (termed DN1a/b, respectively) possess the most efficient capacity to develop into T-cells, with proliferation and differentiation kinetics consistent with a T-cell progenitor, and thus are regarded as containing true ETPs (Porritt et al. 2004). Once they enter the thymus via the CMJ, it is estimated that ETPs spend approximately

10 days in this area prior to their differentiation into DN2 cells (Petrie and Zuniga-Pflucker 2007). During these 10 days, ETPs proliferate and expand approximately 1000-fold in response to CD117 signaling. It appears that CD117 establishes the required starting number of cells destined for T-lymphopoiesis, as mutations in CD117 show T-cell development but are compromised with reductions in cellularity. The reduction in T-cell numbers is also accompanied by abrogated thymic architecture and generation of a less diverse TCR repertoire (Rodewald et al. 1997).

In addition to survival and expansion, ETPs constantly undergo Notch signaling and this may be important for them to retain strong T-lineage potential (Petrie and Zuniga-Pflucker 2007; Schmitt et al. 2004). Although ETPs are not yet fully committed to the T-cell fate, repression of alternative lineages must be maintained to ensure efficient T-lymphopoiesis. One of the many important roles of Notch during the ETP stage is to constrain the critical regulator of myeloid development, *Cebpa* (De Obaldia et al. 2013). Compared to BM progenitor cells, ETPs have increased *Hes1* expression and this coincides with reduced *Cebpa* expression. *Hes1* itself was shown to bind to the *Cebpa* promoter to mediate its transcriptional repression. Within the ETP population, cells with lower *Cebpa* expression gave rise to T-cells more efficiently compared to those with higher expression. Additionally, Notch function in ETPs was also shown to limit their DC potential (Feyerabend et al. 2009). Using a reporter mouse, Feyerabend *et al.* showed that ETPs that were deleted for *Notch1* had a greater propensity to develop into all DC lineages (lymphoid, myeloid, and plasmacytoid DCs) in the thymus compared to control ETPs. Thus, although having distinct functions in either maintaining the ETP pool or retaining T-cell bias in ETPs, CD117 and Notch signaling act in concert during this early stage to provide the most optimal T-cell developmental outcome.

4.1.2.2   Enabling Early T-Cell Progenitors for T-Cell Differentiation

Although ETP cell division is important, eventually they must make the decision to differentiate to continue along the T-cell developmental pathway. Regulation of PU.1 (encoded by *Spi1*), which ETPs inherit from their fetal liver or BM precursors, is critical during this process (Rothenberg et al. 2010). Among thymocytes, PU.1 expression levels are highest in ETP and DN2a cells and thus may serve an important role prior to T-cell commitment (Anderson et al. 1999). PU.1 expression in BM precursors is required for their T-lineage potential, yet must be eventually turned off for full T-cell differentiation to ensue (Anderson et al. 2002a; Scott et al. 1994). A key study by Champhekar *et al.* revealed insight as to why stage-specific regulation of PU.1 is important for the process of T-lymphopoiesis (Champhekar et al. 2015). Compared to control ETPs, those deleted for *Spi1* showed less proliferation when placed on OP9-DL1 cultures. However, *Spi1*-deleted ETPs did show a faster progression toward the T-lineage, as marked by a rapid transition to DN2 and DN3 stages. The accelerated T-cell developmental kinetics by *Spi1*-deficient ETPs is further supported by the fact that these cells showed up-regulated expression of Notch-target genes and T-lineage associated genes, including *Hes1*, *Deltex1*, *Nrarp* and *HEBCan*, *HEBAlt*, *Tcf7*, respectively.

Since PU.1 can antagonize Notch signaling in ETPs, it would seem that the balance between these two factors dictates ETP cell cycling versus transition towards the T-lineage. Notch signaling itself does not directly down-regulate PU.1 expression in ETPs (Del Real and Rothenberg 2013). Therefore, Notch must override the effects of PU.1, or perhaps weaken its activity to tip the balance towards T-lineage progression. By fractioning ETPs into Flt3$^+$ (early) and Flt3$^-$ (late) subsets, one group observed that Flt3$^-$ cells progressed toward the DN2 stage faster than Flt3$^+$ cells (Manesso et al. 2013). This suggested that as ETPs remain longer in the thymus and are thus exposed to more Notch signals, they are better poised for T-cell differentiation. In summary, PU.1 sets a threshold requirement for Notch activation, and could possibly explain why low levels of Notch signaling are sufficient to inhibit B and myeloid cell fates, but higher levels of Notch are needed for induction of T-cell development (Pui et al. 1999; Wilson et al. 2001).

4.1.3 Intrathymic Notch Signaling Reinforces T-Cell Specification

T-cell development initiates in full when BM-derived TSPs enter the thymus and encounter the cues delivered by the thymic microenvironment. These TSPs enter in waves and colonize the thymus depending on niche availability, possibly through coordinated feedback events by the thymus and BM that synchronize release of these progenitors (Donskoy et al. 2003). Once TSPs enter the thymus at the CMJ (now termed ETPs), they encounter the Notch ligand Dll4 that is expressed on the surface of thymic epithelial cells (TECs) and initiate Notch signaling, a highly conserved signaling pathway that is required for the development of many organ systems with great importance in T-cell development (Liu et al. 2010; Mohtashami et al. 2010). ETPs encounter Dll4 ligands through their Notch1 receptors that results in the cleavage of the Notch receptor on both extracellular and intracellular regions, ultimately resulting in the release of the intracellular domain of Notch (ICN), which translocates into the nucleus and interacts with a transcription factor called CSL (CBF1/Suppressor of hairless/Lag-1), or the mouse equivalent RBPJκ (recombination binding protein joining-kappa) (Andersson et al. 2011). For Notch signaling to initiate T-cell specification (the turning ON of T-lineage genes), there must be a recruitment of transcriptional coactivators to the complex (Kovall 2007). The importance of Notch in inducing T-cell development cannot be overstated, as coculturing HSCs with the OP9 BM stromal cell line ectopically expressing Delta-like Notch ligands (OP9-DL) is sufficient to induce their differentiation into T-cells (Holmes and Zuniga-Pflucker 2009; Schmitt and Zuniga-Pflucker 2002). The ability to differentiate T-cells *in vitro* has provided an accelerated understanding of T-cell developmental biology and may have clinical consequences in fields of engineering antigen-specific T-cells and regenerative medicine.

4.1.4 Induction of T-Lineage Genes: Specification and Commitment

Once Notch signaling is initiated in ETPs there is a shift in the gene regulatory network from stem-like and progenitor genes, which support the proliferation of ETPs, to Notch-induced T-cell specification genes (Yui and Rothenberg 2014). Notch signaling induces a cascade of genes responsible for the specification of progenitors into the T-lineage and have been discussed in various reviews (Ciofani and Zuniga-Pflucker 2007; Rothenberg 2014; Rothenberg et al. 2008, 2016; Thompson and Zuniga-Pflucker 2011; Yui and Rothenberg 2014). Nevertheless, there are several key transcription factors induced by the Notch signaling that play instrumental roles in specifying ETPs towards the T-lineage, and generating committed T-cells. These transcription factors will be the focus of this section and highlighted further.

4.1.4.1 Notch-Target Genes

Notch signaling induces the expression of various genes including but not limited to *Il2ra, Deltex1, Ptcra, Nrarp, Notch3, Notch1*, and transcription factors *cMyc, Hes1*, and the E protein HEBAlt (*TCF12*-Alt) (Georgescu et al. 2008; Wang et al. 2006; Yui and Rothenberg 2014). The induced genes all have distinct and important roles throughout specification. The expression of *cMyc* fosters the proliferative burst ETPs undergo as they progress through T-lineage specification (Weng et al. 2006). *Nrarp* is a factor that negatively regulates Notch1 signaling and may help to dampen the Notch signaling pathway once the T-lineage program is initiated (Yun and Bevan 2003). *Nrarp* expression is downregulated once T-cells become committed and may represent an increased need for Notch signaling to induce key T-lineage genes once progenitors are committed to the T-lineage program (Yui and Rothenberg 2014).

Commitment is a crucial process during differentiation as it signifies the achievement of a robust cellular state in which a cell secures its fate. In the case of developing T-lineage cells, this means absolute repression of alternative immune lineages such as the myeloid, DC, and other innate lymphoid fates. Notch signaling induces several transcriptional regulators that function to shut off access to various alternative lineage genes, and thus commits developing T-cells into the T-lineage (Yui and Rothenberg 2014). The basic helix loop helix (bHLH) repressor *Hes1* is induced by Notch signaling until the DN3a stage and is important for repressing the myeloid fate in early developing T-cells by repressing *Cebpa*, a factor that is necessary for the development of myeloid and DCs (Welner et al. 2013; Zhang et al. 1997). Notch signaling also induces the expression of several other critical regulators of T-cell specification and commitment such as *Tcf7, Gata3, Bcl11b* and E proteins, which are discussed in detail below.

4.1.4.2   TCF1

Notch signaling directly activates *Tcf7* (T-cell factor 1; TCF1), which is important for the earliest specification of ETPs into developing T-lineage cells by upregulating genes involved in the survival and proliferation of ETPs (Germar et al. 2011). Overexpression of TCF1 in BM progenitors induces T-lineage development by the upregulation of key T-lineage genes *Gata3* and *Bcl11b* in the absence of Notch signals (Weber et al. 2011). Although TCF1 is not sufficient to activate all Notch-induced genes, it can activate *Gata3*, *Bcl11b*, *Il2ra*, *CD3g*, *Lat*, *Lck*, and endogenous *Tcf7* without Notch signals (Weber et al. 2011). TCF1 also appears not to be dependent on Notch for its expression once it is activated (Del Real and Rothenberg 2013), and given that it is able to upregulate itself, this suggests a positive feedback loop in which TCF1 is able to maintain its expression and the T-lineage developmental program once its expression is initiated. However, TCF1 can be expressed independently of Notch in the later stages of T-cell development and its expression is associated with GATA3 expression levels.

4.1.4.3   GATA3

GATA3 (GATA-binding protein 3) is a zinc-finger transcription factor that is induced by Notch signaling and has important roles in survival, specification and commitment of developing T-cells in a dose-dependent manner (Ho et al. 2009; Yui and Rothenberg 2014). GATA3 collaborates with Notch1 in committing ETPs into the T-lineage by repressing B-cell potential and creating a transcriptional landscape that is required for thymocyte differentiation throughout the early T-cell stages (Garcia-Ojeda et al. 2013). Notch signals are not required for maintaining *Gata3* expression, however they are required to allow for *Gata3* expression in ETPs when *Spi1* is also highly expressed (Del Real and Rothenberg 2013). After progenitors initiate T-lineage specification, *Gata3* expression is regulated in DN2 cells by the E protein family member E2A, a necessary regulator in the proper differentiation of T-lineage cells, as overexpression of *Gata3* leads to prolonged self-renewal of DN2 cells and an arrest of further T-cell development (Xu et al. 2013).

The binding, and thus function, of GATA3 across the genome is different at each stage of T-cell development, even for stages with similar expression levels of GATA3 (Zhang et al. 2012), suggesting GATA3 functions in a stage-specific manner, possibly in collaboration with other factors to promote T-cell development.

4.1.4.4   Bcl11b

Bcl11b is a zinc-finger transcription factor that is critical for inducing T-lineage commitment in developing T-cells (Ikawa et al. 2010; Li et al. 2010a, b). During T-cell development, the Bcl11b enhancer repositions itself from the lamina to the nuclear interior to form a single-loop domain that brings together the Bcl11b promoter and enhancer, which is accomplished with aid from a non-coding RNA called ThymoD (thymocyte differentiation factor) (Isoda et al. 2017). Mice that are conditionally deleted for *Bcl11b* are blocked at the DN2 stage that resemble the DN2a uncommitted T-cells, suggesting it is required for the DN2a-DN2b transition (Ikawa et al. 2010). The upregulation of Bcl11b is triggered by lower levels of IL-7, implying that commitment to the T-lineage may be instructed by signals provided by the thymic microenvironment. Interestingly, IL-7Ra is also greatly downregulated from the DN2–DN3 stage, possibly allowing for the low IL-7 requirement for the commitment step in conjunction with less IL-7 in the thymic microenvironment.

Bcl11b may directly repress genes that promote the myeloid fate, such as *Spi1* and *Cebpa*, which is important for the differentiation to committed T-cells (Ikawa et al. 2010). It is also required to repress NK cell-associated genes and progenitor genes at the commitment checkpoint (Li et al. 2010a). Bcl11b is indispensable for a robust T-cell fate, as deletion of Bcl11b led to all stages of developing T-cells to lose or decrease T-cell-associated gene expression and acquire NK cell properties (Li et al. 2010a). These "induced T-to-NK cells" are genetically and morphologically like conventional NK cells and can exert anti-tumor properties *in vitro* and *in vivo* (Li et al. 2010a). This suggests that the Bcl11b factor is extremely important for promoting and maintaining the committed T-cell fate. Efforts to identify functional targets of Bcl11b during commitment shows its function appears to overlap with E2A-dependent processes (Longabaugh et al. 2017), further illustrating the concerted efforts of core regulatory factors throughout T-cell development.

4.1.4.5 E Proteins

E proteins (*Tcf3* (E2A), *Tcf4* (E2-2), *Tcf12* (HEB)) are a family of bHLH transcription factors that have important roles in various processes throughout T-cell development. E2A and HEB have roles in regulating proliferation and facilitating T-lineage differentiation, with each having several isoforms that are produced by alternative splicing and alternative transcriptional start sites (Belle and Zhuang 2014). E proteins function as dimers by binding other bHLH factors through their HLH domains and the dimerization partners largely determine which genes are regulated. To illustrate this, E2A homodimers are the most prominent E protein dimer in developing B-cells, whereas E2A-HEB heterodimers function primarily in developing T-cells. Subtle differences in preferred DNA-binding sequences are dictated by the specific E protein dimer, and thus differences in gene regulation and developmental consequences are seen as a result (Belle and Zhuang 2014; Hsu et al. 1994; Hu et al. 1992).

E proteins have well described roles in regulating proliferation and survival (Belle and Zhuang 2014). E2A-deficient thymocytes have increased proliferation and develop T-cell leukemia, and overexpression of E2A isoforms in these cancerous cells lead to cell death (Bain et al. 1997; Engel and Murre 1999). HEB-deficient mice have reduction in thymic cellularity and E2A- and HEB-double conditional knockout (KO) mice have a more pronounced reduction in thymic cellularity and increased proliferation at the DN3 stage shown by dysregulation of cell-cycle regulators (Barndt et al. 1999; Wojciechowski et al. 2007).

E proteins have also been shown to regulate the *Rag* genes responsible for rearranging the TCR alleles (Hsu et al. 2003), and thus, in conjunction with inhibiting proliferation, these proteins may be responsible for slowing down proliferation and allowing for RAG-mediated TCR rearrangements before the β- and positive-selection checkpoints. E2A and HEB are required for activating germline transcription of TCRβ, TCRγ and TCRδ loci, and may create a permissive chromatin landscape at these loci to allow for RAG-mediated recombination to occur (Ghosh et al. 2001; Jia et al. 2008; Langerak et al. 2001).

The activity of E proteins is regulated by Id (Inhibitor of DNA-binding) proteins throughout various stages of T-cell development (Belle and Zhuang 2014; Benezra et al. 1990). Id proteins lack the basic DNA-binding domain and bind to E proteins, preventing DNA-binding, effectively inhibiting E-protein-mediated gene expression. The dynamic balance between E and Id proteins and the consequences on the transcriptional landscape can be thought of as a transcriptional switch, which occurs throughout T-cell development at various critical stages (Belle and Zhuang 2014).

The Id2 and Id3 proteins are the main Id family members to function in T-cell development and their expression is upregulated in response to activating signals like that of TCR activation during thymocyte selection, where Id3 is upregulated rapidly and Id2 upregulation is slower but persistent (Bain et al. 2001; Rivera et al. 2000). This suggests that there may be redundant or unique roles for the Id proteins in reversal of E protein-mediated gene expression that may be required for passing through these differentiation steps.

The strength of TCR signaling dictates the levels of Id3 expression and are important for downstream changes in gene expression (Carpenter and Bosselut 2010). High Id3 expression like that in developing γδTCR$^+$ bearing cells is enough to overcome E-protein mediated differentiation blocks at the DN3 stage, however lower Id3 levels in response to pre-TCR signaling would only progress in conjunction with Notch1-induced downregulation of E proteins (Ciofani et al. 2006; Lauritsen et al. 2009; Taghon et al. 2006). Successful TCR allelic rearrangements lead to TCR signaling which reverses the E protein transcriptional network through Id upregulation and allows for cells to proliferate and expand before Id expression decreases and the E protein transcriptional network is once again enforced (Belle and Zhuang 2014; Yashiro-Ohtani et al. 2009).

4.1.4.6 β-Selection

As cells progress through T-lineage development and reach the DN3 stage, they slow in proliferation and undergo RAG-mediated rearrangements of the TCRβ alleles (Carpenter and Bosselut 2010). Successfully rearranged TCRβ chains assemble with an invariant pre-Tα chain and CD3 chains at the cell surface to assemble the pre-TCR complex and undergo pre-TCR signaling (Carpenter and Bosselut 2010). The pre-TCR extracellular portion does not appear to recognize any ligands (Irving et al. 1998), but instead the pre-TCR spontaneously oligomerizes which is mediated by specific charged residues located in the extracellular domain of the pre-Tα chain (Das et al. 2016; Kim et al. 2009; Wang et al. 1998; Yamasaki et al. 2006). Notch1 signaling leads to the enhanced activation of the PI-3 kinase and Akt pathways to reverse the differentiation and proliferative block imposed by E proteins at the DN3 stage, suggesting a role at the β-selection checkpoint independent of the pre-TCR (Carpenter and Bosselut 2010; Ciofani and Zuniga-Pflucker 2005; Kee 2009; Maillard et al. 2006; Wong et al. 2012).

Thymocytes that have successfully rearranged their TCRβ chains undergo pre-TCR signaling which results in intense proliferation as they transit through the DN4 and CD8$^+$ ISP stage to the CD4$^+$CD8$^+$ DP stage (Carpenter and Bosselut 2010; Egawa et al. 2007). DPs become unresponsive to cytokine signals such as IL-7 by downregulation of their IL-7Rα chains and expression of SOCS-1 (suppressor of cytokine signaling) (Carpenter and Bosselut 2010).

4.1.4.7 Positive Selection

At the DP stage, thymocytes begin rearranging their TCRα chains and once successful rearrangements have been made, these DP thymocytes express an αβTCR on the cell surface. To select for developing T-cells with appropriate affinity for peptide antigen, αβTCR$^+$ T-cells undergo a selection process through engagement of TECs and other antigen presenting cells (APCs) (Petrie and Zuniga-Pflucker 2007). TECs and APCs express MHC I and II on their surface, and present self-peptide antigens in the context of MHC to αβTCR$^+$ DP thymocytes. Only the DP thymocytes with TCRs that have appropriate affinity with peptide:MHC survive and proceed to differentiate along the T-lineage in a process called positive selection (Singer et al. 2008). The majority of DP thymocytes do not receive TCR survival signals and undergo death, as DP thymocytes are resistant to the pro-survival cytokine IL-7 (Singer et al. 2008). No matter which peptide:MHC type (class I or lass II) the αβTCR-bearing thymocytes recognize, the CD8 co-receptor is initially downregulated and the developing T-cell senses whether there are persistent signals generated through the αβTCR, which is indicative of continued engagement of the peptide:MHC complex. Since CD8 surface expression is downregulated, this signal must be received through engagement with the peptide:MHC II and the CD4 coreceptor and as a result, GATA3 upregulates ThPOK which functions to inhibit CD8-lineage genes, including *Runx3* and *Cd8*, and specify these cells into the CD4 helper lineage (Kappes et al. 2006; Luckey et al. 2014; Park et al. 2010; Wang et al. 2008). ThPOK is a critical transcription factor in CD4 helper lineage differentiation. Forced expression of ThPOK completely represses Runx3 and inhibits CD8 differentiation and results in re-direction of MHC I selected DP thymocytes into CD4$^+$ T-cells (Sun et al. 2005; Xiong and Bosselut 2012). This may be explained by the ability of ThPOK to repress CD8-lineage genes and antagonize Runx3 function in cells where both factors are expressed (Wildt et al. 2007). In the opposite manner, deletion of the ThPOK gene leads to an absence of CD4$^+$ T-cells and redirection of MHC II-selected DP thymocytes into CD8$^+$ T-cells (Muroi et al. 2008).

As such, ThPOK is a critical specification factor, which has the capability to commit positively selected DP thymocytes into the CD4 lineage.

DP thymocytes that are positively selected and recognize peptide:MHC, but after initial downregulation of the CD8 coreceptor, fail to continue to receive signals through their TCR and are influenced by cytokine signaling, primarily through IL-7 and IL-15 as well as IL-6, IFNγ, TSLP, and TGFβ (Etzensperger et al. 2017). Cytokine signaling upregulates Runx3, which functions to downregulate the CD4 coreceptor permanently, represses the CD4 and ThPOK genes, and specifies these cells into the CD8 lineage (Kohu et al. 2005; Naito and Taniuchi 2010). The CD4 gene contains an intronic silencer that has two Runx-binding sites, which are important for silencing it, as mutating the two binding sites results in de-repression of CD4 in mature CD8$^+$ T-cells (Sawada et al. 1994; Taniuchi et al. 2002a). In CD4$^-$CD8$^-$ DN thymocytes, Runx1 is required for active repression of the CD4 through the silencer, whereas Runx3 is required for epigenetic silencing of CD4 in the mature CD8 lineage (Taniuchi et al. 2002b). Even though overexpression of Runx3 results in increased CD8 SP cells, in part through redirection of MHC II-selected DP thymocytes (Kohu et al. 2005), it is insufficient to repress ThPOK and thus CD4 differentiation (Xiong and Bosselut 2012). Since cytokine signaling was required for the generation of MHC II-selected CD8 SP cells in ThPOK-deficient mice, it can be argued that the coupled importance of both Runx3 expression induced by cytokine signaling, as well as the additional parallel effects of cytokine signaling, are two crucial elements in CD8 lineage differentiation (Etzensperger et al. 2017). Indeed, removal of all six CD8 lineage specifying cytokines during positive selection eliminated Runx3 expression completely and failed to produce CD8 SP thymocytes.

In mice deficient for both ThPOK and Runx3, there was an inability to redirect MHC-selected DP thymocytes into inappropriate CD4/CD8 lineages and CD4$^+$ lineage cells were specified, demonstrating MHC II-selected DP thymocytes may specify into the CD4 lineage independently of ThPOK, but require it to prevent Runx-mediated CD8 lineage differentiation (Egawa and Littman 2008). Collectively, this demonstrates the uncoupled nature of positive selection and CD4/CD8 lineage decision and the interplay of lineage specifying transcription factors to promote robust CD4 and CD8 lineage specification. The processes outlined above are best encompassed by the "kinetic signaling model" (Singer 2002; Singer et al. 2008).

4.1.4.8   NKT and MAIT Cell Differentiation

When thymocytes reach the DP stage of T-cell development, they begin rearranging their TCRα chains. A subset of T-cells called NKT-cells develop at the DP stage that possess distal-most TCRα rearrangements such that the TCRα chains are invariant (Vα14-Jα18 in mice, Vα24-Jα18 in humans) (Gapin 2016). These invariant TCRα chains pair with specific TCRβ chains (Vβ8.2,7 or Vβ2 in mice, and Vβ11 in humans) to recognize lipid antigens α-Galcer and α-Glucer that are presented by a non-polymorphic MHC class I-like molecule called CD1d. Precursors of the NKT lineage must recognize self lipids presented on CD1d complexes that are present on

DP thymocytes through their NKT TCR, and as such the DP thymocytes depend on themselves to undergo positive selection into the NKT lineage (Bendelac 1995). It is precisely this DP-DP engagement that initiates the NKT-cell developmental program (Gapin 2016). The perceived strength of TCR signals during positive selection is higher in NKT-cells than conventional T-cells, even though NKT-cells do not continue to receive strong TCR signals in the periphery, suggesting this increased TCR signal strength may represent a critical component in their developmental program (Moran et al. 2011). Upon the engagement of NKT TCR and CD1d, secondary signals must be received by NKT-cell precursors that involves at least two members of the SLAM (Signaling lymphocytic-activation molecule) family: Slamf1 and Slamf6. These interactions lead to the recruitment of the adaptor SAP (SLAM-associated protein) and the Src kinase Fyn, which are important for the expansion and differentiation of the NKT lineage (Gapin 2016). Although NKT-cells primarily arise from DP thymocytes as described above, recent findings report Vα14$^+$ NKT-cells that develop directly from CD4$^-$CD8$^-$ DN thymocytes, bypassing the DP stage (Dashtsoodol et al. 2017). These DN NKT-cells mainly give rise to a specific subset of NKT-cells called NKT1 cells with strong cytotoxic potential in the periphery. These findings provide further clarification into the developmental origin of DN NKT-cells and demonstrate the intricacies of NKT differentiation.

The differentiation of NKT precursors are characterized into distinct stages which correspond to surface and intracellular marker expression. Stage 0 is characterized by surface expression of CD69, CD24, and the transcription factor Egr2 (Benlagha et al. 2005). The cells progress to stage 1 when they have downregulated both CD69 and CD24 and express high levels of the transcription factor PLZF. Stage 1 cells appear to be otherwise equivalent to naïve mature CD4$^+$ SP thymocytes. At stage 2, the differentiating NKT-cells acquire surface expression of CD44 and within this population of stage 2 cells exist the potential for multiple NKT-cell outcomes. These stage 2 cells may continue to differentiate into the NKT1 sublineage which acquire T-bet expression and the NK-like program (CD161$^+$(NK1.1) CD122^{hi}) and downregulate PLZF and GATA3 and secrete primarily IFNγ upon stimulation (Gapin 2016). They may differentiate into the NKT2 sublineage by retaining PLZF expression and expressing high levels of GATA3 and secrete IL-4 and IL-13, or they may differentiate into the IL-17-secreting NKT17 sublineage by upregulating *Rorc* (RORγt) and maintaining intermediate levels of PLZF. There are two more subsets of NKT-cells: iNKTfh that express *Bcl6* and aid the B-cell response through IL-21 secretion and resemble T-follicular helper (Tfh) subsets; and, NKT10 that express *Nfil3* and do not express PLZF and create an immunosuppressive environment by IL-10 secretion and resemble T-regulatory subsets. The location and signals required for the differentiation of NKTfh and NKT10 subsets are not well understood (Gapin 2016).

E and Id proteins play important roles throughout the development of NKT-cells. HEB has been shown to upregulate RORγt and downstream pro-survival molecule Bcl-XL to allow for enhanced survival in DP thymocytes and distal rearrangements of their TCRα chains (D'Cruz et al. 2010). In Id-deficient mice, E2A activity upregulates genes required for NKT-cell development and function, and Id-deficient DPs show a bias to promote selection into the NKT lineage (Roy et al. 2018). Id proteins Id2 and Id3 redundantly promote NKT lineage specification by inducing PLZF, and Id3 is important for NKT1 differentiation (Verykokakis et al. 2013). Id3 is present in NKT2 subsets, and loss of Id3 increased NKT2 cell expansion (D'Cruz et al. 2014). Interestingly, E proteins were found to directly bind to the PLZF promoter and are required for PLZF expression, as well as expansion and maturation of thymic NKT-cells (D'Cruz et al. 2014). Maintaining E protein activity by blocking Id proteins during positive selection inhibits NKT1 cell differentiation while increasing NKT2 and NKT17 subsets. However, considering the unchanged expression levels of transcription factors that are important for the specification of these NKT sublineages (GATA3, RORγT, Tbet, and Runx3), E proteins may control the number of NKT precursors that differentiate along each NKT subsets (Hu et al. 2013). Although our understanding of E and Id proteins in NKT biology is not clearly defined, it should be appreciated that the interplay between E and Id proteins is not as binary as it appears during early T-lineage specification.

MAIT cells are another type of invariant T-cell that develop in the thymus and respond to bacterially infected cells in the periphery by recognizing MHC class I-related protein-1 (MR1), an antigen-presenting molecule that presents microbial vitamin B metabolites (Le Bourhis et al. 2011). MAIT cell development does not require MR1 expression on TECs (Treiner et al. 2003) or developing T-cells like that of CD1d on DP thymocytes for NKT-cell development. MAIT cells contain an invariant TCRα chain and rearrange the distal-most segment iVα19-Jα33 at the DP stage, and this distal rearrangement is dependent on prolonged survival at the DP stage (Huang and Kanagawa 2001; Le Bourhis et al. 2011). MAIT cells exit the thymus as naïve cells where they acquire MAIT cell characteristics (Le Bourhis et al. 2011) and appear to require MR1 expression on B-cells for their presence in the periphery (Martin et al. 2009). Further work on sites of MAIT cell differentiation as well as signals and transcription factors involved in programming this cell fate will further our understanding of MAIT biology.

4.1.4.9 Negative Selection

Once DP thymocytes are positively selected and begin CD4 or CD8 lineage specification, they encounter another checkpoint, called negative selection that removes potentially autoreactive T-cells, which are T-cells with TCRs that have high affinity for self-antigen presented on MHC. Positively selected thymocytes upregulate the chemokine receptor CCR7 and migrate from the cortex into the medulla to interact with CCR7 ligands CCL19 and CCL21, which are expressed on medullary TECs (mTECs) (Forster et al. 2008). An important feature of negative selection attributed to mTECs is to present peptide antigens from proteins that are expressed in a variety of tissues outside of the thymus to developing T-cells, so that the possibility of T-cells encountering these antigens in the periphery is dramatically lessened (Derbinski et al. 2001). Two key factors are responsible for mediating tissue-specific-antigen (TSA) expression in mTECs: Autoimmune Regulator (Aire) and Forebrain Embryonic Zinc Finger-like 2 (Fezf2) (Takaba and Takayanagi 2017).

Aire nucleates a large transcription factor complex that is required for ectopic expression of peripheral antigens in mTECs (Takaba and Takayanagi 2017). Humans and mice defective in Aire develop multiorgan autoimmune diseases (Anderson et al. 2002b). Aire-dependent genes contain promoter regions that are enriched in repressive chromatin markers, such as methylated H3K27, and a lack of activation markers, such as methylated H3K4 (Takaba and Takayanagi 2017). Subunits of Aire promote a permissive chromatin landscape by removing heterochromatin to promote transcription by recruiting RNA polymerase II and supporting transcription elongation. As such, Aire induces the transcription of many TSA genes by interacting with additional transcriptional regulators.

Although Aire has been prominent in our understanding of TSA expression in mTECs, it is only expressed in 30% of mTECs and is thought to only control approximately 40% of TSAs (Takaba and Takayanagi 2017). Fezf2 is a transcription factor that has also been shown to directly regulate TSA expression in mTECs independently of Aire and is detected at the protein level in more than 80% of mTECs (Takaba et al. 2015). Fezf2-deficient mice do not have defects in positive selection but have altered usage of TCR Vβ chains in both CD4 and CD8 T-cells suggesting Fezf2 is involved in negative selection in some capacity. These mice also have abnormal mTEC distribution and a reduction in mTEC number (Takaba et al. 2015; Takaba and Takayanagi 2017). Mice that were deleted for Fezf2 conditionally in TECs using a Foxn1-Cre showed increases in autoantibody in serum and in the number of activated T-cells in the periphery, as well as increases in the tissues that were infiltrated by immune cells. Although some tissues that were infiltrated by immune cells were common in both Aire- and Fezf2-deficient mice, the autoimmune responses differed suggesting that Fezf2 and Aire may allow for autoreactive T-cells of different specificities to be deleted. Furthermore, gene expression microarray analysis demonstrates differences in Fezf2- and Aire-regulated genes in the mTEC compartments, suggesting that these two factors operate non-redundantly to achieve a relatively complete and thorough removal process of autoreactive T-cells. Additionally, *Fezf2* expression does not appear to be dependent on RANK and CD40-Ligand like that of Aire. Although it is thought that *Fezf2* expression appears to be regulated by the Lymphotoxin β Receptor (LTβR) signaling pathway, since *Ltbr*−/− mice have reduced *Fezf2* mRNA and protein expression in the mTEC compartment, TEC-specific deletion of LTβR in another study did not affect both *Fezf2* and *Aire* expression levels (Cosway et al. 2017). The reduced *Fezf2* expression in *Ltbr*−/− mice could be explained by other factors that are affected by a loss in LTβR signaling that may contribute to dysregulation of *Fezf2*. Nevertheless, there are no differences in Fezf2 gene expression in Aire-deficient mice suggesting Aire and Fezf2 are controlled by distinct inputs and do not regulate each other. It is currently not known which molecules interact with Fezf2 to induce TSA expression.

Only 60% of TSA genes are regulated by Aire and/or Fezf2, which suggests the possibility that other transcriptional regulators may be involved in promoting TSA expression and contributing to the removal of autoreactive T-cells (Takaba et al. 2015; Takaba and Takayanagi 2017). The remaining source of TSA presentation to developing T-cells in enforcing central tolerance could be due to thymus-homing DCs, which have been proposed to transport peripheral TSAs to the thymus and induce clonal T-cell deletion and promote Treg development (Bonasio et al. 2006; Gallegos and Bevan 2004; Hadeiba and Butcher 2013; Leventhal et al. 2016; Mbongue et al. 2014; Proietto et al. 2008). Further clarification into the roles of Fezf2 and uncovering more potential TSA-regulators will greatly benefit our understanding of the negative selection process.

4.1.5 Thymic Regulatory T-Cell Differentiation

During negative selection, CD4 T-cells that have high affinity to self-peptide:MHC II on mTECs or APCs are clonally deleted (Lee and Lee 2018). However, certain CD4 T-cells with high affinity to tissue-restricted self-antigen begin to express CD25 on the cell surface and are identified as progenitor T-regulatory cells (Tregs), an immunoregulatory subset of the CD4 helper lineage. These progenitor Tregs are later stimulated by IL-2 and IL-15 cytokines and are induced to express Foxp3 (Lio and Hsieh 2008). Although the transcription factor Foxp3 is critical for Tregs, its upregulation is followed by cytokine signaling through CD25, which is upregulated due to strong TCR activation. Thus, TCR activation appears to be a prerequisite for Treg lineage commitment along with Foxp3 induction (Lee and Lee 2018).

Cells destined towards the Treg lineage receive a stronger TCR signal during thymocyte development than conventional T-cells that successfully pass negative selection and make it to the periphery (Moran et al. 2011). However, according to the principles of negative selection, T-cells with high affinity for self-antigen are clonally deleted (Lee and Lee 2018). Although CD4 T-cells that have TCRs with high affinity towards peptides that are broadly expressed by mTECs and APCs are clonally deleted, cells that recognize tissue-restricted self-antigens may not undergo deletion and instead proceed through Treg differentiation (Malhotra et al. 2016). The costimulatory molecule B7 is also required for Treg induction in the thymus and their maintenance in the periphery (Lohr et al. 2004; Paust et al. 2004).

Although Treg differentiation is predicated on strong TCR signals in CD4 T-cells from TSAs presented by mTECs and thymic APCs, indicating a quantitative difference in signaling required for specification, it may still be the case that the signals are also qualitative in nature (Lee and Lee 2018). One report shows inverted orientation of a Treg TCR-peptide:MHC where there was a 180 degree rotation of the binding between a TCR and insulin peptide:MHC on an *in vitro*-induced Treg (Beringer et al. 2015). This interaction created a signal through the TCR-self-peptide:MHC complex, and thus demonstrate a possibility that Tregs, and perhaps even other subsets of conventional T-cells, may respond in a qualitatively different manner, which may influence their development or their activation.

4.1.6 Notch Signaling in Late T-Cell Development: CD4/CD8 Lineage Decision

The question of whether Notch signaling is vital only for the initial specification of T-cells or whether it continues to influence T-cell development in the later stages remains open. Although the importance of Notch signaling in T-lineage specification and commitment throughout the early DN stages of T-cell development is well understood, it has been shown to be involved in the decision between CD4 versus CD8 lineages. Overexpression of an activated Notch1 transgene, as well as an activated Notch2 transgene in a different study, analyzed before the mice develop tumors show an increase in CD8 SPs at the expense of CD4 SPs, representing a skew in the CD4 versus CD8 lineage decision (Fowlkes and Robey 2002; Witt et al. 2003). However, inactivation of Notch1 before positive selection showed no changes in the development of SP thymocytes (Wolfer et al. 2001). This is in contrast with two independent studies that used a pharmacological Notch signaling inhibitor, γ-secretase inhibitor (GSI), at various stages of T-cell development in fetal thymic organ cultures (FTOCs) and saw reduced development of CD8 SP cells, attributing functions of Notch signaling to the promotion of the CD8 SP lineage (Doerfler et al. 2001; Hadland et al. 2001). Studies involving *in vitro* generated CD8 T-cells from BM-derived HSCs revealed a Notch-dependent role in the efficient differentiation and selection of mature CD8 SP cells (Dervovic et al. 2013). Notch signaling also appears to have roles in promoting the maturation of CD4 and CD8 SP thymocytes in the absence of TCR and MHC-ligand interactions on TECs and suggests TCR-independent roles in differentiation into mature SP thymocytes (Deftos et al. 2000). This notion is strengthened by the observation that TCR signaling is necessary but not sufficient to promote positive selection of DP to SP thymocytes (Huang et al. 2004). Comparing distinct transcriptional programs in DP thymocytes undergoing positive selection, TCR stimulation, and Notch activation revealed genes that are induced by Notch activation that overlap with those induced by positive selection but not by TCR stimulation, suggesting Notch activity may contribute during positive selection in a manner independent to TCR-induced selection (Huang et al. 2004). Increased Notch1 signaling is also detrimental to DP thymocyte development, as constitutive Notch1 expression at the DP stage impaired thymocyte responses to TCR stimulation and thus inhibited the maturation of DP thymocytes into mature SP T-cells (Izon et al. 2001). Oppositely, deletions of Presenilin1 (PS1) and PS2, genes involved in cleavage of ICN, which block Notch signaling in DP thymocytes, showed reduced TCR signaling and suggests Notch may have an involvement in TCR signal transduction during positive selection (Laky and Fowlkes 2007). Collectively this suggests that Notch signaling is influential for the maturation of DP thymocytes into mature SP cells and may contribute to the regulation of TCR signaling during positive selection. The Notch and TCR signaling pathways, although distinct yet parallel, may be linked functionally in the context of T-lymphocyte development.

Interestingly, peripheral T-cells also possess a functional co-dependence of both Notch and TCR signaling. Activated peripheral T-cells showed decreased proliferation when Notch activation was inhibited resulting in lower IL-2 production and CD25 expression (Adler et al. 2003; Laky et al. 2015). Complementary to this observation, forced expression of a constitutively active Notch1 in primary T-cells led to increased sensitivity to cognate antigen, IL-2-mediated proliferation and CD25 surface expression (Adler et al. 2003). TCR signaling also appears to induce Notch expression, and inhibition of Notch activation decreases both CD4 and CD8 T-cell proliferation, and blocks NFκB activity and IFNγ production in peripheral T-cells (Palaga et al. 2003). Thus, the Notch signaling pathway continues to exert its influence in T-cell function long after its vital requirement in the development of T-cells within the thymus.

4.1.7   Beyond T-Lineage Commitment: αβ- vs. γδ-Lineage Bifurcation

4.1.7.1   TCR Signaling

T-lineage committed DN3 thymocytes further differentiate into two fundamentally distinct T-cell subsets, αβ T-cells or γδ T-cells, which have differential tissue distributions, and thus functions, with respect to tissue homeostasis and immune responses (Vantourout and Hayday 2013). Therefore, understanding the mechanisms that govern their lineage bifurcation remains an important area of investigation. The prevailing and current understanding of how DN3 thymocytes make the decision to enter either the αβ-lineage or γδ-lineage is the "TCR signal strength" model (Zarin et al. 2015). This model proposes that DN3 cells that have successfully rearranged γ and δ chains versus those that have successfully rearranged a β chain, commit to their respective lineages in a complex process that is not merely dictated by the TCR type they express. Evidence for this comes from the observation that in TCRβ$^{-/-}$ mice, small but detectable numbers of CD4$^+$CD8$^+$ "αβ-lineage" DPs were found in the thymus (Hayes et al. 2005). Conversely, αβTCR transgenic mice were still able to generate CD4$^-$CD8$^-$ T-cells that have phenotypic and functional similarities to conventional γδ T-cells. In the KN6 TCR (Vγ4Vδ5) transgenic system, in which the ligand for KN6 is the non-classical MHC Ib molecule T10/T22, deletion of the MHC Ib subunit, β$_2$M, led to the absence of ligand surface expression and consequently, many KN6 thymocytes failed to mature into γδ-lineage cells and instead develop into DP cells (Haks et al. 2005). To reinforce the TCR signal strength argument, Zarin *et al.* demonstrated that forced expression of a pre-TCR in KN6 thymocytes potentiated, rather than inhibited, their commitment to the γδ-lineage (Zarin et al. 2014). In the presence of weaker ligand signals, expression of a pre-TCR led to fewer KN6 thymocytes developing into DPs, and more of them upregulating CD73, downregulating CD24, and having higher levels of γδ-lineage genes, which all represent hallmarks of the γδ-lineage. Thus, these experiments point to TCR signal strength, rather than the intrinsic nature of the TCR type, as the deciding factor for αβ- versus γδ-lineage choice.

The mechanism by which differences in the TCR signal strength dictate the $\alpha\beta$/$\gamma\delta$-lineage decision is possibly through differential induction of Id3 by the ERK-Egr signaling cascade, which in turn leads to differential inhibition of E protein (E2A, E2-2, HEB) activity (Lauritsen et al. 2009; Wong and Zuniga-Pflucker). Thus, it is likely that the $\gamma\delta$TCR leads to higher activation of the ERK-Egr-Id3 pathway, such that there is an enforced inhibition of E protein activity. Weaker signals delivered by the pre-TCR, in turn, lead to modest activation of the ERK-Egr-Id3 pathway and incomplete inhibition of E protein activity, promoting development along the $\alpha\beta$-lineage. Even in the presence of strong ligand, deletion of *Erk1* and *Erk2* in KN6 thymocytes led to a block in their maturation to the $\gamma\delta$-lineage and instead they developed into DP cells. Conversely, forced expression of Egr1 rescued the commitment of KN6 cells to the $\gamma\delta$-lineage, despite the absence of a ligand with β_2M deletion. Similar to deletion of the *Erk* genes, deletion of *Id3* has been shown to divert KN6 cells towards the $\alpha\beta$-lineage (Lauritsen et al. 2009). More recent evidence demonstrated that it is not only the intensity of ERK signaling, but also its duration, that affects this lineage fate (Lee et al. 2014). KN6 thymocytes from a $\gamma\delta$-inducing environment (presence of ligand) had more intense ERK phosphorylation that lasted over 60 min, while KN6 thymocytes from an $\alpha\beta$-inducing environment (absence of ligand) had less intense ERK phosphorylation that lasted only 20 min. This led to increased association of Egr1 with ERK and thus increased stability in KN6 cells adopting the $\gamma\delta$ T-cell fate. Outside of the KN6 TCR transgenic context, deletion of *Id3* also led to fewer V$\gamma4^+$ cells in the spleen and V$\gamma5^+$ cells in the epidermis, suggesting that a normal $\gamma\delta$ T-cell repertoire also requires strong TCR signals for commitment and maturation to the $\gamma\delta$-lineage (Lauritsen et al. 2009). Taken together, these experiments highlight that higher and increased duration of TCR signaling promotes $\gamma\delta$-lineage development.

4.1.7.2 Notch Signaling

Several studies have demonstrated a differential dependency on Notch for $\alpha\beta$- versus $\gamma\delta$-lineage bifurcation (Ciofani et al. 2006; Tanigaki et al. 2004; Washburn et al. 1997). In a study by Ciofani *et al.*, $\alpha\beta$-lineage development was shown to require Notch signaling, as in the presence of low concentrations of GSI, DP cells failed to develop on OP9-DL1 cultures while $\gamma\delta$TCR$^+$ cells only declined slightly in number. The authors later deduced that the timing of this differential dependency on Notch occurs at the DN3 stage. Isolated DN2 or DN3 cells placed on OP9-DL1 were able to differentiate into TCRβ^+ and $\gamma\delta$TCR$^+$ cells. However, when DN2 cells were placed on OP9-Control without Notch ligands, both cell types failed to develop. Conversely, DN3 cells on OP9-Control did not generate TCRβ^+ cells but did generate $\gamma\delta$TCR$^+$ cells, with some of them down-regulating CD24, indicating their commitment and maturation to the $\gamma\delta$ T-cell lineage. In further support of this idea, Lauritsen *et al.* correlated the induction of Id3 to be sufficient for $\gamma\delta$TCR-bearing cells to no longer require Notch for further differentiation (Lauritsen et al. 2009).

Their study revealed that even without a γδTCR, DN3 cells forced to express Id3 survived and downregulated CD24 in OP9-Control cultures. A possible explanation for the differential dependency on Notch is that although αβ T-cells develop through incomplete inhibition of E protein activity, a certain level of inhibition must still be met for β-selection to proceed, and the pre-TCR alone provides insufficient signals for this to occur (Wong and Zuniga-Pflucker 2010). It is proposed that Notch can then provide the additional signals needed to meet this level of E protein inhibition (Nie et al. 2003, 2011). Indeed, pre-TCR-expressing cells are only able to reduce E2A levels in the presence of Notch signaling, while γδTCR-expressing cells can do so irrespective of the presence of Notch ligands (Lauritsen et al. 2009). However, whether HEB is under similar regulation remains to be determined.

The studies described so far showed that Notch signaling is not required for γδ T-cell linage commitment. However, they do not necessarily indicate whether Notch signals can positively or negatively regulate the generation of γδ T-cells. Using mice in which RBPJ is conditionally deleted in Lck-expressing cells (RBPJ$^{f/f}$Lck-Cre mice), Tanigaki et al. showed that in the absence of Notch responsiveness, there is enhanced γδ T-cell generation in the thymus (Tanigaki et al. 2004). Compared to control mice, RBPJ$^{f/f}$Lck-Cre mice had more BrdU$^+$ γδ T-cells in the thymus. The RBPJ-deficient thymocytes did not have decreased expression of TCRβ or pre-Tα, but failed to progress beyond the DN3 to the DP stage. Therefore, the enhanced generation of γδ T-cells was not due to an impaired ability of thymocytes to form the pre-TCR, but rather Notch may limit γδ T-cell expansion or survival. The authors also noted that in the absence of Notch responsiveness, there was accelerated thymic egress of γδ T-cells, which culminated in an increase of these cells in the spleen and lymph nodes.

Other studies implicate Notch as being directly involved with αβ- versus γδ-lineage choice (Garbe et al. 2006; Washburn et al. 1997), rather than by exclusively limiting the expansion or survival of γδ T-cells. Comparing Notch1$^{+/+}$ to Notch1$^{+/-}$ cells, one study showed that changes in *Notch1* gene dosage can affect this lineage decision, as Notch1$^{+/+}$ progenitor cells had a greater propensity to generate αβ T-cells, while Notch1$^{+/-}$ progenitor cells had a greater propensity to generate γδ T-cells (Washburn et al. 1997). To further illustrate this point, the authors demonstrated that having an activated form of Notch signaling favored the αβ T-cell fate. Furthermore, ICN transgenic mice, even in a G8 γδTCR transgenic background, had diversion of the γδTCR-bearing cells towards the αβ-lineage. These cells were expressing either CD4 or CD8αβ, as opposed to being CD4$^-$CD8$^-$ in the case of G8 γδTCR transgenic alone. In ICN transgenic mice, even in a TCRβ mutant background, there was restoration of the CD4$^+$CD8$^+$ DP population.

4.1.8 Effector Function Programming of γδ T-Cells

4.1.8.1 TCR Signaling

Acquisition of CD73 expression by developing γδ T-cells marks their commitment to the γδ-lineage, however this is still considered an intermediate stage at which they must still mature (marked by CD24 downregulation) (Coffey et al. 2014). This intermediate stage is a point at which developing γδ T-cells still have not resolved their effector fate. Like conventional αβ T-cells, γδ T-cells have the capacity to produce diverse cytokines, including IFNγ, IL-4, and IL-17 (Prinz et al. 2013). However, unlike αβ T-cells that exit the thymus as naïve cells and differentiate into specific T-helper subsets in the periphery, γδ T-cells can acquire their effector fate during their thymic differentiation (Zarin et al. 2015). Due to this preprogramming, γδ T-cells can produce cytokines rapidly during immunological challenge, and thus are often labeled "innate" T-cells. The same signaling factors that govern αβ- vs. γδ-lineage choice are involved in effector differentiation of γδ T-cells as well, namely TCR signaling (Turchinovich and Pennington 2011; Zarin et al. 2015). The effector function of γδ T-cells is dictated by their Vγ usage, which is temporally determined during ontogeny. In mice, there are five Vγ subsets with a known bias towards particular cytokine productions—Vγ1⁺ cells produce IFNγ or IL-4, Vγ4⁺ cells produce IFNγ or IL-17: Vγ5⁺ cells produce IFNγ, Vγ6⁺ cells produce IL-17, and Vγ7⁺ cells produce IFNγ (Carding and Egan 2002; Prinz et al. 2013). In addition, the generation of some Vγ, and thus effector, subsets is thought to be dependent on either fetal or postnatal periods of production. The reason for this could be due to differences in fetal versus postnatal thymic environments that express different selecting ligands (Havran and Allison 1990; Nitta et al. 2015; Turchinovich and Hayday 2011). Thus, having different γδTCRs and ligands can influence γδ T-cell effector fates, pointing towards a "TCR signal strength" model once again in dictating this outcome.

The KN6 transgenic model has been extensively used to address the role of γδTCR in directing γδ T-cell effector outcome. The advantage of this system stems from the fact that, although KN6 can select and mature on both T10 and T22, it recognizes T22 at a tenfold higher affinity compared to T10 (Ito et al. 1990). While B6 mice have functional expression of both T10 and T22, BALB/c mice harbor a mutation that does not allow T22 to be expressed. Using this to their advantage, Zarin *et al.* demonstrated that KN6 cells cultured with BALB/c embryonic fibroblasts were predominantly directed towards the IL-17 fate, compared to cultures where T22 was present (Zarin et al. 2018). Consequently, these cells had higher expression of RORγt and greater production of IL-17. The authors noted that the expression of T-bet (*Tbx21*) and production of IFNγ were not significantly affected by differences in TCR signal strength. This observation leads to a question of how weak γδTCR signaling can drive the IL-17 fate. A clue may come from a study that investigated how the E protein, HEB, programs IL-17 producing γδ T-cells (In et al. 2017). In this study, HEB was shown to directly interact with the *Sox4* and *Sox13* loci, factors important for the IL-17 program (Malhotra et al. 2013). In keeping with this, HEB deficiency caused an impaired ability of fetal T-cell progenitors to generate RORγt⁺ IL-17 producing γδ T-cells. The authors proposed that if Id3, which is an inhibitor of E protein activity, is a downstream consequence of γδTCR signaling, this may explain why weak γδTCR signals are optimal for IL-17 programming, such that in this setting, the function of HEB can persist. Indeed, they showed that when uncommitted, immature γδ T-cells were forced to express Id3, this led to decreased expression levels of *Sox4*, *Sox13*, and *Rorc*.

The experiments performed thus far in the field agree that lower γδTCR signals favor the IL-17 fate. However, there is continued debate about whether generating IL-17 producers requires ligand engagement. In keeping with the KN6 system, one would argue that experiencing ligand is obligatory to even commit to the γδ-lineage, and IL-17 producing KN6 cells lie within the CD73 fraction, a marker of ligand-experienced cells (Coffey et al. 2014). However, in the HEB study mentioned previously, the authors reported that IL-17 producers could be found within both the CD73⁺ and CD73⁻ fractions, whereby the CD73⁺ cells were mostly Vγ4⁺ and the CD73⁻ cells were mostly Vγ6⁺ (In et al. 2017). Indeed, a study by Jensen *et al.* demonstrated that in the thymus of C57BL/6 mice, CD122 expression on γδ T-cells marked those that had encountered antigen, and those that were CD122⁺ were not able to produce IL-17 (Jensen et al. 2008). Only within the CD122⁻ fraction were there γδ T-cells that contained this function. An explanation for the discrepancy that antigen-naïve γδ T-cells exist but that TCR signaling is critical for γδ-lineage commitment, is that other signaling factors can mediate induction of Id3 in place of TCR signaling. An example of such a factor is Skint1, which is expressed by TECs and aids in the proper development of fetal Vγ5⁺ cells (Roberts et al. 2012; Turchinovich and Hayday 2011). In mice with a *Skint1* mutation, Vγ5⁺ cells have reduced expression of *Egr3* and became IL-17 producing cells instead (Turchinovich and Hayday 2011). Therefore, although not a γδTCR ligand, Skint1 can mediate the effects of γδTCR signaling by activating the pathway that leads to Id3 induction. Thus, development of antigen-naïve IL-17 producing γδ T-cells could first involve engagement of non-TCR ligands that can mediate γδ-lineage commitment, followed by IL-17 effector differentiation through delivery of weak γδTCR signals.

While the programming of IFNγ producers is not influenced by differences in γδTCR signaling strength and IL-17 production requires a weak γδTCR signaling environment, it is still unclear what the requirements are for programming IL-4 production in γδ T-cells. Lauritsen *et al.* observed that although Id3 deficiency led to an impaired development of Vγ4⁺ and Vγ5⁺ cells in mice, the Vγ1⁺ population was increased (Lauritsen et al. 2009). A subset of the Vγ1⁺ population (which also expresses Vδ6.3) is thought to be comprised of γδ NKT-cells, as they express PLZF and NK1.1 (Alonzo et al. 2010). Their function includes simultaneous production of both IFNγ and IL-4. The authors reasoned that these γδ NKT-cells might normally be selected on strong signals that would cause them to be deleted. However, in the absence of Id3, their TCR signals might be dampened to allow them to survive (Lauritsen et al. 2009). Thus, although reduced TCR signals would allow γδ NKT-cells to survive and expand, the hypothesis put forward is that strong ligand selection is required for IL-4 programming. This observation has been confirmed in work done by Alonzo *et al.*, in which they further showed that Id3 KO mice have increased percentage and number of Vγ1⁺Vδ6.3⁺ cells in both the thymus and spleen (Alonzo et al. 2010). Within the Vγ1⁺Vδ6.3⁺ population, those from Id3 KO mice also displayed enhanced expression of PLZF and increased production of IL-4. Other KO mouse models involving components of the TCR signaling pathway such as Itk-deficient mice also showed an increased population of IL-4 producing Vγ1⁺Vδ6.3⁺ cells (Felices et al. 2009; Qi et al. 2009). Interestingly, the Itk KO model demonstrated the immunological consequence of having an increased generation of IL-4 producing γδ T-cells, which is elevated production of IgE by B-cells. Thus, regulation of TCR signaling in γδ T-cells can partake in the development of allergic responses.

4.1.8.2   Notch and Cytokine Signaling

Notch1 and *Hes1* expression have been found to be higher in thymic and peripheral γδ T-cells that were CD27⁻ (marking IL-17 producers) compared to those that were CD27⁺ (marking IFNγ producers) (Shibata et al. 2011). This indicated that Notch signaling may be important for the development of IL-17 producing γδ T-cells. *Hes1*-deficient mice do not have alterations in the γδTCR repertoire but do show a severe reduction in IL-17 producers in the E18 fetal thymus. This observation was strengthened by the fact that fetal-derived γδ T-cells cultured *in vitro* without Notch ligands displayed no IL-17 producing capacity. Zarin *et al.* also showed that in combination with weak γδTCR signaling, the presence of Notch ligands *in vitro* enhanced the ability of KN6 T-cells to make IL-17 (Zarin et al. 2018). These authors further described the importance of other critical signaling factors that augment the IL-17 program, including cytokines such as IL-1β, IL-21, and IL-23. FTOCs treated with anti-IL-1β, IL-21, and IL-23 antibodies failed to generate IL-17 producing γδ T-cells. While these cytokines were initially shown to stimulate peripheral γδ T-cells to produce IL-17 (Sutton et al. 2009), it seems that they play an integral role in the generation of these cells also.

4.2   Overview of Human T-Cell Development

T-cell development in the human thymus, like its mouse counterpart, is contingent on BM-derived cells, which seed the thymus as TSPs, and become gradually differentiated into fully mature and functionally programmed T-lymphocytes, as outlined in Fig. 4.1 (lower panels). The earliest intrathymic cells are CD34⁺, and they lack CD4 and CD8 expression, as well as TCR rearrangements. Notch signaling subsequently induces expression of the cell surface marker CD7, and later CD5, which together mark specification towards the T-lineage in humans (Brauer et al. 2016). Subsequent acquisition of CD1a indicates T-lineage commitment as thymocytes lose the ability to generate alternative lineage types (Payne and Crooks 2002), including B-cells and myeloid cells. A unique property of human precursors is that they pass through a CD4 ISP stage instead of a CD8 ISP stage. At this stage, there is a bifurcation towards either the αβ or γδ T-cell lineage. Cells proceeding towards the αβ T-cell lineage subsequently progress through a CD4$^+$CD8α$^+$β$^-$ early DP stage, followed by a CD4$^+$CD8α$^+$β$^+$ late DP stage. Following productive expression of the αβTCR, positive and negative selection events occur in the thymic cortex and medulla, respectively. CD4 SP or CD8 SP cells are then able to migrate to the periphery. These steps, as well as key signals initiating their generation are described in further detail below. As is the case in mice, it is likely that the distinct developmental steps are coordinated with migration through different thymic niches that provide stage-specific environmental factors and signals for maturation.

4.2.1   Notch Signaling During Human T-Cell Development

Notch receptor-ligand interactions in the thymus are crucial in the T-cell developmental process in mice (Schmitt et al. 2004), and the same signaling pathway appears to be equally critical in humans (La Motte-Mohs et al. 2005; Van de Walle et al. 2009, 2011). Several groups have studied the expression patterns of Notch ligands within the human fetal and postnatal thymus, providing insights into the Notch ligands that are capable of supporting early and late stages of T-cell maturation (Van de Walle et al. 2011). Based on these reports, *JAG2* is predominantly expressed in both cortical TEC (cTEC) and mTEC subsets. DLL1 is also expressed in both cTEC and mTEC subsets in the postnatal thymus (Garcia-Leon et al. 2018; Van de Walle et al. 2011), consistent with previously described reports for the mouse thymus (Garcia-Leon et al. 2018; Hozumi et al. 2008; Koch et al. 2008; Schmitt et al. 2004). In contrast to DLL1 expression, a recent study revealed that the critical Notch ligand DLL4 appears to be absent on human postnatal cTECs and expressed only at low levels on mTECs (Garcia-Leon et al. 2018). High DLL4-expressing populations were rare, but were predominantly comprised of mesenchymal perivascular cells, endothelium and myeloid medullary cells within the human postnatal thymus. From these studies, one would expect that Jagged2 might play a major role during human T-cell development, and that the discrepancies in DLL4 expression between mouse and human highlight a species-specific role of DLL4 in postnatal T-cell development, implying a role for DLL1 in supporting human T-cell development postnatally. Additionally, JAG2 protein overexpression allowed for generation of human T-cells at the expense of B-cells, and initiation of the human T-lineage program *in vitro* (Van de Walle et al. 2011); however, the signaling events initiated by JAG2 are much weaker than those initiated by either DLL1 or DLL4 and while JAG2 could replace DLL4 function in humans, it cannot replace this function in mice (Hozumi et al. 2008; Koch et al. 2008). Furthermore, although DLL4 is unexpectedly observed at low levels or absent in the postnatal cortex (Garcia-Leon et al. 2018), high Dll4 expression in the cortical endothelium and perivascular regions, which are the first point of contact for BM-derived TSPs, is consistent with the notion of Notch1-dependent T-cell priming. Importantly, DLL4 expression was also significantly higher in the human fetal thymus, indicating that it is still functionally relevant for human T-cell development before birth, and consistent with the fact that human T-cell development declines with age (Brauer et al. 2016; Fry and Mackall 2002; Goronzy and Weyand 2005; Mackall and Gress 1997). Taken together, it remains well established that Notch1-Dll4 interactions are required for T-cell commitment and differentiation of TSPs in both mice and humans.

4.2.2   The Role of IL-7

The cytokine IL-7 induces a proliferative response in $CD34^+$ thymic T-cell progenitors. Similar to the mouse context, IL-7 is critical for T-cell development in humans, as seen in patients who lack the γc cytokine receptor subunit (i.e. SCID-X1 patients) as well as patients deficient for IL-7Rα or JAK3, a downstream signaling component of the IL-7R complex (DiSanto et al. 1994; Nosaka et al. 1995; Peschon et al. 1994). In patients, these deficiencies lead to a complete abrogation of T-cells, while mice with the same defects have only a reduction in the number of T-cells. These observations suggest a differential role for IL-7 in humans, where it is more than just a mere survival factor for precursor cells, as it is in mice.

Establishing a further role for IL-7 signaling, and explaining this species difference, SCID-X1 patients fail to rearrange TCRβ V-D-J segments (Sleasman et al. 1994), whereas in mice, IL-7 is only partially involved in TCRγ rearrangements, with no impact on TCRβ (Durum et al. 1998; Maki et al. 1996). This implies potential involvement of IL-7 in initiation of human TCRβ rearrangements, and this is further supported by observed enhancement of TCRβ rearrangements in FTOC cultures through addition of recombinant human IL-7 (Okamoto et al. 2002).

4.2.3   The Human Thymus-Seeding Progenitor

In humans, BM HSC-derived TSPs settle within the thymus by weeks 8–9 of gestation (Farley et al. 2013; Haddad et al. 2006). Numerous groups have made efforts towards the characterization of the earliest human T-lineage progenitor, although interpretation of these findings has been unclear given the rarity of human fetal and postnatal organs for analysis.

In recent years, tremendous progress has been made with respect to the identification of early ETPs in human. It is well established that the earliest intrathymic cells lack intracellular TCR rearrangements and cell surface CD4 and CD8 expression, consistent with what is seen in mice. These cells are represented within the $CD34^+CD1a^-$ T-cell progenitor population (Blom and Spits 2006; Haynes et al. 1988; Kurtzberg et al. 1989), and they can differentiate into T-cells but are not committed to the T-lineage, given lack of VDJ rearrangement and the ability to give rise to alternative lineages. Thus, human T-cell commitment occurs after thymic entry. Upon induction of Notch signaling, CD7 is the first marker to be upregulated on developing cells within the thymus (Awong et al. 2009; Galy et al. 1993; Haddad et al. 2006; Spits et al. 1995; Terstappen et al. 1992). In their analysis of CD34$^+$CD1a$^-$ progenitors based on CD7 expression, Hao *et al.* identified that CD34$^+$CD1a$^-$CD7$^-$ progenitors, which exist as a rare population (0.2%) within the human thymus, are the most immature subset, capable of generating lymphoid, myeloid and erythroid fates (Hao et al. 2008; Weerkamp et al. 2006). The ability to generate erythroid cells highlights the more primitive nature of human ETPs compared to mouse ETPs, suggesting that this population warrants further characterization, as it may include features of HSCs or MPPs. While HSCs have been shown to effectively differentiate into T-cells when directly placed within a thymic microenvironment (Spangrude and Scollay 1990), HSCs are devoid of thymus-seeding properties, as they have low to absent CCR9 and CCR7 cell surface expression and therefore are unable to respond to the major chemokine axis (CCL25 and CCL21) guiding progenitors towards the thymus (Schwarz et al. 2007).

Cells that begin to upregulate CD7 (CD34$^+$CD1a$^-$CD7^{int}) are devoid of myeloid and erythroid potentials and appear to possess the features of LMPPs (Hao et al. 2008; Hoebeke et al. 2007). Indeed, CD7^{int} cells are capable of making both T- and B-cells *in vitro*. Given that continuous Notch signaling is required to suppress B-cell development (Krueger et al. 2006; Taghon et al. 2005), it is plausible that these cells have only begun specification towards the T-cell lineage and it is CD7$^+$ cells that have undergone extended Notch signaling and specification towards the T-lineage.

Interestingly, both CD7$^-$ and CD7^{int} subsets have been found in human umbilical cord blood and adult bone marrow. Furthermore, these subsets express CD10, a marker that denotes early differentiation towards the lymphoid lineage. Therefore, it is conceivable that the human thymus can be colonized by cells possessing either of these phenotypes (Doulatov et al. 2010; Galy et al. 1995; Six et al. 2007). Further supporting this notion and suggesting a potential pre-thymic event during human T-cell development, CD34$^+$CD7^{int}CD10$^+$ cells were identified in human bone marrow by Klein and colleagues (Klein et al. 2003). Thus, numerous human ETP candidates, as well as TSP candidates at the level of the bone marrow and circulating blood, have been observed.

While the majority of these studies were largely based on observational evidence, the first functional technique that began to test the robustness of the human TSP was the *in vitro* xenograft approach using FTOCs (Fisher et al. 1990). FTOCs revealed that both CD7^{int} cells (Haddad et al. 2006) and CD7$^-$ cells were able to develop into DP cells (Hoebeke et al. 2007; Taghon et al. 2002). These findings suggested that few, but physiologically relevant and sufficient, CD7$^-$ cells can enter the thymus. The identity of the human TSP became clearer from a functional standpoint when CD34$^+$CD7$^+$ cells were shown to engraft the thymus in immunodeficient mice (Awong et al. 2009; Haddad et al. 2006; Hao et al. 2008; Hoebeke et al. 2007; Reimann et al. 2012). Furthermore, the precursor-progeny relationship between CD7$^-$, CD7^{int} and CD7$^+$ cells became more apparent, where induction of CD7 expression by Notch was consistent with loss of multipotency (De Smedt et al. 2002; Jaleco et al. 2001; Van de Walle et al. 2009, 2011), with CD7$^+$ cells appearing to be T/NK-cell progenitors, similar to mouse DN2a cells (Yui et al. 2010). These studies highlighted a potential caveat in the technical approaches used to define TSP function in FTOC, with a shift towards a more rigorous methodology, such as *in vivo* humanized mice, to define functional TSPs. Based on these and subsequent studies that have engrafted CD34$^+$CD7$^+$ cells into humanized mice (Reimann et al. 2012), there seems to be a contradiction with the earlier experiments that suggested that TSPs were CD7$^-$ or CD7^{int}. While it seems unlikely that CD7$^-$ cells would engraft *in vivo* given the rarity of T-cell reconstitution with these cells in FTOC, it is possible that the lymphoid-primed CD7^{int} cell population could contain cells with thymus-homing potential.

4.2.4 Human T-Lineage Specification and Commitment

The transition of CD34$^+$CD7$^+$CD5$^+$CD1a$^-$ cells to a CD1a$^+$ phenotype is strongly associated with T-lineage commitment and an inability to generate alternative lineages including NK cells, DCs, or plasmacytoid DCs (pDCs) (Blom and Spits 2006; Spits et al. 1998, 2000). Furthermore, in the mouse, Notch1/Dll4 interactions serve as the pivotal force behind Notch-dependent initiation of the T-lineage program and the same holds true in humans. *NOTCH1* and *NOTCH2* are expressed at high levels in human TSPs and uncommitted human postnatal thymocytes (CD34$^+$CD1a$^-$), and this expression persists into the DP stages of T-cell development before downregulation in mature CD4 SP and CD8 SP cells (Ghisi et al. 2011; Van de Walle et al. 2009). While the expression of these receptors is similar to the expression of *Notch1* and *Notch2* in the mouse thymus, *NOTCH3* expression in human reaches significant levels in uncommitted CD34$^+$CD1a$^-$ thymocytes, and appears earlier during human T-cell development compared to the mouse. However, *NOTCH3* expression strongly overlaps with *NOTCH1* in human precursors, suggesting that Notch1 may still be the predominant receptor during these early stages of T-cell development, similar to the mouse (Shi et al. 2011; Suliman et al. 2011).

ICN overexpression in human CD34$^+$ UCB cells showed a block in CD19$^+$ B-cells, and an increase in CD7$^+$ T/NK progenitor cells and CD1a$^+$ committed T-cells (De Smedt et al. 2002). Culture of these transduced cells in FTOC generated more γδ T-cells, whereas the other subpopulations were largely unaffected. In contrast, overexpression of dominant negative Mastermind-like 1 (DNMAML1) (Taghon et al. 2009) or GSI experiments (De Smedt et al. 2005; Van de Walle et al. 2009) showed the inhibition of T-cell development and the induction of B- and NK cell fates. Stromal co-culture data have also shown that Notch is equally important in human HSCs to induce T-cell development at the expense of alternative lineages (Awong et al. 2009, 2013; De Smedt et al. 2004; La Motte-Mohs et al. 2005; Van de Walle et al. 2011), and acquisition of the cell surface markers CD5 followed by CD1a mark specification and commitment to the T-lineage (Awong et al. 2009; Spits 2002; Taghon et al. 2012).

The combination of these approaches has also yielded insights at the gene regulatory level, suggesting that strong levels of Notch signaling are required for driving

T-lineage specification—*DTX1*, *HES1*, and *MYC* upregulation—in uncommitted CD34$^+$CD1a$^-$ thymocytes. Given that NOTCH1-JAG2 signaling is not as strong as NOTCH1-DLL4 interactions, it is more likely that NOTCH1-DLL4 is also responsible for induction of human T-lineage specification, despite the abundant JAG2 expression by human TECs (Van de Walle et al. 2011).

Following the initial T-cell specification program and as progenitors start to undergo T-lineage commitment, the expression of NOTCH-target genes, *DTX1* and *NRARP*, is reduced, while expression levels of *HES1* and *MYC* are maintained, suggesting that there is a reduction in NOTCH signaling during this developmental stage, possibly due to direct negative feedback by GATA3 on Notch-dependent genes (Van de Walle et al. 2016). Consistent with this notion, investigation of Notch signal strength in OP9-DL co-cultures of uncommitted human progenitors using GSI (Van de Walle et al. 2009) or weaker Notch ligands such as OP9-JAG1 (Dontje et al. 2006), revealed enhanced differentiation of uncommitted T-lineage progenitors into DP thymocytes. Remarkably, even without Notch signaling, uncommitted human thymocytes could generate CD8β-expressing DP cells, albeit inefficiently, given that Notch is involved in proliferation and survival of DPs (Taghon et al. 2009). This suggests a less important role for Notch signaling during αβ T-cell development in humans, and a role for other molecular mechanisms beyond Notch signaling, such as TCF-1, GATA3 and BCL11b, for further T-cell differentiation (Taghon et al. 2012). These mechanisms, similar to the mouse context, could either be recruited to regulate Notch target genes, however their function during human T-cell development remains unclear.

GATA3 is necessary for T-lineage specification and commitment in mouse thymocytes (Garcia-Ojeda et al. 2013), and appears to similarly have a role in human T-cell development. GATA3 has been implicated in repressing stem cell and B-lineage genes (Van de Walle et al. 2016), as well as restricting the pDC fate of human cells by antagonizing SPIB (Dontje et al. 2006) and NK cell fate through repression of NK-signature genes, including *ID2* and *RUNX3* (Van de Walle et al. 2016). Furthermore, GATA-3 overexpression in human thymocytes resulted in aberrant T-lymphopoiesis and enhanced apoptosis in the presence of Notch ligands in FTOC (Taghon et al. 2001), possibly by antagonizing the effects of Notch signaling in promoting the T-lineage (Van de Walle et al. 2016). To reveal a direct role for GATA3 during human T-lineage commitment, a chromatin immunoprecipitation (ChIP)-sequencing experiment was performed in human postnatal thymocytes, which revealed that GATA3 binds motifs at several critical T-lineage specific genes including *TCF7*, *BCL11B*, *TCF12* and the Notch-target gene *HES1* (Van de Walle et al. 2016). Taken together, these data suggest that GATA3 is an important T-lineage driver in human cells.

Finally, E-proteins also play a role in human T-cell commitment. Enforced expression of the E-protein antagonists ID2 or ID3 in human CD34$^+$ precursors from fetal liver, cord blood or thymus resulted in a block in human T-cell development and promotion of the innate lymphoid cell (ILC) or NK cell fates (Blom et al. 1999; Heemskerk et al. 1997). Similarly, human pluripotent stem cell-derived CD34$^+$ precursors that were HEB deficient failed to develop into T-cells on OP9-DL cultures

(Li et al. 2017). These findings are consistent with the critical role of Id2 in NK cell development in the mouse (Ikawa et al. 2001; Yokota et al. 1999) and it is therefore likely that one or more E proteins are also important for human T-cell commitment.

4.2.5   Beyond T-Lineage Commitment

After T-lineage commitment, CD34⁺CD1a⁺ cells begin to express CD4, but not CD8, and are termed CD4 ISP cells (Galy et al. 1993; Kraft et al. 1993). At this point, TCR rearrangements at the *TCRD*, *TCRG* and *TCRB* loci are fully underway (Dik et al. 2005), thus this subset is upstream of the β-selection checkpoint in humans, and can become both γδ and αβ T-cells (Blom et al. 1999; Ramiro et al. 1996). It is unclear why the appearance of ISP is dominated by CD4 ISP in human compared to CD8 ISP in mouse, but it is possible that this difference is cytokine-mediated or mediated by the strength of pre-TCR signals. As illustrated by high levels of *RAG* and *PTCRA* expression, CD4 ISP cells are similar to mouse DN3a cells (Taghon et al. 2006).

Productive rearrangement of the TCRδ and TCRγ chains yields γδTCR⁺CD3⁺ T-cells. In contrast, a functionally rearranged TCRβ chain will pair with the surrogate pre-Tα to induce β-selection. Once cells acquire cell surface expression of CD28, they have completed β-selection and are comparable to mouse DN3b cells (Blom and Spits 2006). At this stage, cells undergo extensive Notch-dependent selective survival and proliferation, and differentiation into CD4⁺CD8α⁺β⁺ DP thymocytes. Failure to rearrange a functional TCRβ results in cell death (Spits 2002). Importantly, the precise stage at which β-selection occurs in the human thymus remains unclear and represents a key distinction from mouse T-cell development. Blom *et al.* have reported that β-selection is initiated at the CD4 ISP stage, as they were able to detect a small proportion of cells (~5%) that expressed cytoplasmic TCRβ, and overexpression of LCK in FTOC led to an accumulation of CD4 ISP cells (Blom et al. 1999). The same study showed that the subsequent stage of development, early DP cells (CD4⁺CD8α⁺β⁻), have a significant proportion of cells expressing cytoplasmic TCRβ. Based on their work and the work done by Toribio and colleagues (Carrasco et al. 1999), it is possible that β-selection can take place at the CD4 ISP stage, but mainly occurs when cells express CD8α and CD8β.

After selection of TCRβ-expressing cells, there is subsequent rearrangement of the TCRα chain. The first step of this rearrangement involves deletion of the TCRδ locus, and thus completion of V-Jα rearrangements mainly leads to cells expressing functional a αβTCR (Spits et al. 1998; Verschuren et al. 1997). Finally, positive and negative selection in the thymus determine whether cells mature to CD4 SP or CD8 SP αβTCR⁺ T-cells.

4.2.6   Human αβ Versus γδ T-Lineage Choice

In humans, three main Vδ segments, Vδ1, Vδ2 and Vδ3, are most frequently used in rearrangement of the δ chain, while seven functional Vγ gene segments, Vγ2, Vγ3, Vγ4, Vγ5, Vγ8, Vγ9 and Vγ11 are used for rearrangement of the γ chain (Adams et al. 2015). Similar to the mouse, the restricted repertoire of Vδ and Vγ gene segments available has led to the speculation that these γδ TCRs recognize conserved ligand motifs (Adams et al. 2015). In the mouse, pairing biases of particular Vγ and Vδ chains occur, thus supporting this idea, although this has yet to be observed in humans. Nevertheless, it is well established that γδ T-cells expressing Vδ1 constitute >50% of fetal blood γδ T-cells at birth, and in adulthood, these same cells are nearly absent in the blood, but instead populate epithelial tissues including the intestine (Dimova et al. 2015). The major subset of γδ cells found in adult human blood, comprising 1–10% of total healthy T-cells in the blood in humans (Adams et al. 2015; Morita et al. 2000), are the Vγ9Vδ2 T-cells (Dimova et al. 2015). These cells are also found at high frequencies in the gut, liver and other mucosal sites, and potentially respond to tumor cells and microbial pathogens (Adams et al. 2015). Thus, understanding their development in the human context has become increasingly important for harnessing these cells for immunotherapeutic application.

As mentioned earlier, γδ T-cells bifurcate from αβ T-cells at the CD34+CD1a+CD4+ ISP stage of development, as the TCRD, TCRG and TCRB segments begin rearrangement (Joachims et al. 2006; Spits 2002; Van de Walle et al. 2009). However, the γδ-lineage choice remains an option during key stages of αβ T-lineage differentiation as well, given that early human DPs (CD4+CD8α+) still contain γδ T-lineage potential, and absolute commitment to the αβ-lineage reaches completion at the late DP stage (CD4+CD8α+β+) (Carrasco et al. 1999; Joachims et al. 2006; Taghon and Rothenberg 2008). Early DP γδ T-cells were also observed in the human postnatal thymus (Offner et al. 1997). These findings show that although the CD4+CD8α+ DP subset dictates an αβ T-cell fate in the mouse, this association is not apparent in humans.

Interestingly, human γδ T-cells are favored over αβ T-cells under conditions of high Notch activation, in contrast to the paradigm for mouse γδ T-cell development (Robey et al. 1996; Washburn et al.). A series of studies including ICN overexpression (De Smedt et al. 2002; Garcia-Peydro et al. 2003) and OP9-DL co-culture experiments (Van de Walle et al. 2009) using CD34+ thymic progenitors revealed that high levels of Notch signaling favor γδ T-cell differentiation at the expense of the αβ T-cell lineage (Garcia-Peydro et al. 2003; Van de Walle et al. 2011, 2013). However, it is worth noting that these studies were done with intrathymically-derived cells, and this would not necessarily hold true for progenitor cells derived from the BM. Nevertheless, when human T-lineage specified or committed progenitors were cultured in the absence of Notch, virtually no γδ T-cells developed (Taghon et al. 2009). Consistent with the differential Notch signaling requirements between human and mouse, slightly higher levels of expression of key Notch target genes (*DTX1*, *HES1*, *NRARP* and *MYC*) were observed in γδ compared to αβ T-cells (Taghon et al. 2012; Van de Walle et al. 2009). Work by Taghon and colleagues has proposed that in the context of human γδ T-cell development, these differences may be due to potential regulation of RUNX3, as *RUNX3* gene expression is extremely high in γδ T-lineage cells in both mice (Rothenberg et al. 2008; Taghon et al. 2006) and humans (Van de Walle et al. 2009). Given that Runx1 is critical for repression of CD4 in DN and Runx3 is required for CD8 SP development in the mouse, the precise role that Runx3 plays during the αβ versus γδ decision is not yet clear. While GSI titration experiments hint that *RUNX3* regulation may be controlled by Notch itself (Van de Walle et al. 2009), further experimentation is required. Arguably, Sox13 is the transcription factor that most clearly separates γδ cells from αβ cells that have undergone β-selection. Similar to the sustained expression of Sox13 in γδ T-cells in the mouse, this is also true in humans (Taghon and Rothenberg 2008). Thus, some of the transcriptional profiles associated with mouse γδ T-cells may be conserved in humans, with a potential discrepancy in the requirements for Notch.

While high levels of Notch favor γδ T-cell differentiation, it is not entirely clear how this fits with the TCR signal strength model, as defined in the mouse context. It is known that maturation of human γδ T-cells can occur equally well in the presence or absence of Notch signaling (Van Coppernolle et al. 2012). Furthermore, ID3 overexpression in T-lineage committed human cells reduces αβ T-cell differentiation while γδ T-cells are unaffected (Blom et al. 1999), suggesting that the signal strength model may hold true in humans. Nevertheless, the combination of Notch signaling with TCR signal strength in γδ versus αβ bifurcation requires further analysis such as transduction experiments with a specific γδTCR. Furthermore, manipulating Notch signaling thresholds by harnessing different Notch receptor-ligand interactions, as well as understanding the differential availability of Notch ligands in particular intrathymic niches, may provide critical insight into this lineage bifurcation. Both Toribio and Taghon's groups have suggested that JAG2-mediated signaling could foster human γδ T-cell development (Garcia-Leon et al. 2018; Van de Walle et al. 2013), which is perhaps surprising, given the weaker Notch activation initiated by NOTCH1-JAG2 interactions, and the clear requirement for stronger Notch activation for human γδ T-cell fate.

4.2.7 Intrathymic Development of Alternative Lineages

Similar to the supportive role that the mouse thymus plays towards generation of unconventional intrathymic T-cell lineages, there appears to be a role for the human thymus to support NKT-cell and Treg development. Human type 1 NKT-cells express the Vα24-Jα18 and Vβ11 αβTCR chains, which like the mouse counterparts have the ability to bind to lipid antigens such as α-GalCer (Dellabona et al. 1994; Lantz and Bendelac 1994; Porcelli et al. 1993). While the intrathymic intricacies of NKT-cell development are unclear (Das et al. 2010), mature human NKT-cells within the blood express CD161, similar to the mouse NK1.1, and are either CD4+, CD8+ or DN (Baev et al. 2004; Berzins et al. 2005). While CD8 is not expressed on mouse NKT-cells, and therefore it's role still remains elusive, it is important to note that all three human NKT subsets have unique profiles with respect to gene expression, cytokine production capacity and the ability to activate neighboring cells, and human NKT-cells differ in their tissue localization and abundance compared to mouse NKT-cells (Lin et al. 2006, 2004).

MAIT cells also exist in humans and comprise ~10% of all peripheral T-cells (Gherardin et al. 2018). Their development is also intrathymically determined, consisting of Vα7.2 joined to Jα33, Jα12 or Jα20 (Lepore et al. 2014). While human MAIT cells can express a range of TCRβ chains, they are biased toward Vβ2 and Vβ13 (Rahimpour et al. 2015), enabling MAIT cells to detect MR1. Both human NKT-cells and MAIT cells are discussed in more detail by others (Godfrey et al. 2004; Le Bourhis et al. 2011; Martin et al. 2009; Treiner et al. 2005; Treiner and Lantz 2006). Nevertheless, an expanding number of NKT- and MAIT cell subsets are being identified in humans; thus, it remains important to investigate their precise developmental requirements for manipulating or enhancing normal immunity.

The thymic origin of human Treg cells is also unclear. However, observed autoimmunity in patients with immunodysregulation, polyendocrinopathy, enteropathy, X-linked (IPEX) syndrome, has established a key role for FOXP3$^+$ Treg cells in the maintenance of self-tolerance (Bennett et al. 2001; Wildin et al. 2001). Mouse thymus-derived Tregs are described as CD4$^+$Foxp3$^+$CD25$^+$ T-cells. While FOXP3$^+$ thymocytes do exist within the human thymus (Lee and Hsieh 2009; Tuovinen et al. 2008a, b), little is known about the developmental requirements needed for their generation. The use of FOXP3 and CD25 to define human Tregs has been widely disputed as transient FOXP3 expression was observed on CD4$^+$CD25$^-$ effector T-cells in the presence or absence of TGFβ, and these cells did not necessarily contain suppressive function (Roncarolo and Gregori 2008). Furthermore, in peripheral-derived Treg cells, called adaptive T-regulatory cells type 1 (TR1 cells), suppressive function was shown to be independent from FOXP3 expression. CD25 expression cannot be used in human studies, as only 1–2% of CD25-expressing cells have been shown to be functionally suppressive and considered Treg cells (Baecher-Allan et al. 2001). Several groups have refined alternative schemes for purifying human Treg cells including lack of cell surface CD127 (Seddiki et al. 2006) and a CD62L$^+$ phenotype (Hamann et al. 2000) to distinguish these cells from naïve CD4$^+$ T-cells, or the presence of co-stimulatory molecules such as ICOS (Ito et al. 2008). Given the known similarities described between mouse and human T-cell development, groups have postulated that the requirements for human Tregs may closely resemble those in the mouse. However, the medulla of the human thymus contains a unique structure, Hassall's corpuscles, which may provide a distinct niche for supporting human Treg cells given the secretion of TSLP, which has a known ability to activate CD11c$^+$ DCs to upregulate their co-stimulatory molecules (Hanabuchi et al. 2010; Watanabe et al. 2004). These DCs, in turn, appear to induce FOXP3 expression in immature CD4$^+$CD8$^-$CD25$^-$ thymocytes (Watanabe et al. 2005). DCs that are activated by TSLP may also contribute to positive selection of extremely self-reactive thymocytes, supporting their differentiation into mature FOXP3-expressing Treg cells.

4.3   Conclusion

The intricate process of T-cell development enables the production of functionally diverse mature T-cells from the few BM-derived progenitors that enter the thymus each day (summarized in Fig. 4.1). Equipped with potentials for various lineages, early progenitors in the thymus receive the appropriate signaling cues to lose alternative lineage potentials and secure their T-lineage fate. Numerous studies have yielded insights into how Notch signaling, in particular, plays a critical role in the stage-specific induction of factors that dictate T-lineage specification and commitment. Once committed to the T-lineage fate, T-cell progenitors receive further instructional cues for differentiation into either γδ T-cells, conventional CD4 or CD8 SP αβ T-cells, or unconventional T-cell subsets including NKT, MAIT or Treg cells, which themselves are programmed by unique combinations of lineage-specific transcription factors. Despite the advances that have been made in the field, numerous questions in the field of T-cell biology remain unanswered. In particular, it remains unclear in both mice and humans whether BM-derived progenitors entering the thymus already possess a strong T-lineage bias. Furthermore, the precise mechanism by which Notch-target genes shut off alternative lineages for induction of T-cell commitment is still being refined. More recent studies have focused on evaluation and construction of a global gene regulatory network capable of driving T-cell fate. While our knowledge of intrathymic T-cell development has been centered around the progress in mice, we anticipate that technical advances such as knock-out and knock-in approaches, which are already being explored, as well as increased access to primary human tissues and humanized mouse models, will generate important insights for these processes in humans. Nevertheless, the human processes closely follow many of the same fundamental principles of T-lineage specification/commitment and lineage bifurcation as in the mouse, but with clear distinctions in the developmental stages and checkpoints that arise during development. Additional insights into the signaling factors that the human thymic microenvironment provides, as well as a more comprehensive understanding of the cellular changes that occur during each stage of T-cell development, may reveal the mechanisms behind these differences. Continuing to tackle these questions and refining our understanding of the intricate process of T-cell development will enable our understanding of normal and malignant processes in both mice and human.

References

Adams EJ, Gu S, Luoma AM (2015) Human gamma delta T cells: evolution and ligand recognition. Cell Immunol 296:31–40. https://doi.org/10.1016/j.cellimm.2015.04.008

Adler SH et al (2003) Notch signaling augments T cell responsiveness by enhancing CD25 expression. J Immunol 171:2896–2903

Alonzo ES et al (2010) Development of promyelocytic zinc finger and ThPOK-expressing innate gamma delta T cells is controlled by strength of TCR signaling and Id3. J Immunol 184:1268–1279. https://doi.org/10.4049/jimmunol.0903218

Anderson MK, Hernandez-Hoyos G, Diamond RA, Rothenberg EV (1999) Precise developmental regulation of Ets family transcription factors during specification and commitment to the T cell lineage. Development 126:3131–3148

Anderson MK, Weiss AH, Hernandez-Hoyos G, Dionne CJ, Rothenberg EV (2002a) Constitutive expression of PU.1 in fetal hematopoietic progenitors blocks T cell development at the pro-T cell stage. Immunity 16:285–296

Anderson MS et al (2002b) Projection of an immunological self shadow within the thymus by the aire protein. Science 298:1395–1401. https://doi.org/10.1126/science.1075958

Andersson ER, Sandberg R, Lendahl U (2011) Notch signaling: simplicity in design, versatility in function. Development 138:3593–3612. https://doi.org/10.1242/dev.063610

Awong G, Herer E, Surh CD, Dick JE, La Motte-Mohs RN, Zuniga-Pflucker JC (2009) Characterization in vitro and engraftment potential in vivo of human progenitor T cells generated from hematopoietic stem cells. Blood 114:972–982. https://doi.org/10.1182/blood-2008-10-187013

Awong G et al (2013) Human proT-cells generated in vitro facilitate hematopoietic stem cell-derived T-lymphopoiesis in vivo and restore thymic architecture. Blood 122:4210–4219

Baecher-Allan C, Brown JA, Freeman GJ, Hafler DA (2001) CD4+CD25high regulatory cells in human peripheral blood. J Immunol 167:1245–1253

Baev DV, Peng XH, Song L, Barnhart JR, Crooks GM, Weinberg KI, Metelitsa LS (2004) Distinct homeostatic requirements of CD4+ and CD4- subsets of Valpha24-invariant natural killer T cells in humans. Blood 104:4150–4156. https://doi.org/10.1182/blood-2004-04-1629

Bain G et al (1997) E2A deficiency leads to abnormalities in alphabeta T-cell development and to rapid development of T-cell lymphomas. Mol Cell Biol 17:4782–4791

Bain G, Cravatt CB, Loomans C, Alberola-Ila J, Hedrick SM, Murre C (2001) Regulation of the helix-loop-helix proteins, E2A and Id3, by the Ras-ERK MAPK cascade. Nat Immunol 2:165–171. https://doi.org/10.1038/84273

Barndt R, Dai MF, Zhuang Y (1999) A novel role for HEB downstream or parallel to the pre-TCR signaling pathway during alpha beta thymopoiesis. J Immunol 163:3331–3343

Belle I, Zhuang Y (2014) E proteins in lymphocyte development and lymphoid diseases. Curr Top Dev Biol 110:153–187. https://doi.org/10.1016/B978-0-12-405943-6.00004-X

Bendelac A (1995) Positive selection of mouse NK1+ T cells by CD1-expressing cortical thymocytes. J Exp Med 182:2091–2096

Benezra R, Davis RL, Lockshon D, Turner DL, Weintraub H (1990) The protein Id: a negative regulator of helix-loop-helix DNA binding proteins. Cell 61:49–59

Benlagha K, Wei DG, Veiga J, Teyton L, Bendelac A (2005) Characterization of the early stages of thymic NKT cell development. J Exp Med 202:485–492. https://doi.org/10.1084/jem.20050456

Bennett CL et al (2001) The immune dysregulation, polyendocrinopathy, enteropathy, X-linked syndrome (IPEX) is caused by mutations of FOXP3. Nat Genet 27:20–21. https://doi.org/10.1038/83713

Beringer DX et al (2015) T cell receptor reversed polarity recognition of a self-antigen major histocompatibility complex. Nat Immunol 16:1153–1161. https://doi.org/10.1038/ni.3271

Berzins SP, Cochrane AD, Pellicci DG, Smyth MJ, Godfrey DI (2005) Limited correlation between human thymus and blood NKT cell content revealed by an ontogeny study of paired tissue samples. Eur J Immunol 35:1399–1407. https://doi.org/10.1002/eji.200425958

Blom B, Spits H (2006) Development of human lymphoid cells. Annu Rev Immunol 24:287–320. https://doi.org/10.1146/annurev.immunol.24.021605.090612

Blom B et al (1999) TCR gene rearrangements and expression of the pre-T cell receptor complex during human T-cell differentiation. Blood 93:3033–3043

Bonasio R, Scimone ML, Schaerli P, Grabie N, Lichtman AH, von Andrian UH (2006) Clonal deletion of thymocytes by circulating dendritic cells homing to the thymus. Nat Immunol 7:1092–1100. https://doi.org/10.1038/ni1385

Brauer PM, Singh J, Xhiku S, Zuniga-Pflucker JC (2016) T cell genesis: in vitro veritas est? Trends Immunol 37:889–901. https://doi.org/10.1016/j.it.2016.09.008

Carding SR, Egan PJ (2002) Gammadelta T cells: functional plasticity and heterogeneity. Nat Rev Immunol 2:336–345. https://doi.org/10.1038/nri797

Carpenter AC, Bosselut R (2010) Decision checkpoints in the thymus. Nat Immunol 11:666–673. https://doi.org/10.1038/ni.1887

Carrasco YR, Trigueros C, Ramiro AR, de Yebenes VG, Toribio ML (1999) Beta-selection is associated with the onset of CD8beta chain expression on CD4(+)CD8alphaalpha(+) pre-T cells during human intrathymic development. Blood 94:3491–3498

Champhekar A, Damle SS, Freedman G, Carotta S, Nutt SL, Rothenberg EV (2015) Regulation of early T-lineage gene expression and developmental progression by the progenitor cell transcription factor PU.1. Genes Dev 29:832–848. https://doi.org/10.1101/gad.259879.115

Ciofani M, Zuniga-Pflucker JC (2005) Notch promotes survival of pre-T cells at the beta-selection checkpoint by regulating cellular metabolism. Nat Immunol 6:881–888. https://doi.org/10.1038/ni1234

Ciofani M, Zuniga-Pflucker JC (2007) The thymus as an inductive site for T lymphopoiesis. Annu Rev Cell Dev Biol 23:463–493. https://doi.org/10.1146/annurev.cellbio.23.090506.123547

Ciofani M, Knowles GC, Wiest DL, von Boehmer H, Zuniga-Pflucker JC (2006) Stage-specific and differential notch dependency at the alphabeta and gammadelta T lineage bifurcation. Immunity 25:105–116. https://doi.org/10.1016/j.immuni.2006.05.010

Coffey F et al (2014) The TCR ligand-inducible expression of CD73 marks gammadelta lineage commitment and a metastable intermediate in effector specification. J Exp Med 211:329–343. https://doi.org/10.1084/jem.20131540

Cosway EJ et al (2017) Redefining thymus medulla specialization for central tolerance. J Exp Med 214:3183–3195. https://doi.org/10.1084/jem.20171000

D'Cruz LM, Knell J, Fujimoto JK, Goldrath AW (2010) An essential role for the transcription factor HEB in thymocyte survival, Tcra rearrangement and the development of natural killer T cells. Nat Immunol 11:240–249. https://doi.org/10.1038/ni.1845

D'Cruz LM, Stradner MH, Yang CY, Goldrath AW (2014) E and Id proteins influence invariant NKT cell sublineage differentiation and proliferation. J Immunol 192:2227–2236. https://doi.org/10.4049/jimmunol.1302904

Das R, Sant'Angelo DB, Nichols KE (2010) Transcriptional control of invariant NKT cell development. Immunol Rev 238:195–215. https://doi.org/10.1111/j.1600-065X.2010.00962.x

Das DK et al (2016) Pre-T cell receptors (Pre-TCRs) leverage vbeta complementarity determining regions (CDRs) and hydrophobic patch in mechanosensing thymic self-ligands. J Biol Chem 291:25292–25305. https://doi.org/10.1074/jbc.M116.752865

Dashtsoodol N et al (2017) Alternative pathway for the development of Valpha14(+) NKT cells directly from CD4(−)CD8(-) thymocytes that bypasses the CD4(+)CD8(+) stage. Nat Immunol 18:274–282. https://doi.org/10.1038/ni.3668

De Obaldia ME et al (2013) T cell development requires constraint of the myeloid regulator C/EBP-alpha by the Notch target and transcriptional repressor Hes1. Nat Immunol 14:1277–1284. https://doi.org/10.1038/ni.2760

De Smedt M et al (2002) Active form of Notch imposes T cell fate in human progenitor cells. J Immunol 169:3021–3029

De Smedt M, Hoebeke I, Plum J (2004) Human bone marrow CD34+ progenitor cells mature to T cells on OP9-DL1 stromal cell line without thymus microenvironment. Blood Cells Mol Dis 33:227–232. https://doi.org/10.1016/j.bcmd.2004.08.007

De Smedt M, Hoebeke I, Reynvoet K, Leclercq G, Plum J (2005) Different thresholds of Notch signaling bias human precursor cells toward B-, NK-, monocytic/dendritic-, or T-cell lineage in thymus microenvironment. Blood 106:3498–3506. https://doi.org/10.1182/blood-2005-02-0496

Deftos ML, Huang E, Ojala EW, Forbush KA, Bevan MJ (2000) Notch1 signaling promotes the maturation of CD4 and CD8 SP thymocytes. Immunity 13:73–84

Del Real MM, Rothenberg EV (2013) Architecture of a lymphomyeloid developmental switch controlled by PU.1, Notch and Gata3. Development 140:1207–1219. https://doi.org/10.1242/dev.088559

Dellabona P, Padovan E, Casorati G, Brockhaus M, Lanzavecchia A (1994) An invariant V alpha 24-J alpha Q/V beta 11 T cell receptor is expressed in all individuals by clonally expanded CD4-8- T cells. J Exp Med 180:1171–1176

Derbinski J, Schulte A, Kyewski B, Klein L (2001) Promiscuous gene expression in medullary thymic epithelial cells mirrors the peripheral self. Nat Immunol 2:1032–1039. https://doi.org/10.1038/ni723

Dervovic DD et al (2013) Cellular and molecular requirements for the selection of in vitro-generated CD8 T cells reveal a role for Notch. J Immunol 191:1704–1715. https://doi.org/10.4049/jimmunol.1300417

Dik WA et al (2005) New insights on human T cell development by quantitative T cell receptor gene rearrangement studies and gene expression profiling. J Exp Med 201:1715–1723. https://doi.org/10.1084/jem.20042524

Dimova T et al (2015) Effector Vgamma9Vdelta2 T cells dominate the human fetal gammadelta T-cell repertoire. Proc Natl Acad Sci U S A 112:E556–E565. https://doi.org/10.1073/pnas.1412058112

DiSanto JP, Rieux-Laucat F, Dautry-Varsat A, Fischer A, de Saint Basile G (1994) Defective human interleukin 2 receptor gamma chain in an atypical X chromosome-linked severe combined immunodeficiency with peripheral T cells. Proc Natl Acad Sci U S A 91:9466–9470

Doerfler P, Shearman MS, Perlmutter RM (2001) Presenilin-dependent gamma-secretase activity modulates thymocyte development. Proc Natl Acad Sci U S A 98:9312–9317. https://doi.org/10.1073/pnas.161102498

Donskoy E, Foss D, Goldschneider I (2003) Gated importation of prothymocytes by adult mouse thymus is coordinated with their periodic mobilization from bone marrow. J Immunol 171:3568–3575

Dontje W et al (2006) Delta-like1-induced Notch1 signaling regulates the human plasmacytoid dendritic cell versus T-cell lineage decision through control of GATA-3 and Spi-B. Blood 107:2446–2452. https://doi.org/10.1182/blood-2005-05-2090

Doulatov S, Notta F, Eppert K, Nguyen LT, Ohashi PS, Dick JE (2010) Revised map of the human progenitor hierarchy shows the origin of macrophages and dendritic cells in early lymphoid development. Nat Immunol 11:585–593. https://doi.org/10.1038/ni.1889

Durum SK, Candeias S, Nakajima H, Leonard WJ, Baird AM, Berg LJ, Muegge K (1998) Interleukin 7 receptor control of T cell receptor gamma gene rearrangement: role of receptor-associated chains and locus accessibility. J Exp Med 188:2233–2241

Egawa T, Littman DR (2008) ThPOK acts late in specification of the helper T cell lineage and suppresses Runx-mediated commitment to the cytotoxic T cell lineage. Nat Immunol 9:1131–1139. https://doi.org/10.1038/ni.1652

Egawa T, Tillman RE, Naoe Y, Taniuchi I, Littman DR (2007) The role of the Runx transcription factors in thymocyte differentiation and in homeostasis of naive T cells. J Exp Med 204:1945–1957. https://doi.org/10.1084/jem.20070133

Engel I, Murre C (1999) Ectopic expression of E47 or E12 promotes the death of E2A-deficient lymphomas. Proc Natl Acad Sci U S A 96:996–1001

Etzensperger R et al (2017) Identification of lineage-specifying cytokines that signal all CD8(+)-cytotoxic-lineage-fate 'decisions' in the thymus. Nat Immunol 18:1218–1227. https://doi.org/10.1038/ni.3847

Farley AM et al (2013) Dynamics of thymus organogenesis and colonization in early human development. Development 140:2015–2026. https://doi.org/10.1242/dev.087320

Felices M, Yin CC, Kosaka Y, Kang J, Berg LJ (2009) Tec kinase Itk in gammadeltaT cells is pivotal for controlling IgE production in vivo. Proc Natl Acad Sci U S A 106:8308–8313. https://doi.org/10.1073/pnas.0808459106

Feyerabend TB et al (2009) Deletion of Notch1 converts pro-T cells to dendritic cells and promotes thymic B cells by cell-extrinsic and cell-intrinsic mechanisms. Immunity 30:67–79. https://doi.org/10.1016/j.immuni.2008.10.016

Fisher AG, Larsson L, Goff LK, Restall DE, Happerfield L, Merkenschlager M (1990) Human thymocyte development in mouse organ cultures. Int Immunol 2:571–578

Forster R, Davalos-Misslitz AC, Rot A (2008) CCR7 and its ligands: balancing immunity and tolerance. Nat Rev Immunol 8:362–371. https://doi.org/10.1038/nri2297

Foss DL, Donskoy E, Goldschneider I (2001) The importation of hematogenous precursors by the thymus is a gated phenomenon in normal adult mice. J Exp Med 193:365–374

Fowlkes BJ, Robey EA (2002) A reassessment of the effect of activated Notch1 on CD4 and CD8 T cell development. J Immunol 169:1817–1821

Fry TJ, Mackall CL (2002) Current concepts of thymic aging. Springer Semin Immunopathol 24:7–22

Gallegos AM, Bevan MJ (2004) Central tolerance to tissue-specific antigens mediated by direct and indirect antigen presentation. J Exp Med 200:1039–1049. https://doi.org/10.1084/jem.20041457

Galy A, Verma S, Barcena A, Spits H (1993) Precursors of CD3+CD4+CD8+ cells in the human thymus are defined by expression of CD34. Delineation of early events in human thymic development. J Exp Med 178:391–401

Galy A, Travis M, Cen D, Chen B (1995) Human T, B, natural killer, and dendritic cells arise from a common bone marrow progenitor cell subset. Immunity 3:459–473

Gapin L (2016) Development of invariant natural killer T cells. Curr Opin Immunol 39:68–74. https://doi.org/10.1016/j.coi.2016.01.001

Garbe AI, Krueger A, Gounari F, Zuniga-Pflucker JC, von Boehmer H (2006) Differential synergy of Notch and T cell receptor signaling determines alphabeta versus gammadelta lineage fate. J Exp Med 203:1579–1590. https://doi.org/10.1084/jem.20060474

Garcia-Leon MJ, Fuentes P, de la Pompa JL, Toribio ML (2018) Dynamic regulation of Notch1 activation and Notch ligand expression in human thymus development. Development 145:pii: dev165597. https://doi.org/10.1242/dev.165597

Garcia-Ojeda ME et al (2013) GATA-3 promotes T-cell specification by repressing B-cell potential in pro-T cells in mice. Blood 121:1749–1759. https://doi.org/10.1182/blood-2012-06-440065

Garcia-Peydro M, de Yebenes VG, Toribio ML (2003) Sustained Notch1 signaling instructs the earliest human intrathymic precursors to adopt a gammadelta T-cell fate in fetal thymus organ culture. Blood 102:2444–2451. https://doi.org/10.1182/blood-2002-10-3261

Georgescu C et al (2008) A gene regulatory network armature for T lymphocyte specification. Proc Natl Acad Sci U S A 105:20100–20105. https://doi.org/10.1073/pnas.0806501105

Germar K et al (2011) T-cell factor 1 is a gatekeeper for T-cell specification in response to Notch signaling. Proc Natl Acad Sci U S A 108:20060–20065. https://doi.org/10.1073/pnas.1110230108

Gherardin NA et al (2018) Human blood MAIT cell subsets defined using MR1 tetramers. Immunol Cell Biol 96:507–525. https://doi.org/10.1111/imcb.12021

Ghisi M et al (2011) Modulation of microRNA expression in human T-cell development: targeting of NOTCH3 by miR-150. Blood 117:7053–7062. https://doi.org/10.1182/blood-2010-12-326629

Ghosh JK, Romanow WJ, Murre C (2001) Induction of a diverse T cell receptor gamma/delta repertoire by the helix-loop-helix proteins E2A and HEB in nonlymphoid cells. J Exp Med 193:769–776

Godfrey DI, MacDonald HR, Kronenberg M, Smyth MJ, Van Kaer L (2004) NKT cells: what's in a name? Nat Rev Immunol 4:231–237. https://doi.org/10.1038/nri1309

Goldschneider I, Komschlies KL, Greiner DL (1986) Studies of thymocytopoiesis in rats and mice. I. Kinetics of appearance of thymocytes using a direct intrathymic adoptive transfer assay for thymocyte precursors. J Exp Med 163:1–17

Goronzy JJ, Weyand CM (2005) T cell development and receptor diversity during aging. Curr Opin Immunol 17:468–475. https://doi.org/10.1016/j.coi.2005.07.020

Haddad R et al (2006) Dynamics of thymus-colonizing cells during human development. Immunity 24:217–230

Hadeiba H, Butcher EC (2013) Thymus-homing dendritic cells in central tolerance. Eur J Immunol 43:1425–1429. https://doi.org/10.1002/eji.201243192

Hadland BK et al (2001) Gamma -secretase inhibitors repress thymocyte development. Proc Natl Acad Sci U S A 98:7487–7491. https://doi.org/10.1073/pnas.131202798

Haks MC et al (2005) Attenuation of gammadeltaTCR signaling efficiently diverts thymocytes to the alphabeta lineage. Immunity 22:595–606. https://doi.org/10.1016/j.immuni.2005.04.003

Hamann A, Klugewitz K, Austrup F, Jablonski-Westrich D (2000) Activation induces rapid and profound alterations in the trafficking of T cells. Eur J Immunol 30:3207–3218. https://doi.org/10.1002/1521-4141(200011)30:11<3207::AID-IMMU3207>3.0.CO;2-L

Hanabuchi S et al (2010) Thymic stromal lymphopoietin-activated plasmacytoid dendritic cells induce the generation of FOXP3+ regulatory T cells in human thymus. J Immunol 184:2999–3007. https://doi.org/10.4049/jimmunol.0804106

Hao QL et al (2008) Human intrathymic lineage commitment is marked by differential CD7 expression: identification of CD7- lympho-myeloid thymic progenitors. Blood 111:1318–1326

Havran WL, Allison JP (1990) Origin of Thy-1+ dendritic epidermal cells of adult mice from fetal thymic precursors. Nature 344:68–70. https://doi.org/10.1038/344068a0

Hayes SM, Li L, Love PE (2005) TCR signal strength influences alphabeta/gammadelta lineage fate. Immunity 22:583–593. https://doi.org/10.1016/j.immuni.2005.03.014

Haynes BF, Martin ME, Kay HH, Kurtzberg J (1988) Early events in human T cell ontogeny. Phenotypic characterization and immunohistologic localization of T cell precursors in early human fetal tissues. J Exp Med 168:1061–1080

Heemskerk MH et al (1997) Inhibition of T cell and promotion of natural killer cell development by the dominant negative helix loop helix factor Id3. J Exp Med 186:1597–1602

Ho IC, Tai TS, Pai SY (2009) GATA3 and the T-cell lineage: essential functions before and after T-helper-2-cell differentiation. Nat Rev Immunol 9:125–135. https://doi.org/10.1038/nri2476

Hoebeke I, De Smedt M, Stolz F, Pike-Overzet K, Staal FJ, Plum J, Leclercq G (2007) T-, B- and NK-lymphoid, but not myeloid cells arise from human CD34(+)CD38(-)CD7(+) common lymphoid progenitors expressing lymphoid-specific genes. Leukemia 21:311–319. https://doi.org/10.1038/sj.leu.2404488

Holmes R, Zuniga-Pflucker JC (2009) The OP9-DL1 system: generation of T-lymphocytes from embryonic or hematopoietic stem cells in vitro. Cold Spring Harb Protoc 2009:pdb prot5156. https://doi.org/10.1101/pdb.prot5156

Hozumi K et al (2008) Delta-like 4 is indispensable in thymic environment specific for T cell development. J Exp Med 205:2507–2513. https://doi.org/10.1084/jem.20080134

Hsu HL et al (1994) Preferred sequences for DNA recognition by the TAL1 helix-loop-helix proteins. Mol Cell Biol 14:1256–1265

Hsu LY, Lauring J, Liang HE, Greenbaum S, Cado D, Zhuang Y, Schlissel MS (2003) A conserved transcriptional enhancer regulates RAG gene expression in developing B cells. Immunity 19:105–117

Hu JS, Olson EN, Kingston RE (1992) HEB, a helix-loop-helix protein related to E2A and ITF2 that can modulate the DNA-binding ability of myogenic regulatory factors. Mol Cell Biol 12:1031–1042

Hu T et al (2013) Increased level of E protein activity during invariant NKT development promotes differentiation of invariant NKT2 and invariant NKT17 subsets. J Immunol 191:5065–5073. https://doi.org/10.4049/jimmunol.1301546

Huang C, Kanagawa O (2001) Ordered and coordinated rearrangement of the TCR alpha locus: role of secondary rearrangement in thymic selection. J Immunol 166:2597–2601

Huang YH, Li D, Winoto A, Robey EA (2004) Distinct transcriptional programs in thymocytes responding to T cell receptor, Notch, and positive selection signals. Proc Natl Acad Sci U S A 101:4936–4941. https://doi.org/10.1073/pnas.0401133101

Ikawa T, Fujimoto S, Kawamoto H, Katsura Y, Yokota Y (2001) Commitment to natural killer cells requires the helix-loop-helix inhibitor Id2. Proc Natl Acad Sci U S A 98:5164–5169. https://doi.org/10.1073/pnas.091537598

Ikawa T et al (2010) An essential developmental checkpoint for production of the T cell lineage. Science 329:93–96. https://doi.org/10.1126/science.1188995

In TSH et al (2017) HEB is required for the specification of fetal IL-17-producing gammadelta T cells. Nat Commun 8:2004. https://doi.org/10.1038/s41467-017-02225-5

Irving BA, Alt FW, Killeen N (1998) Thymocyte development in the absence of pre-T cell receptor extracellular immunoglobulin domains. Science 280:905–908

Isoda T et al (2017) Non-coding transcription instructs chromatin folding and compartmentalization to dictate enhancer-promoter communication and T cell fate. Cell 171:103–119 e118. https://doi.org/10.1016/j.cell.2017.09.001

Ito K, Van Kaer L, Bonneville M, Hsu S, Murphy DB, Tonegawa S (1990) Recognition of the product of a novel MHC TL region gene (27b) by a mouse gamma delta T cell receptor. Cell 62:549–561

Ito T et al (2008) Two functional subsets of FOXP3+ regulatory T cells in human thymus and periphery. Immunity 28:870–880. https://doi.org/10.1016/j.immuni.2008.03.018

Izon DJ et al (2001) Notch1 regulates maturation of CD4+ and CD8+ thymocytes by modulating TCR signal strength. Immunity 14:253–264

Jaleco AC et al (2001) Differential effects of Notch ligands Delta-1 and Jagged-1 in human lymphoid differentiation. J Exp Med 194:991–1002

Jensen KD et al (2008) Thymic selection determines gammadelta T cell effector fate: antigen-naive cells make interleukin-17 and antigen-experienced cells make interferon gamma. Immunity 29:90–100. https://doi.org/10.1016/j.immuni.2008.04.022

Jia J, Dai M, Zhuang Y (2008) E proteins are required to activate germline transcription of the TCR Vbeta8.2 gene. Eur J Immunol 38:2806–2820. https://doi.org/10.1002/eji.200838144

Joachims ML, Chain JL, Hooker SW, Knott-Craig CJ, Thompson LF (2006) Human alpha beta and gamma delta thymocyte development: TCR gene rearrangements, intracellular TCR beta expression, and gamma delta developmental potential--differences between men and mice. J Immunol 176:1543–1552

Kappes DJ, He X, He X (2006) Role of the transcription factor Th-POK in CD4:CD8 lineage commitment. Immunol Rev 209:237–252. https://doi.org/10.1111/j.0105-2896.2006.00344.x

Kee BL (2009) E and ID proteins branch out. Nat Rev Immunol 9:175–184. https://doi.org/10.1038/nri2507

Kim ST et al (2009) The alphabeta T cell receptor is an anisotropic mechanosensor. J Biol Chem 284:31028–31037. https://doi.org/10.1074/jbc.M109.052712

Klein F et al (2003) T lymphoid differentiation in human bone marrow. Proc Natl Acad Sci U S A 100:6747–6752. https://doi.org/10.1073/pnas.1031503100

Koch U et al (2008) Delta-like 4 is the essential, nonredundant ligand for Notch1 during thymic T cell lineage commitment. J Exp Med 205:2515–2523. https://doi.org/10.1084/jem.20080829

Kohu K et al (2005) Overexpression of the Runx3 transcription factor increases the proportion of mature thymocytes of the CD8 single-positive lineage. J Immunol 174:2627–2636

Kondo M, Weissman IL, Akashi K (1997) Identification of clonogenic common lymphoid progenitors in mouse bone marrow. Cell 91:661–672

Kovall RA (2007) Structures of CSL, Notch and Mastermind proteins: piecing together an active transcription complex. Curr Opin Struct Biol 17:117–127. https://doi.org/10.1016/j.sbi.2006.11.004

Kraft DL, Weissman IL, Waller EK (1993) Differentiation of CD3-4-8- human fetal thymocytes in vivo: characterization of a CD3-4+8- intermediate. J Exp Med 178:265–277

Krueger A, Garbe AI, von Boehmer H (2006) Phenotypic plasticity of T cell progenitors upon exposure to Notch ligands. J Exp Med 203:1977–1984. https://doi.org/10.1084/jem.20060731

Krueger A, Willenzon S, Lyszkiewicz M, Kremmer E, Forster R (2010) CC chemokine receptor 7 and 9 double-deficient hematopoietic progenitors are severely impaired in seeding the adult thymus. Blood 115:1906–1912. https://doi.org/10.1182/blood-2009-07-235721

Kurtzberg J, Denning SM, Nycum LM, Singer KH, Haynes BF (1989) Immature human thymocytes can be driven to differentiate into nonlymphoid lineages by cytokines from thymic epithelial cells. Proc Natl Acad Sci U S A 86:7575–7579

La Motte-Mohs RN, Herer E, Zuniga-Pflucker JC (2005) Induction of T-cell development from human cord blood hematopoietic stem cells by Delta-like 1 in vitro. Blood 105:1431–1439. https://doi.org/10.1182/blood-2004-04-1293

Lai AY, Kondo M (2007) Identification of a bone marrow precursor of the earliest thymocytes in adult mouse. Proc Natl Acad Sci U S A 104:6311–6316. https://doi.org/10.1073/pnas.0609608104

Laky K, Fowlkes BJ (2007) Presenilins regulate alphabeta T cell development by modulating TCR signaling. J Exp Med 204:2115–2129. https://doi.org/10.1084/jem.20070550

Laky K, Evans S, Perez-Diez A, Fowlkes BJ (2015) Notch signaling regulates antigen sensitivity of naive CD4+ T cells by tuning co-stimulation. Immunity 42:80–94. https://doi.org/10.1016/j.immuni.2014.12.027

Langerak AW, Wolvers-Tettero IL, van Gastel-Mol EJ, Oud ME, van Dongen JJ (2001) Basic helix-loop-helix proteins E2A and HEB induce immature T-cell receptor rearrangements in nonlymphoid cells. Blood 98:2456–2465

Lantz O, Bendelac A (1994) An invariant T cell receptor alpha chain is used by a unique subset of major histocompatibility complex class I-specific CD4+ and CD4-8- T cells in mice and humans. J Exp Med 180:1097–1106

Lauritsen JP et al (2009) Marked induction of the helix-loop-helix protein Id3 promotes the gammadelta T cell fate and renders their functional maturation Notch independent. Immunity 31:565–575. https://doi.org/10.1016/j.immuni.2009.07.010

Le Bourhis L, Guerri L, Dusseaux M, Martin E, Soudais C, Lantz O (2011) Mucosal-associated invariant T cells: unconventional development and function. Trends Immunol 32:212–218. https://doi.org/10.1016/j.it.2011.02.005

Lee HM, Hsieh CS (2009) Rare development of Foxp3+ thymocytes in the CD4+CD8+ subset. J Immunol 183:2261–2266. https://doi.org/10.4049/jimmunol.0901304

Lee W, Lee GR (2018) Transcriptional regulation and development of regulatory T cells. Exp Mol Med 50:e456. https://doi.org/10.1038/emm.2017.313

Lee SY et al (2014) Noncanonical mode of ERK action controls alternative alphabeta and gammadelta T cell lineage fates. Immunity 41:934–946. https://doi.org/10.1016/j.immuni.2014.10.021

Lepore M et al (2014) Parallel T-cell cloning and deep sequencing of human MAIT cells reveal stable oligoclonal TCRbeta repertoire. Nat Commun 5:3866. https://doi.org/10.1038/ncomms4866

Leventhal DS et al (2016) Dendritic cells coordinate the development and homeostasis of organ-specific regulatory T cells. Immunity 44:847–859. https://doi.org/10.1016/j.immuni.2016.01.025

Li L, Leid M, Rothenberg EV (2010a) An early T cell lineage commitment checkpoint dependent on the transcription factor Bcl11b. Science 329:89–93. https://doi.org/10.1126/science.1188989

Li P et al (2010b) Reprogramming of T cells to natural killer-like cells upon Bcl11b deletion. Science 329:85–89. https://doi.org/10.1126/science.1188063

Li Y, Brauer PM, Singh J, Xhiku S, Yoganathan K, Zuniga-Pflucker JC, Anderson MK (2017) Targeted disruption of TCF12 reveals HEB as essential in human mesodermal specification and hematopoiesis. Stem Cell Rep 9:779–795. https://doi.org/10.1016/j.stemcr.2017.07.011

Lin H, Nieda M, Nicol AJ (2004) Differential proliferative response of NKT cell subpopulations to in vitro stimulation in presence of different cytokines. Eur J Immunol 34:2664–2671. https://doi.org/10.1002/eji.200324834

Lin H, Nieda M, Hutton JF, Rozenkov V, Nicol AJ (2006) Comparative gene expression analysis of NKT cell subpopulations. J Leukoc Biol 80:164–173. https://doi.org/10.1189/jlb.0705421

Lio CW, Hsieh CS (2008) A two-step process for thymic regulatory T cell development. Immunity 28:100–111. https://doi.org/10.1016/j.immuni.2007.11.021

Liu J, Sato C, Cerletti M, Wagers A (2010) Notch signaling in the regulation of stem cell self-renewal and differentiation. Curr Top Dev Biol 92:367–409. https://doi.org/10.1016/S0070-2153(10)92012-7

Lohr J, Knoechel B, Kahn EC, Abbas AK (2004) Role of B7 in T cell tolerance. J Immunol 173:5028–5035

Longabaugh WJR et al (2017) Bcl11b and combinatorial resolution of cell fate in the T-cell gene regulatory network. Proc Natl Acad Sci U S A 114:5800–5807. https://doi.org/10.1073/pnas.1610617114

Luckey MA, Kimura MY, Waickman AT, Feigenbaum L, Singer A, Park JH (2014) The transcription factor ThPOK suppresses Runx3 and imposes CD4(+) lineage fate by inducing the SOCS suppressors of cytokine signaling. Nat Immunol 15:638–645. https://doi.org/10.1038/ni.2917

Mackall CL, Gress RE (1997) Thymic aging and T-cell regeneration. Immunol Rev 160:91–102

Maillard I et al (2006) The requirement for Notch signaling at the beta-selection checkpoint in vivo is absolute and independent of the pre-T cell receptor. J Exp Med 203:2239–2245. https://doi.org/10.1084/jem.20061020

Maki K, Sunaga S, Ikuta K (1996) The V-J recombination of T cell receptor-gamma genes is blocked in interleukin-7 receptor-deficient mice. J Exp Med 184:2423–2427

Malhotra N et al (2013) A network of high-mobility group box transcription factors programs innate interleukin-17 production. Immunity 38:681–693. https://doi.org/10.1016/j.immuni.2013.01.010

Malhotra D et al (2016) Tolerance is established in polyclonal CD4(+) T cells by distinct mechanisms, according to self-peptide expression patterns. Nat Immunol 17:187–195. https://doi.org/10.1038/ni.3327

Manesso E, Chickarmane V, Kueh HY, Rothenberg EV, Peterson C (2013) Computational modelling of T-cell formation kinetics: output regulated by initial proliferation-linked deferral of developmental competence. J R Soc Interface 10:20120774. https://doi.org/10.1098/rsif.2012.0774

Martin E et al (2009) Stepwise development of MAIT cells in mouse and human. PLoS Biol 7:e54. https://doi.org/10.1371/journal.pbio.1000054

Mbongue J, Nicholas D, Firek A, Langridge W (2014) The role of dendritic cells in tissue-specific autoimmunity. J Immunol Res 2014:857143. https://doi.org/10.1155/2014/857143

Michie AM, Zuniga-Pflucker JC (2002) Regulation of thymocyte differentiation: pre-TCR signals and beta-selection. Semin Immunol 14:311–323

Mohtashami M, Shah DK, Nakase H, Kianizad K, Petrie HT, Zuniga-Pflucker JC (2010) Direct comparison of Dll1- and Dll4-mediated Notch activation levels shows differential lympho-myeloid lineage commitment outcomes. J Immunol 185:867–876. https://doi.org/10.4049/jimmunol.1000782

Moran AE, Holzapfel KL, Xing Y, Cunningham NR, Maltzman JS, Punt J, Hogquist KA (2011) T cell receptor signal strength in Treg and iNKT cell development demonstrated by a novel fluorescent reporter mouse. J Exp Med 208:1279–1289. https://doi.org/10.1084/jem.20110308

Morita CT, Mariuzza RA, Brenner MB (2000) Antigen recognition by human gamma delta T cells: pattern recognition by the adaptive immune system. Springer Semin Immunopathol 22:191–217

Muroi S et al (2008) Cascading suppression of transcriptional silencers by ThPOK seals helper T cell fate. Nat Immunol 9:1113–1121. https://doi.org/10.1038/ni.1650

Naito T, Taniuchi I (2010) The network of transcription factors that underlie the CD4 versus CD8 lineage decision. Int Immunol 22:791–796. https://doi.org/10.1093/intimm/dxq436

Nie L, Xu M, Vladimirova A, Sun XH (2003) Notch-induced E2A ubiquitination and degradation are controlled by MAP kinase activities. EMBO J 22:5780–5792. https://doi.org/10.1093/emboj/cdg567

Nie L, Zhao Y, Wu W, Yang YZ, Wang HC, Sun XH (2011) Notch-induced Asb2 expression promotes protein ubiquitination by forming non-canonical E3 ligase complexes. Cell Res 21:754–
769. https://doi.org/10.1038/cr.2010.165

Nitta T et al (2015) The thymic cortical epithelium determines the TCR repertoire of IL-17-
producing gammadeltaT cells. EMBO Rep 16:638–653. https://doi.org/10.15252/
embr.201540096

Nosaka T et al (1995) Defective lymphoid development in mice lacking Jak3. Science 270:800–802

Offner F, Van Beneden K, Debacker V, Vanhecke D, Vandekerckhove B, Plum J, Leclercq G
(1997) Phenotypic and functional maturation of TCR gammadelta cells in the human thymus.
J Immunol 158:4634–4641

Okamoto Y, Douek DC, McFarland RD, Koup RA (2002) Effects of exogenous interleukin-7 on
human thymus function. Blood 99:2851–2858

Palaga T, Miele L, Golde TE, Osborne BA (2003) TCR-mediated Notch signaling regulates proliferation and IFN-gamma production in peripheral T cells. J Immunol 171:3019–3024

Park K, He X, Lee HO, Hua X, Li Y, Wiest D, Kappes DJ (2010) TCR-mediated ThPOK induction
promotes development of mature (CD24-) gammadelta thymocytes. EMBO J 29:2329–2341.
https://doi.org/10.1038/emboj.2010.113

Paust S, Lu L, McCarty N, Cantor H (2004) Engagement of B7 on effector T cells by regulatory
T cells prevents autoimmune disease. Proc Natl Acad Sci U S A 101:10398–10403. https://doi.
org/10.1073/pnas.0403342101

Payne KJ, Crooks GM (2002) Human hematopoietic lineage commitment. Immunol Rev
187:48–64

Perry SS, Wang H, Pierce LJ, Yang AM, Tsai S, Spangrude GJ (2004) L-selectin defines a bone
marrow analog to the thymic early T-lineage progenitor. Blood 103:2990–2996. https://doi.
org/10.1182/blood-2003-09-3030

Peschon JJ et al (1994) Early lymphocyte expansion is severely impaired in interleukin 7 receptor-
deficient mice. J Exp Med 180:1955–1960

Petrie HT, Zuniga-Pflucker JC (2007) Zoned out: functional mapping of stromal signaling microenvironments in the thymus. Annu Rev Immunol 25:649–679. https://doi.org/10.1146/annurev.
immunol.23.021704.115715

Porcelli S, Yockey CE, Brenner MB, Balk SP (1993) Analysis of T cell antigen receptor (TCR)
expression by human peripheral blood CD4-8- alpha/beta T cells demonstrates preferential use
of several V beta genes and an invariant TCR alpha chain. J Exp Med 178:1–16

Porritt HE, Rumfelt LL, Tabrizifard S, Schmitt TM, Zuniga-Pflucker JC, Petrie HT (2004)
Heterogeneity among DN1 prothymocytes reveals multiple progenitors with different capacities to generate T cell and non-T cell lineages. Immunity 20:735–745. https://doi.org/10.1016/j.
immuni.2004.05.004

Prinz I, Silva-Santos B, Pennington DJ (2013) Functional development of gammadelta T cells. Eur
J Immunol 43:1988–1994. https://doi.org/10.1002/eji.201343759

Proietto AI et al (2008) Dendritic cells in the thymus contribute to T-regulatory cell induction. Proc
Natl Acad Sci U S A 105:19869–19874. https://doi.org/10.1073/pnas.0810268105

Pui JC et al (1999) Notch1 expression in early lymphopoiesis influences B versus T lineage determination. Immunity 11:299–308

Qi Q, Xia M, Hu J, Hicks E, Iyer A, Xiong N, August A (2009) Enhanced development of CD4+
gammadelta T cells in the absence of Itk results in elevated IgE production. Blood 114:564–
571. https://doi.org/10.1182/blood-2008-12-196345

Radtke F, Wilson A, Stark G, Bauer M, van Meerwijk J, MacDonald HR, Aguet M (1999) Deficient
T cell fate specification in mice with an induced inactivation of Notch1. Immunity 10:547–558

Rahimpour A et al (2015) Identification of phenotypically and functionally heterogeneous mouse
mucosal-associated invariant T cells using MR1 tetramers. J Exp Med 212:1095–1108. https://
doi.org/10.1084/jem.20142110

Ramiro AR, Trigueros C, Marquez C, San Millan JL, Toribio ML (1996) Regulation of pre-T cell
receptor (pT alpha-TCR beta) gene expression during human thymic development. J Exp Med
184:519–530

Ramond C et al (2014) Two waves of distinct hematopoietic progenitor cells colonize the fetal thymus. Nat Immunol 15:27–35. https://doi.org/10.1038/ni.2782

Reimann C et al (2012) Human T-lymphoid progenitors generated in a feeder-cell-free Delta-like-4 culture system promote T-cell reconstitution in NOD/SCID/gammac(-/-) mice. Stem Cells 30:1771–1780. https://doi.org/10.1002/stem.1145

Rivera RR, Johns CP, Quan J, Johnson RS, Murre C (2000) Thymocyte selection is regulated by the helix-loop-helix inhibitor protein, Id3. Immunity 12:17–26

Roberts NA et al (2012) Rank signaling links the development of invariant gammadelta T cell progenitors and aire(+) medullary epithelium. Immunity 36:427–437. https://doi.org/10.1016/j.immuni.2012.01.016

Robey E et al (1996) An activated form of Notch influences the choice between CD4 and CD8 T cell lineages. Cell 87:483–492

Rodewald HR, Ogawa M, Haller C, Waskow C, DiSanto JP (1997) Pro-thymocyte expansion by c-kit and the common cytokine receptor gamma chain is essential for repertoire formation. Immunity 6:265–272

Roncarolo MG, Gregori S (2008) Is FOXP3 a bona fide marker for human regulatory T cells? Eur J Immunol 38:925–927. https://doi.org/10.1002/eji.200838168

Rossi FM et al (2005) Recruitment of adult thymic progenitors is regulated by P-selectin and its ligand PSGL-1. Nat Immunol 6:626–634. https://doi.org/10.1038/ni1203

Rothenberg EV (2014) Transcriptional control of early T and B cell developmental choices. Annu Rev Immunol 32:283–321. https://doi.org/10.1146/annurev-immunol-032712-100024

Rothenberg EV, Moore JE, Yui MA (2008) Launching the T-cell-lineage developmental programme. Nat Rev Immunol 8:9–21. https://doi.org/10.1038/nri2232

Rothenberg EV, Zhang J, Li L (2010) Multilayered specification of the T-cell lineage fate. Immunol Rev 238:150–168. https://doi.org/10.1111/j.1600-065X.2010.00964.x

Rothenberg EV, Ungerback J, Champhekar A (2016) Forging T-lymphocyte identity: intersecting networks of transcriptional control. Adv Immunol 129:109–174. https://doi.org/10.1016/bs.ai.2015.09.002

Roy S et al (2018) Id proteins suppress E2A-driven invariant natural killer T cell development prior to TCR selection. Front Immunol 9:42. https://doi.org/10.3389/fimmu.2018.00042

Sambandam A et al (2005) Notch signaling controls the generation and differentiation of early T lineage progenitors. Nat Immunol 6:663–670. https://doi.org/10.1038/ni1216

Sawada S, Scarborough JD, Killeen N, Littman DR (1994) A lineage-specific transcriptional silencer regulates CD4 gene expression during T lymphocyte development. Cell 77:917–929

Schmitt TM, Zuniga-Pflucker JC (2002) Induction of T cell development from hematopoietic progenitor cells by delta-like-1 in vitro. Immunity 17:749–756

Schmitt TM, Ciofani M, Petrie HT, Zuniga-Pflucker JC (2004) Maintenance of T cell specification and differentiation requires recurrent notch receptor-ligand interactions. J Exp Med 200:469–479. https://doi.org/10.1084/jem.20040394

Schwarz BA, Bhandoola A (2004) Circulating hematopoietic progenitors with T lineage potential. Nat Immunol 5:953–960. https://doi.org/10.1038/ni1101

Schwarz BA, Sambandam A, Maillard I, Harman BC, Love PE, Bhandoola A (2007) Selective thymus settling regulated by cytokine and chemokine receptors. J Immunol 178:2008–2017

Scott EW, Simon MC, Anastasi J, Singh H (1994) Requirement of transcription factor PU.1 in the development of multiple hematopoietic lineages. Science 265:1573–1577

Seddiki N et al (2006) Expression of interleukin (IL)-2 and IL-7 receptors discriminates between human regulatory and activated T cells. J Exp Med 203:1693–1700. https://doi.org/10.1084/jem.20060468

Serwold T, Ehrlich LI, Weissman IL (2009) Reductive isolation from bone marrow and blood implicates common lymphoid progenitors as the major source of thymopoiesis. Blood 113:807–815. https://doi.org/10.1182/blood-2008-08-173682

Shi J, Fallahi M, Luo JL, Petrie HT (2011) Nonoverlapping functions for Notch1 and Notch3 during murine steady-state thymic lymphopoiesis. Blood 118:2511–2519. https://doi.org/10.1182/blood-2011-04-346726

Shibata K et al (2011) Notch-Hes1 pathway is required for the development of IL-17-producing gammadelta T cells. Blood 118:586–593. https://doi.org/10.1182/blood-2011-02-334995

Singer A (2002) New perspectives on a developmental dilemma: the kinetic signaling model and the importance of signal duration for the CD4/CD8 lineage decision. Curr Opin Immunol 14:207–215

Singer A, Adoro S, Park JH (2008) Lineage fate and intense debate: myths, models and mechanisms of CD4- versus CD8-lineage choice. Nat Rev Immunol 8:788–801. https://doi.org/10.1038/nri2416

Six EM et al (2007) A human postnatal lymphoid progenitor capable of circulating and seeding the thymus. J Exp Med 204:3085–3093. https://doi.org/10.1084/jem.20071003

Sleasman JW, Harville TO, White GB, George JF, Barrett DJ, Goodenow MM (1994) Arrested rearrangement of TCR V beta genes in thymocytes from children with X-linked severe combined immunodeficiency disease. J Immunol 153:442–448

Spangrude GJ, Scollay R (1990) Differentiation of hematopoietic stem cells in irradiated mouse thymic lobes. Kinetics and phenotype of progeny. J Immunol 145:3661–3668

Spits H (2002) Development of alphabeta T cells in the human thymus. Nat Rev Immunol 2:760–772

Spits H, Lanier LL, Phillips JH (1995) Development of human T and natural killer cells. Blood 85:2654–2670

Spits H et al (1998) Early stages in the development of human T, natural killer and thymic dendritic cells. Immunol Rev 165:75–86

Spits H, Couwenberg F, Bakker AQ, Weijer K, Uittenbogaart CH (2000) Id2 and Id3 inhibit development of CD34(+) stem cells into predendritic cell (pre-DC)2 but not into pre-DC1. Evidence for a lymphoid origin of pre-DC2. J Exp Med 192:1775–1784

Suliman S et al (2011) Notch3 is dispensable for thymocyte beta-selection and Notch1-induced T cell leukemogenesis. PLoS One 6:e24937. https://doi.org/10.1371/journal.pone.0024937

Sun G et al (2005) The zinc finger protein cKrox directs CD4 lineage differentiation during intrathymic T cell positive selection. Nat Immunol 6:373–381. https://doi.org/10.1038/ni1183

Sutton CE, Lalor SJ, Sweeney CM, Brereton CF, Lavelle EC, Mills KH (2009) Interleukin-1 and IL-23 induce innate IL-17 production from gammadelta T cells, amplifying Th17 responses and autoimmunity. Immunity 31:331–341. https://doi.org/10.1016/j.immuni.2009.08.001

Taghon T, Rothenberg EV (2008) Molecular mechanisms that control mouse and human TCR-alphabeta and TCR-gammadelta T cell development. Semin Immunopathol 30:383–398. https://doi.org/10.1007/s00281-008-0134-3

Taghon T, De Smedt M, Stolz F, Cnockaert M, Plum J, Leclercq G (2001) Enforced expression of GATA-3 severely reduces human thymic cellularity. J Immunol 167:4468–4475

Taghon T, Stolz F, De Smedt M, Cnockaert M, Verhasselt B, Plum J, Leclercq G (2002) HOX-A10 regulates hematopoietic lineage commitment: evidence for a monocyte-specific transcription factor. Blood 99:1197–1204

Taghon TN, David ES, Zuniga-Pflucker JC, Rothenberg EV (2005) Delayed, asynchronous, and reversible T-lineage specification induced by Notch/Delta signaling. Genes Dev 19:965–978. https://doi.org/10.1101/gad.1298305

Taghon T, Yui MA, Pant R, Diamond RA, Rothenberg EV (2006) Developmental and molecular characterization of emerging beta- and gammadelta-selected pre-T cells in the adult mouse thymus. Immunity 24:53–64. https://doi.org/10.1016/j.immuni.2005.11.012

Taghon T, Van de Walle I, De Smet G, De Smedt M, Leclercq G, Vandekerckhove B, Plum J (2009) Notch signaling is required for proliferation but not for differentiation at a well-defined beta-selection checkpoint during human T-cell development. Blood 113:3254–3263. https://doi.org/10.1182/blood-2008-07-168906

Taghon T, Waegemans E, Van de Walle I (2012) Notch signaling during human T cell development. Curr Top Microbiol Immunol 360:75–97

Takaba H, Takayanagi H (2017) The mechanisms of T cell selection in the thymus. Trends Immunol 38:805–816. https://doi.org/10.1016/j.it.2017.07.010

Takaba H et al (2015) Fezf2 orchestrates a thymic program of self-antigen expression for immune tolerance. Cell 163:975–987. https://doi.org/10.1016/j.cell.2015.10.013

Tan JB, Visan I, Yuan JS, Guidos CJ (2005) Requirement for Notch1 signals at sequential early stages of intrathymic T cell development. Nat Immunol 6:671–679. https://doi.org/10.1038/ni1217

Tanigaki K et al (2004) Regulation of alphabeta/gammadelta T cell lineage commitment and peripheral T cell responses by Notch/RBP-J signaling. Immunity 20:611–622

Taniuchi I et al (2002a) Differential requirements for Runx proteins in CD4 repression and epigenetic silencing during T lymphocyte development. Cell 111:621–633

Taniuchi I, Sunshine MJ, Festenstein R, Littman DR (2002b) Evidence for distinct CD4 silencer functions at different stages of thymocyte differentiation. Mol Cell 10:1083–1096

Terstappen LW, Huang S, Picker LJ (1992) Flow cytometric assessment of human T-cell differentiation in thymus and bone marrow. Blood 79:666–677

Thompson PK, Zuniga-Pflucker JC (2011) On becoming a T cell, a convergence of factors kick it up a Notch along the way. Semin Immunol 23:350–359. https://doi.org/10.1016/j.smim.2011.08.007

Treiner E, Lantz O (2006) CD1d- and MR1-restricted invariant T cells: of mice and men. Curr Opin Immunol 18:519–526. https://doi.org/10.1016/j.coi.2006.07.001

Treiner E et al (2003) Selection of evolutionarily conserved mucosal-associated invariant T cells by MR1. Nature 422:164–169. https://doi.org/10.1038/nature01433

Treiner E, Duban L, Moura IC, Hansen T, Gilfillan S, Lantz O (2005) Mucosal-associated invariant T (MAIT) cells: an evolutionarily conserved T cell subset. Microbes Infect 7:552–559. https://doi.org/10.1016/j.micinf.2004.12.013

Tuovinen H, Kekalainen E, Rossi LH, Puntila J, Arstila TP (2008a) Cutting edge: human CD4-CD8- thymocytes express FOXP3 in the absence of a TCR. J Immunol 180:3651–3654

Tuovinen H, Pekkarinen PT, Rossi LH, Mattila I, Arstila TP (2008b) The FOXP3+ subset of human CD4+CD8+ thymocytes is immature and subject to intrathymic selection. Immunol Cell Biol 86:523–529. https://doi.org/10.1038/icb.2008.36

Turchinovich G, Hayday AC (2011) Skint-1 identifies a common molecular mechanism for the development of interferon-gamma-secreting versus interleukin-17-secreting gammadelta T cells. Immunity 35:59–68. https://doi.org/10.1016/j.immuni.2011.04.018

Turchinovich G, Pennington DJ (2011) T cell receptor signalling in gammadelta cell development: strength isn't everything. Trends Immunol 32:567–573. https://doi.org/10.1016/j.it.2011.09.005

Uehara S, Grinberg A, Farber JM, Love PE (2002) A role for CCR9 in T lymphocyte development and migration. J Immunol 168:2811–2819

Ueno T et al (2002) Role for CCR7 ligands in the emigration of newly generated T lymphocytes from the neonatal thymus. Immunity 16:205–218

Van Coppernolle S et al (2012) Notch induces human T-cell receptor gammadelta+ thymocytes to differentiate along a parallel, highly proliferative and bipotent CD4 CD8 double-positive pathway. Leukemia 26:127–138. https://doi.org/10.1038/leu.2011.324

Van de Walle I, De Smet G, De Smedt M, Vandekerckhove B, Leclercq G, Plum J, Taghon T (2009) An early decrease in Notch activation is required for human TCR-alphabeta lineage differentiation at the expense of TCR-gammadelta T cells. Blood 113:2988–2998. https://doi.org/10.1182/blood-2008-06-164871

Van de Walle I et al (2011) Jagged2 acts as a Delta-like Notch ligand during early hematopoietic cell fate decisions. Blood 117:4449–4459. https://doi.org/10.1182/blood-2010-06-290049

Van de Walle I et al (2013) Specific Notch receptor-ligand interactions control human TCR-alphabeta/gammadelta development by inducing differential Notch signal strength. J Exp Med 210:683–697. https://doi.org/10.1084/jem.20121798

Van de Walle I et al (2016) GATA3 induces human T-cell commitment by restraining Notch activity and repressing NK-cell fate. Nat Commun 7:11171. https://doi.org/10.1038/ncomms11171

Vantourout P, Hayday A (2013) Six-of-the-best: unique contributions of gammadelta T cells to immunology. Nat Rev Immunol 13:88–100. https://doi.org/10.1038/nri3384

Verschuren MC, Wolvers-Tettero IL, Breit TM, Noordzij J, van Wering ER, van Dongen JJ (1997) Preferential rearrangements of the T cell receptor-delta-deleting elements in human T cells. J Immunol 158:1208–1216

Verykokakis M, Krishnamoorthy V, Iavarone A, Lasorella A, Sigvardsson M, Kee BL (2013) Essential functions for ID proteins at multiple checkpoints in invariant NKT cell development. J Immunol 191:5973–5983. https://doi.org/10.4049/jimmunol.1301521

Wallis VJ, Leuchars E, Chwalinski S, Davies AJ (1975) On the sparse seeding of bone marrow and thymus in radiation chimaeras. Transplantation 19:2–11

Wang J et al (1998) Atomic structure of an alphabeta T cell receptor (TCR) heterodimer in complex with an anti-TCR fab fragment derived from a mitogenic antibody. EMBO J 17:10–26. https://doi.org/10.1093/emboj/17.1.10

Wang D et al (2006) The basic helix-loop-helix transcription factor HEBAlt is expressed in pro-T cells and enhances the generation of T cell precursors. J Immunol 177:109–119

Wang L et al (2008) Distinct functions for the transcription factors GATA-3 and ThPOK during intrathymic differentiation of CD4(+) T cells. Nat Immunol 9:1122–1130. https://doi.org/10.1038/ni.1647

Washburn T, Schweighoffer E, Gridley T, Chang D, Fowlkes BJ, Cado D, Robey E (1997) Notch activity influences the alphabeta versus gammadelta T cell lineage decision. Cell 88:833–843

Watanabe N, Hanabuchi S, Soumelis V, Yuan W, Ho S, de Waal Malefyt R, Liu YJ (2004) Human thymic stromal lymphopoietin promotes dendritic cell-mediated CD4+ T cell homeostatic expansion. Nat Immunol 5:426–434. https://doi.org/10.1038/ni1048

Watanabe N, Wang YH, Lee HK, Ito T, Wang YH, Cao W, Liu YJ (2005) Hassall's corpuscles instruct dendritic cells to induce CD4+CD25+ regulatory T cells in human thymus. Nature 436:1181–1185. https://doi.org/10.1038/nature03886

Weber BN, Chi AW, Chavez A, Yashiro-Ohtani Y, Yang Q, Shestova O, Bhandoola A (2011) A critical role for TCF-1 in T-lineage specification and differentiation. Nature 476:63–68. https://doi.org/10.1038/nature10279

Weerkamp F, Luis TC, Naber BA, Koster EE, Jeannotte L, van Dongen JJ, Staal FJ (2006) Identification of Notch target genes in uncommitted T-cell progenitors: no direct induction of a T-cell specific gene program. Leukemia 20:1967–1977. https://doi.org/10.1038/sj.leu.2404396

Welner RS et al (2013) C/EBPalpha is required for development of dendritic cell progenitors. Blood 121:4073–4081. https://doi.org/10.1182/blood-2012-10-463448

Weng AP et al (2006) c-Myc is an important direct target of Notch1 in T-cell acute lymphoblastic leukemia/lymphoma. Genes Dev 20:2096–2109. https://doi.org/10.1101/gad.1450406

Wildin RS et al (2001) X-linked neonatal diabetes mellitus, enteropathy and endocrinopathy syndrome is the human equivalent of mouse scurfy. Nat Genet 27:18–20. https://doi.org/10.1038/83707

Wildt KF, Sun G, Grueter B, Fischer M, Zamisch M, Ehlers M, Bosselut R (2007) The transcription factor Zbtb7b promotes CD4 expression by antagonizing Runx-mediated activation of the CD4 silencer. J Immunol 179:4405–4414

Wilson A, MacDonald HR, Radtke F (2001) Notch 1-deficient common lymphoid precursors adopt a B cell fate in the thymus. J Exp Med 194:1003–1012

Witt CM, Hurez V, Swindle CS, Hamada Y, Klug CA (2003) Activated Notch2 potentiates CD8 lineage maturation and promotes the selective development of B1 B cells. Mol Cell Biol 23:8637–8650

Wojciechowski J, Lai A, Kondo M, Zhuang Y (2007) E2A and HEB are required to block thymocyte proliferation prior to pre-TCR expression. J Immunol 178:5717–5726

Wolfer A et al (2001) Inactivation of Notch 1 in immature thymocytes does not perturb CD4 or CD8T cell development. Nat Immunol 2:235–241. https://doi.org/10.1038/85294

Wong GW, Zuniga-Pflucker JC (2010) gammadelta and alphabeta T cell lineage choice: resolution by a stronger sense of being. Semin Immunol 22:228–236. https://doi.org/10.1016/j.smim.2010.04.005

Wong GW, Knowles GC, Mak TW, Ferrando AA, Zuniga-Pflucker JC (2012) HES1 opposes a PTEN-dependent check on survival, differentiation, and proliferation of TCRbeta-selected mouse thymocytes. Blood 120:1439–1448. https://doi.org/10.1182/blood-2011-12-395319

Xiong Y, Bosselut R (2012) CD4-CD8 differentiation in the thymus: connecting circuits and building memories. Curr Opin Immunol 24:139–145. https://doi.org/10.1016/j.coi.2012.02.002

Xu W, Carr T, Ramirez K, McGregor S, Sigvardsson M, Kee BL (2013) E2A transcription factors limit expression of Gata3 to facilitate T lymphocyte lineage commitment. Blood 121:1534–1542. https://doi.org/10.1182/blood-2012-08-449447

Yamasaki S et al (2006) Mechanistic basis of pre-T cell receptor-mediated autonomous signaling critical for thymocyte development. Nat Immunol 7:67 75. https://doi.org/10.1038/ni1290

Yashiro-Ohtani Y et al (2009) Pre-TCR signaling inactivates Notch1 transcription by antagonizing E2A. Genes Dev 23:1665–1676. https://doi.org/10.1101/gad.1793709

Yokota Y, Mansouri A, Mori S, Sugawara S, Adachi S, Nishikawa S, Gruss P (1999) Development of peripheral lymphoid organs and natural killer cells depends on the helix-loop-helix inhibitor Id2. Nature 397:702–706. https://doi.org/10.1038/17812

Yu VW et al (2015) Specific bone cells produce DLL4 to generate thymus-seeding progenitors from bone marrow. J Exp Med 212:759–774. https://doi.org/10.1084/jem.20141843

Yui MA, Rothenberg EV (2014) Developmental gene networks: a triathlon on the course to T cell identity. Nat Rev Immunol 14:529–545. https://doi.org/10.1038/nri3702

Yui MA, Feng N, Rothenberg EV (2010) Fine-scale staging of T cell lineage commitment in adult mouse thymus. J Immunol 185:284–293. https://doi.org/10.4049/jimmunol.1000679

Yun TJ, Bevan MJ (2003) Notch-regulated ankyrin-repeat protein inhibits Notch1 signaling: multiple Notch1 signaling pathways involved in T cell development. J Immunol 170:5834–5841

Zarin P, Wong GW, Mohtashami M, Wiest DL, Zuniga-Pflucker JC (2014) Enforcement of gammadelta-lineage commitment by the pre-T-cell receptor in precursors with weak gammadelta-TCR signals. Proc Natl Acad Sci U S A 111:5658–5663. https://doi.org/10.1073/pnas.1312872111

Zarin P, Chen EL, In TS, Anderson MK, Zuniga-Pflucker JC (2015) Gamma delta T-cell differentiation and effector function programming, TCR signal strength, when and how much? Cell Immunol 296:70–75. https://doi.org/10.1016/j.cellimm.2015.03.007

Zarin P et al (2018) Integration of T-cell receptor, Notch and cytokine signals programs mouse gammadelta T-cell effector differentiation. Immunol Cell Biol 96:994. https://doi.org/10.1111/imcb.12164

Zhang DE, Zhang P, Wang ND, Hetherington CJ, Darlington GJ, Tenen DG (1997) Absence of granulocyte colony-stimulating factor signaling and neutrophil development in CCAAT enhancer binding protein alpha-deficient mice. Proc Natl Acad Sci U S A 94:569–574

Zhang JA, Mortazavi A, Williams BA, Wold BJ, Rothenberg EV (2012) Dynamic transformations of genome-wide epigenetic marking and transcriptional control establish T cell identity. Cell 149:467–482. https://doi.org/10.1016/j.cell.2012.01.056

Chapter 5
Intrathymic Cell Migration: Implications in Thymocyte Development and T-Cell Repertoire Formation

Daniella Arêas Mendes-da-Cruz, Carolina Valença Messias, Julia Pereira Lemos, and Wilson Savino

Abstract During intrathymic T-cell development, differentiating thymocytes migrate through distinct thymic compartments, interacting with the cortical and medullary microenvironments of the thymic lobules. This migration is mainly regulated by adhesion molecules (including extracellular matrix ligands and receptors) as well as soluble factors such as chemokines, which are also crucial for thymocyte differentiation. The migration events orchestrated by these molecules comprise the entry of bone marrow derived lymphoid progenitors, the migration within the thymus and the emigration of mature thymocytes. Importantly, migration of developing T cells can also impact the positive and negative selection processes, which are essential for avoiding the generation of autoreactive T cells. In this chapter we will focus on key molecules involved in thymocyte migration and how their expression pattern can possibly impact T-cell development and repertoire formation.

5.1   Introduction

Developing thymocytes undergo differentiation while migrate throughout distinct thymic niches within the thymic lobules, in a key process for proper T-cell functional maturation (Savino et al. 2002a). Different signals guide precursor cells entrance in the thymus, the migration of developing thymocytes inside the organ and mature thymocytes egress to the periphery of the immune system. During this journey, thymocytes experience distinct types of interactions and receive appropriate signals for cell survival, proliferation and differentiation that culminate in the

D. A. Mendes-da-Cruz (✉) · C. V. Messias · J. P. Lemos · W. Savino
Laboratory on Thymus Research, Oswaldo Cruz Institute, Oswaldo Cruz Foundation, Rio de Janeiro, RJ, Brazil

National Institute of Science and Technology on Neuroimmunomodulation (INCT-NIM), Rio de Janeiro, RJ, Brazil
e-mail: daniella@ioc.fiocruz.br

© Springer Nature Switzerland AG 2019
G. A. Passos (ed.), *Thymus Transcriptome and Cell Biology*,
https://doi.org/10.1007/978-3-030-12040-5_5

generation of a T cell repertoire able to respond to foreign antigens. Those signals can be transmitted by cell-cell interactions, for example, mediated by class I and class II major histocompatibility complex (MHC) and T-cell receptor (TCR), expressed by thymic epithelial cells (TEC) and developing T cells respectively; by soluble moieties as cytokines and chemokines and respective receptors; by classical adhesion molecules as LFA-3/CD2 and ICAM-1/LFA-1, as well as extracellular matrix (ECM) ligands and corresponding integrin-type receptors (Savino et al. 2000; Ayres-Martins et al. 2004). As detailed below, this chapter will be devoted to various cellular interactions involved in the control of the intrathymic T-cell migration as well as their implications for the generation of the T-cell repertoire. Nevertheless, it seemed worthwhile to provide a basic background on the ligand receptor pairs for which consistent evidence upon intrathymic cell migration has been established.

5.2 Intrathymic Expression of Extracellular Matrix Ligands and Receptors: Role Upon Thymocyte Adhesion

The first description of specific ECM proteins in the thymus dates the 1980s when fibronectin, laminin, and various collagens were identified in the human thymic parenchyma and TEC cultures (Henry 1967). Importantly, the patterns of intrathymic distribution of these ECM proteins were found to be very conserved in mammals (Meireles De Souza et al. 1993), suggesting since then an essential role in thymus physiology and thymocyte development. Over the years, a variety of ECM ligands were found to be expressed in the organ (Berrih et al. 1985; Lannes-Vieira et al. 1991), being produced mainly by TEC, but also by fibroblasts and MHC class II+ phagocytic cells, as summarized in Table 5.1.

Fibronectin is a major ECM structural component in the adult thymus, commonly found adjacent to medullary TECs (mTEC), macrophages and fibroblasts, basement membranes bordering the thymic capsule, septa, as well as the perivascular spaces (PVS). The classical fibronectin isoform, recognized by VLA-5 (the $\alpha5\beta1$ integrin, CD49e/CD29) is expressed throughout the thymic parenchyma, whereas the alternative form, resulted from pre-mRNA alternative splicing, and recognized by the integrin VLA-4 ($\alpha4\beta1$, CD49d/CD29), is restricted to the medulla of the thymic lobules (Crisa et al. 1996).

Both thymic microenvironment cells and developing thymocytes express VLA-4 and VLA-5 (Shimizu and Shaw 1991; Savino et al. 2000, 2003). VLA-4 is expressed by almost all thymocyte subpopulations; the membrane expression is higher on DN cells and decreases during thymocyte maturation, being slightly and strongly decreased in DP and SP subpopulations, respectively (Mojcik et al. 1995). VLA-5 is also expressed by the majority of thymocytes, although its expression pattern differs from that observed for VLA-4; it is expressed at high levels by DN cells, decreases at the DP stage and is up-regulated in mature SP thymocytes (Mojcik et al. 1995). Functionally, immature thymocytes adhere more efficiently to fibronec-

Table 5.1 Expression of thymocyte migration-related molecules and corresponding receptors in the thymus

ECM-related molecules	Receptors	References
Fibronectin isoform 1	VLA-5	Crisa et al. (1996)
Fibronectin isoform 2	VLA-4	Crisa et al. (1996)
Laminin-221	α7β1	Kutleša et al. (2002), Savino et al. (2015)
Laminin-211	VLA-6	Iwao et al. (2000), Magner et al. (2000), Kutleša et al. (2002), Ocampo et al. (2008), Golbert et al. (2014)
Laminin-332	VLA-6, α6β4	Vivinus-Nebot et al. (1999), Kim et al. (2000), Kutleša et al. (2002), Golbert et al. (2014)
Laminin-411	VLA-3, VLA-6, α7β1	Kutleša et al. (2002), Nishiuchi et al. (2006), Golbert et al. (2014)
Laminin-421	VLA-6	Kutleša et al. (2002), Golbert et al. (2014)
Laminin-511	VLA-3, VLA-6, α6β4	Kutleša et al. (2002), Nishiuchi et al. (2006), Golbert et al. (2014)
Laminin-521	VLA-3, VLA-6, α6β4	Kutleša et al. (2002), Nishiuchi et al. (2006), Golbert et al. (2014)
Type I Collagen	VLA-1, VLA-2, VLA-3	Berrih et al. (1985), Lannes-Vieira et al. (1991), Villa-Verde et al. (1994), Mori et al. (2007)
Type III Collagen	VLA-1, VLA-2, VLA-3	Berrih et al. (1985), Lannes-Vieira et al. (1991), Villa-Verde et al. (1994), Mori et al. (2007)
Type IV Collagen	VLA-1, VLA-2, CD26	Berrih et al. (1985), Simon et al. (1991), Lannes-Vieira et al. (1991), Goldman et al. (1992), Villa-Verde et al. (1994), Gorski and Kupiec-Weglinski (1995), Mori et al. (2007)
Hyaluronic acid	CD44, RHAMM	Lynch and Ceredig (1988), Gares and Pilarski (1999, 2000)
Tenascin	ND	Hemesath and Stefansson (1994), Freitas et al. (1995)
Netrin-1	Unc5B, α6β4	Hong et al. (1999), Corset et al. (2000), Yebra et al. (2003), Barallobre et al. (2005)
Soluble factors	**Receptors**	**References**
CCL1	CCR8	Zingoni et al. (1998), Kremer et al. (2001)
CCL2	CCR2	Baba et al. (2009)
CCL4	CCR5	Dairaghi et al. (1998)
CCL11	CCR3	Franz-bacon et al. (1999)
CCL17	CCR4	Cowan et al. (2014), Hu et al. (2015)
CCL19	CCR7	Misslitz et al. (2004), Krueger et al. (2010), Zlotoff et al. (2010)
CCL20	CCR6	Bunting et al. (2014, 2014)
CCL21	CCR7	Misslitz et al. (2004), Krueger et al. (2010), Zlotoff et al. (2010)
CCL22	CCR4	Cowan et al. (2014), Hu et al. (2015)

(continued)

Table 5.1 (continued)

CCL25	CCR9	Uehara et al. (2002b), Schwarz et al. (2007), Gossens et al. (2009), Krueger et al. (2010)
CXCL9	CXCR3	Romagnani et al. (2001), Drennan et al. (2009)
CXCL10	CXCR3	Romagnani et al. (2001), Drennan et al. (2009)
CXCL11	CXCR3	Romagnani et al. (2001), Drennan et al. (2009)
CXCL12	CXCR4	Aiuti et al. (1999), Robertson et al. (2006)
Sphingosin-1-phosphate	S1P1	Matloubian et al. (2004), Allende et al. (2011)
Growth hormone	GHR	Baxter et al. (1991), Mertani et al. (1995)
Triiodothyronine	TR	Villa-Verde et al. (1993)
Semaphorin-3A	NRP1[a]/NRP2	Lepelletier et al. (2007)
Semaphorin-3F	NRP1/NRP2[a]	Takahashi et al. (2008), Mendes-da-Cruz et al. (2014)
Membrane-bound proteins	**Receptors**	**References**
Ephrin A1	EphA1	Coulthard et al. (2001), Muñoz et al. (2002)
Ephrin A2	EphA1, A2, A3, A4, A5	Muñoz et al. (2002)
Ephrin A3	EphA1, A2, A3, A4, A5	Muñoz et al. (2002)
Ephrin B2	EphB1, B2, B3, B4, B6	Stimamiglio et al. (2010), García-Ceca et al. (2013)
Ephrin B3	EphB1, B2, B3, B4, B6	García-Ceca et al. (2013), Montero-Herradón et al. (2017)

ND = not determined, RHAMM = receptor for hyaluronan-mediated motility, Unc-5b = Unc-5 Netrin Receptor B, VLA = Very late antigen, GHR = Growth hormone receptor, TR = Triiodothyronine receptors, NRP1 = neuropilin-1, NRP2 = neuropilin-2
[a]Higher affinity

tin and TEC monolayers (Utsumi et al. 1991) and mature SP cells are less adhesive to thymic microenvironmental cells (Sawada et al. 1992).

Blockage of thymocyte adhesion to fibronectin using a trans-dominant inhibitor of β1 integrin in the thymus led to partial blockade of DN to DP differentiation and reduction of CD4 SP cells (Schmeissner et al. 2001). Moreover, it has been shown that DN thymocytes cultured with fibronectin differentiate to DP and CD4 or CD8 SP cells; conversely, the addition of synthetic peptides or anti-fibronectin antibody inhibited T cell differentiation (Utsumi et al. 1991).

Some studies show that immature human thymocytes preferentially use VLA-4 to adhere and migrate on fibronectin, while mature SP cells use both VLA-4 and VLA-5 receptors, which seem to induce migration rather than firm adhesion in these subpopulations (Salomon et al. 1994; Crisa et al. 1996). Moreover, anti-VLA-4 monoclonal antibody or fibronectin LDV sequence-containing synthetic peptides blocked the adhesion of immature DN thymocytes to TEC monolayers or fibronectin-coated plates, whereas anti-VLA-5 or RGD peptide did not (Sawada et al. 1992; Dalmau et al. 1999).

Interestingly, specific small interference RNA gene silencing of CD49e (the α5 integrin chain) in TEC resulted in the modulation of more than 100 genes, some of them coding for other proteins involved in thymocyte adhesion, but also for signaling pathways triggered after integrin activation, as well as the control of F-actin formation (Linhares-Lacerda et al. 2010). These studies show the importance of fibronectin/VLA-5 signaling for thymocytes and TECs physiology that can impact the generation of mature T cells.

A second major ECM molecule constitutively expressed in the thymus is laminin, a heterotrimeric cross-shaped coiled-coil glycoprotein, formed by the combination of three chains generated by the transcription of distinct genes. They are classified according to the corresponding α, β, and γ chains and at least 18 laminin isoforms have been described, with five α, three β, and three γ polypeptide chains chemically characterized (Savino et al. 2015). The current laminin nomenclature uses Arabic numerals based on the chain numbers. So laminin-111 is formed by α1, β1, and γ1 chains, for example.

Many isoforms were described to be heterogeneously distributed within the thymic lobules, being produced by TEC and the cortically-located lymphoepithelial complexes, the thymic nurse cells (TNC), as well as phagocytic cells and fibroblasts (Anderson et al. 1997; Savino et al. 2000; Suniara et al. 2000). Both humans and mice display a dense laminin network in the thymic medulla, compared to the cortex, as well as in septal membranes and perivascular spaces (Berrih et al. 1985; Bofill et al. 1985; Lannes-Vieira et al. 1991; Ayres-Martins et al. 2004).

Laminin-111 is the mostly conspicuous isoform during early embryogenesis, but its expression in adults is limited to certain epithelial basement membranes within the lobules and septa (Savino et al. 1991; Falk et al. 1999; Smyth et al. 1999; Murray and Edgar 2000; Virtanen et al. 2000). Laminin-211 was the first isoform characterized in the mouse thymus (Chang et al. 1993). Immunocytochemistry revealed a selective distribution within the thymic lobules, being restricted to cortical TECs in the subcapsular epithelium and blood vessels of human thymus (Kutleša et al. 2002). Laminin isoforms containing α3 chains were also described in human thymus (Virtanen et al. 1996), with laminin-332 (α3Aβ3γ2) being mainly found in the basement membranes of human thymic medulla, produced by medullary TECs, in blood vessels and in the subcapsular cortical epithelium in mice (Vivinus-Nebot et al. 1999; Kim et al. 2000; Kutleša et al. 2002). Moreover, expression of α4, α5, β1, β2, β3, γ1 and γ2 chains were also detected in human thymus, suggesting the expression of other laminin isoforms, such as 221, 511 and 521 (Kutleša et al. 2002; Savino et al. 2015).

In addition to the intrathymic ligand expression, both thymocytes and TEC constitutively express the laminin receptors VLA-6 (α6β1 integrin, CD49f/CD29), VLA-3 (α3β1, CD49c/CD29) and α6β4 (Wadsworth et al. 1992, 1993; Lannes-Vieira et al. 1993; Ocampo et al. 2008). Functionally, immature thymocytes preferentially adhere to laminin, 2–4-fold higher than mature thymocytes (Savino et al. 2003). These cells use the VLA-6 rather than α6β4 integrin to adhere to the laminin-211 isoform, suggesting that VLA-6 is activated in most thymocytes (Wadsworth et al. 1993; Chang et al. 1993). Thymocyte adhesion to TEC monolayers is also partially mediated by VLA-6 and the adhesion is impaired by α6 monoclonal antibodies or laminin-211 (Lannes-Vieira et al. 1993; Ocampo et al. 2008). Moreover, knocking-down the ITGA6 gene, which encodes CD49f (the integrin α6 chain) or the use of anti-VLA-6 antibodies, led to decreased thymocyte adhesion (Virtanen et al. 1996; Vivinus-Nebot et al. 1999; Kutleša et al. 2002; Golbert et al. 2013).

As mentioned above, laminins-211 and -221 are mainly found in the subcapsulary region and immature thymocytes present higher adhesion capacity to laminin-211 than mature medulla located cells (Wadsworth et al. 1992). Also, early DN cells strongly adhere to laminin-511 ($\alpha5\beta1\gamma1$) and laminin-521 ($\alpha5\beta2\gamma1$) using VLA-6 (Iwao et al. 2000; Magner et al. 2000; Kutleša et al. 2002).

Human CD8 SP cells adhere to laminin-332 expressed by medullary TEC. This interaction is dependent on the $\alpha6\beta4$ integrin, suggesting an involvement of this laminin isoform in mature thymocyte differentiation. Although CD4 SP thymocytes also express VLA-3, VLA-6 and $\alpha6\beta4$, they were not able to adhere to this isoform, suggesting that these integrins, although being expressed are not activated. Also, in human thymus, laminin-332 inhibits human thymocyte proliferation induced by anti-CD3 and IL-2 via $\alpha6\beta4$, and induces mature thymocyte migration (Vivinus-Nebot et al. 1999; Kutleša et al. 2002).

In addition to fibronectin and laminin isoforms, collagens are present at several sites in the murine and human thymus parenchyma. They are predominantly expressed in the thymic capsule and septa, as well as in blood vessels and PVS, being produced by fibroblasts and TEC, including the epithelial component of TNC. In young human thymuses, type I collagen predominates in the capsule, septa and PVS; type III collagen forms reticulum fibers and the type IV collagen is mainly found in the basement membranes adjacent to TEC and PVS. In keeping with this localization, it has been demonstrated that TEC constitutively express type IV collagen (Berrih et al. 1985; Lannes-Vieira et al. 1991; Villa-Verde et al. 1994; Mori et al. 2007).

Tenascin is another ECM protein expressed in both human and murine thymuses, both in the medulla and corticomedullary junction (Saga et al. 1991; Hemesath and Stefansson 1994; Freitas et al. 1995). *In vitro* studies suggest that tenascin-producing cells are not epithelial and probably not fibroblastic microenvironmental elements (Freitas et al. 1995). Also, a tenascin receptor on thymocytes is not yet well characterized, although a possible candidate is an integrin RGD-dependent binding receptor, such as the $\alpha v\beta3$ integrin (Bourdon and Ruoslahti 1989).

A further adhesive protein found in the mouse thymus is netrin-1, a laminin-related molecule originally described in the nervous system (Serafini et al. 1996; Guo et al. 2013). Netrin-1 can induce either attraction or repulsion depending on the cell type and kind of receptor. Several receptors have been described for netrin-1, belonging to the UNC5 family (Unc5a, Unc5b, Unc5c, and Unc5d), the adenosine A2b receptor and the $\alpha6\beta4$ integrin (Hong et al. 1999; Corset et al. 2000; Yebra et al. 2003; Barallobre et al. 2005). In the thymus, netrin-1 is expressed by microenvironmental cells and is inducible in thymocytes, mainly in DP cells, after stimulation with anti-CD3 or IL-7, suggesting a role in thymocyte localization and selection. Concerning the receptors, thymocytes express the Unc5b and $\alpha6\beta4$, and adhesion to netrin-1 is dependent on $\alpha6\beta4$. Netrin-1 alone is not able to induce thymocyte migratory responses, but potentiate chemotaxis induced by the chemokine CXCL12. Interestingly, netrin-1 had no effect on total thymocyte migration induced by the chemokine CCL19, suggesting that this effect in specific for certain chemokines (Guo et al. 2013).

The hyaluronic acid or hyaluronan, is a glycosaminoglycan produced by TEC and phagocytic cells, also capable to modulate TEC/thymocyte adhesion (Villa-Verde et al. 1994; Patel et al. 1995; Oliveira-dos-Santos et al. 1997; Werneck et al. 2000). The distribution of hialuronan is similar to that observed for fibronectin and its first described receptor, CD44, is mainly expressed in immature DN cells (Lynch and Ceredig 1988). In this respect, it has been showed that the generation of DP cells from CD25$^+$CD44$^+$ DN precursors is diminished in the presence of the anti-CD44 blocking antibody, as well as the adhesion of the latter to ECM-associated fibroblasts that reside in the cortical region (Anderson et al. 1997).

Taken together, data on the expression and role of ECM ligands and receptors in the thymus show the importance of these molecules in essential steps necessary for thymocyte differentiation including thymocyte adhesion to microenvironmental cells. As seen below, these interactions also affect thymocyte migration throughout the thymic lobules.

5.3 Chemokines and Other Soluble Factors Binding to G Protein-Coupled Receptors

A number of cell migration related soluble factors are constitutively expressed in the thymus, particularly chemokines and the sphingosine-1-phosphate (S1P).

Chemokines are small chemotactic soluble proteins (8–14 kDa) that regulate cellular trafficking through its interaction with seven-spanning transmembrane G protein-coupled receptors (Hughes and Nibbs 2018). Chemokines can be expressed on the cell surface, associated to ECM molecules or secreted in the microenvironment. These chemotactic molecules and their receptors are classified according to the cysteine residues present in a conserved region in the ligands. To identify the different chemokines, an "L" suffix followed by decimal numbers is added to the subfamily name, while a "R" suffix is added to identify the receptors. To date, 48 chemokines and 23 chemokines receptors have been identified, which indicate that at least some chemokines and receptors have a promiscuous interaction (Barondes et al. 1994; Zlotnik and Yoshie 2000, 2012; Hughes and Nibbs 2018).

Some chemokines can induce the attraction of leukocytes considering an increasing concentration gradient (chemotaxis) or induce the repulsion of these cells considering a decreasing concentration gradient (fugetaxis), as the CXCL12 chemokine. The type of guiding response will depend on the chemokine concentration in the microenvironment (Poznansky et al. 2000; Vianello et al. 2005).

A considerable number of chemokines were detected in the thymus and several studies demonstrated that these molecules are produced in specific thymic compartments and their receptors are expressed by distinct thymocyte subpopulations (Kim et al. 1998; Campbell et al. 1999; Bleul and Boehm 2000; Annunziato et al. 2001; Savino et al. 2002a; Halkias et al. 2014). As detailed below in this chapter, several of them modulate intrathymic cell migration at different stages of thymocyte development.

In addition to chemokines, sphingosine1-phosphate (S1P) bind to G protein-coupled receptors and affect cell migration (Fyrst and Saba 2010), also playing an important role in development during angiogenesis, cardiogenesis, neurogenesis and limbic development (Mendelson et al. 2014). S1P receptors, like chemokines receptors, are seven-spanning transmembrane G protein-coupled receptors (Watterson et al. 2003). However, they are able to interact with different G-proteins (G_i, G_s, G_q and $G_{12/13}$), allowing S1P to act in several physiological processes (Takabe et al. 2008). Despite exerting different actions depending on the G protein activated, S1P interaction with all of the five receptors (S1P1-5) can lead to cell migration in the immune system (Kohno et al. 2003; Allende et al. 2004, 2011; Matloubian et al. 2004; Walzer et al. 2007; Jenne et al. 2009; Mayol et al. 2011; Keul et al. 2011; Green et al. 2011; Ishii and Kikuta 2013; Nussbaum et al. 2015).

Apparently, erythrocytes are the main source of S1P in the blood (Thuy et al. 2014), whereas lymphatic endothelial cells are the main source in lymph (Spiegel and Milstien 2011). A S1P gradient is observed between the vascular (plasma/serum) and extravascular (lymphoid tissues) compartments. Higher concentrations are found in the plasma and serum whereas lower concentrations are found in lymphoid tissues.

It has been demonstrated that murine plasma contains 330 ng/g (860 nM) of S1P, while the thymus, lymph nodes and spleen contain 30, 40 and 150 ng/g of S1P respectively (Schwab et al. 2005). S1P concentrations in human plasma is approximately 400 nM, whereas S1P concentration in human serum is around 800 nM (Murata et al. 2000). So far, the S1P concentrations in human lymphoid organs are not known.

Like cellular motility mediated by chemokines, chemotactic or fugetactic responses induced by S1P will depend on the S1P concentration in the microenvironment (Ishii et al. 2010; Messias et al. 2016). The first evidence on the role of these molecules upon thymocytes came from studies with pertussis toxin (PTX), which impairs signaling through $G_{\alpha i}$ protein-coupled receptors. Thymocytes treated with PTX are unable to leave the thymus and migrate to secondary lymphoid organs (Chaffin and Perlmutter 1991). Moreover, mice treated with PTX present a decrease in thymic weight and cell numbers in the cortex. In this case, SP thymocytes are not found in the thymic medulla, since developing thymocytes remain retained in the cortical region (Person et al. 1992; Suzuki et al. 1999).

5.4   Entry of Hematopoietic Progenitors in the Thymus

T cell progenitors enter the thymus through postcapillary veins present in the corticomedullary junction of the thymic lobules (Lind et al. 2001; Mori et al. 2007). Transmigration of progenitor cells through the endothelium begins with rolling and subsequent firm adhesion of these cells to integrins ligands expressed by endothelial cells. However, this firm adhesion is only possible due to chemokine action that increases integrin affinity for their ligand.

It was demonstrated that T cell progenitors express CCR7 and CCR9 (Schwarz et al. 2007; Krueger et al. 2010; Zlotoff et al. 2010). Their ligands are detected in the thymus: CCL19 and CCL21, both CCR7 ligands, are expressed by medullary thymic epithelial cells and thymic endothelial cells (Ueno et al. 2002), whereas CCL25, a CCR9 ligand, is expressed mainly by cortical epithelial cells (Buono et al. 2016). Some studies demonstrated that CCR7-deficient mice have increased frequency of T cell progenitors (Krueger et al. 2010) and morphological changes in the organ, with smaller but more numerous medullary regions (Misslitz et al. 2004). Conversely, smaller frequencies of T cell progenitors were detected in the thymus of CCR9-deficient mice (Krueger et al. 2010). Furthermore, competitive intravenous transfer of normal plus CCR9-deficient bone marrow cells revealed that the lack of this chemokine receptor impairs T cell progenitor entry (Uehara et al. 2002a).

Importantly, practically no T cell progenitors were observed in the thymus of CCR7-CCR9-double-deficient mice (Krueger et al. 2010; Zlotoff et al. 2010), suggesting that together these receptors play a major role in the recruitment of T cell progenitors.

Other chemokine receptors are also implicated in the entrance of T cell progenitors in the thymus. For example, CXCR4 blockage partially inhibits T cell progenitor homing into the thymus (Robertson et al. 2006). Accordingly, *in vitro* assays have shown that human fetal CD34$^+$ T cell progenitors express CXCR4 and migrate towards CXCL12 (Aiuti et al. 1999), which is widely expressed in the thymus parenchyma (Annunziato et al. 2001; Hernandez-Lopez et al. 2002; Plotkin et al. 2003; Trampont et al. 2010).

In summary, apparently both CCR7 and CCR9 have a critical role for the recruitment of T cell precursors to the thymus, although CXCR4 also seems to contribute to this process.

In addition to chemokines, entry of T-cell precursors into the organ is also dependent of laminin, and its interaction with the VLA-6 laminin receptor. Blockage of the ligand or its receptor by neutralizing antibodies abrogates the entrance of these precursors (Savagner et al. 1986; Stimamiglio et al. 2010). Furthermore, several data show that the interaction with fibronectin may be important to direct cell migration and to stabilize the interactions between developing thymocytes and thymic microenvironmental cells, such as TEC. During embryonic development, fibronectin secreted by TEC and VLA-5 interactions seems to be involved in early hematopoietic precursor cells entry into the thymus, since this process was blocked by anti-fibronectin antibody or synthetic peptides containing the VLA-5 RGD-binding sequence (Savagner et al. 1986). Likewise, adhesion to microenvironmental cells through fibronectin was shown to be required for thymocyte differentiation *in vitro*.

Laminin-111 mediated interactions involving integrin-type laminin receptors also participate in precursor cells homing to the thymus during embryogenesis (Ruiz et al. 1995; Stimamiglio et al. 2010). Additionally, *in vitro* experiments revealed a role of hyaluronic acid in these early events, since that the entrance of T cells progenitors is dependent on CD44 and the transmigration through blood vessels can be abrogated with anti-CD44 antibody (Ruiz et al. 1995).

5.5   Migration from the Corticomedullary Junction
to the Outer Cortex

Once inside the thymus, DN thymocytes receive different survival and proliferation signals that culminate with the T cell lineage commitment. These cells encounter several chemokine gradients spread across different niches and it is believed that their migration pattern in each stages of development depends on the chemokine receptors expressed by each subpopulation.

The CXCL12/CXCR4 ligand-receptor pair is an example. All DN subpopulations (DN1–4) express CXCR4, although DN1 and DN4 cells express lower membrane expression levels, as compared with DN2 and DN3 (Kim et al. 1998; Suzuki et al. 1999; Plotkin et al. 2003; Trampont et al. 2010).

In addition to CXCR4, DN thymocytes express low amounts of CCR7 and CCR9, with CCR9 being upregulated in DN3 to DN4 cells (Kim et al. 1998; Misslitz et al. 2004; Wurbel et al. 2006).

DN1 thymocytes migrate towards CXCL12 (Kim et al. 1998; Bleul and Boehm 2000; Plotkin et al. 2003), but are not able to respond to CCL25 and CCL19 chemotactic stimuli (Kim et al. 1998; Wurbel et al. 2006). Indeed, it was proposed that the low amounts of chemokine receptors expressed by this subpopulation may explain their sluggish transit from the corticomedullary junction towards the outer cortex, facilitating initial differentiation events, such as T cell lineage commitment (Bunting et al. 2011). The importance of CXCL12/CXCR4 axis in the early stages of thymocyte development became even clearer when it was demonstrated that conditional depletion of CXCR4 in these cells blocks their differentiation. Transfer of CXCR4-deficient bone marrow-derived T cell progenitors led to the retention of DN1 cells in this differentiation stage and accumulation of these cells at the corticomedullary junction in the thymus (Plotkin et al. 2003).

A small group of DN1 thymocytes also expresses CCR6, a CCL20 receptor, which is predominately expressed in the thymic medulla, although lower levels were detected in the cortical region. These cells migrate in response to CCL20, and CCR6-deficent mice were shown to exhibit reduced relative and absolute numbers of DN2 and DN3 subpopulations, suggesting that CCR6 regulates DN1 migration. However, DN4 subpopulation in CCR6-deficient mice is normal and no other significant changes in T-cell development were detected in these animals (Bunting et al. 2014).

DN2 thymocytes are localized up to the central region of the cortex (Petrie and Zúñiga-Pflücker 2007), also known as mid-cortex, and as DN1 thymocytes, they migrate towards a CXCL12 gradient, but do not respond to the chemotactic stimulus of CCL19 (Kim et al. 1998; Bleul and Boehm 2000; Plotkin et al. 2003).

Albeit DN thymocytes express low amounts of CCR7, it seems that this receptor is important in the early stages of T-cell development. Besides the altered thymic architecture, demonstrated by smaller and more numerous medullary regions, the thymus of CCR7-deficient mice also exhibit reduced cellularity. Yet, global T cell development is apparently normal in young mice, since DP and SP thymocytes are found in normal ratios. In contrast, increased proportion of DN1/2 thymocytes was found in the mutant mice, whereas DN3 and DN4 cells were reduced. In fact, DN1/2 thymocytes accumulate in the corticomedullary junction, suggesting that CCR7 is also important for DN2 migration to the mid-cortex (Misslitz et al. 2004).

DN3 thymocytes are found from the mid-cortex to the outer cortex (Petrie and Zúñiga-Pflücker 2007) and migrate towards CXCL12 but do not respond to CCL19 and CCL21 (Kim et al. 1998; Bleul and Boehm 2000; Plotkin et al. 2003). As expected, they respond to CCL25, which is in keeping with the fact that during this differentiation stage there is an increase in CCR9 expression levels (Wurbel et al. 2006). CCR9-deficient mice present normal numbers of DN3 thymocytes, although these cells are found spread throughout the whole cortex, rather than compartmentalized between the mid and outer cortex. Despite their aberrant location, these mutant thymocytes present normal T cell development and repertoire. These data provide evidence that CCR9/CCL25 axis is important for DN3 thymocyte localization in the cortical region but are not essential for the further differentiation developmental stages (Uehara et al. 2002a; Benz et al. 2004).

The role of CXCR4/CXCL12 axis in the DN3 stage of T cell development was also demonstrated. Conditional CXCR4 depletion in DN2 and DN3 subpopulation induced an increase in DN3 subpopulation, with a decrease in total thymocytes and in DN4 and DP subpopulations being detected. DN3 thymocytes were found distributed around the cortex, where apoptotic cells were detected. In fact, DN2/3 CXCR4-deficient thymocytes presented reduced expression of Bcl2-A1, a pro-survival molecule that protect early thymocytes from cell death and impaired proliferation. As these results were not restored by molecules induced by pre-TCR signaling, it was proposed that CXCR4 act as a costimulatory molecule during β-selection (Trampont et al. 2010).

DN4 thymocytes, which are also found in outer cortex (Petrie and Zúñiga-Pflücker 2007), as the DN3 subpopulation, migrate towards CXCL12 and CCL25 but do not respond to CLCL19 and CCL21 (Kim et al. 1998; Bleul and Boehm 2000; Plotkin et al. 2003; Wurbel et al. 2006).

Taken together, these data suggest that migration of DN thymocytes from the corticomedullary junction to the outer cortex is coordinated by CXCR4/CXCL12 and CCR9/CCL25 interactions. Although further evidence is needed to reveal a role for CCR7/CCL19/CCL21 axis in this process, this may be mediated by a chemorepulsive response, since these chemokines are mainly described in the thymic medulla.

It should be pointed out that migration through TNCs (lymphoepithelial complexes that essentially harbor DP and to a minor extent DN thymocytes) largely depend on ECM-mediated interactions involving laminin, fibronectin and type IV collagen and their integrin-type corresponding receptors. In this respect, entrance and exit of developing thymocytes are enhanced by the ligands and can be impaired by treating TNCs with antibodies specific for the ligands or respective receptors (Lannes-Vieira et al. 1993; Villa-Verde et al. 1994; Ocampo et al. 2008).

5.6   Migration from the Outer Cortex to the Medulla

Following β-selection in the DN4 differentiation stage, developing thymocytes gain CD4 and CD8 expression, being named as double-positive (DP) cells. With the exception of the outer cortex, these thymocytes are found throughout the cortex and interact with cortical epithelial cells (Petrie and Zúñiga-Pflücker 2007).

DP thymocytes express CXCR4 and CCR9, but low amounts of CCR7, similar to the DN4 subpopulation, and present chemotactic responses towards CXCL12 and CCL25 (Kim et al. 1998; Schabath et al. 1999; Bleul and Boehm 2000; Wurbel et al. 2006). As described above, CXCR4 deletion in DN2/3 thymocytes impacts the DP subpopulation, due to its role in the β-selection (Trampont et al. 2010). Furthermore, human DP thymocytes express CCR5 and migrate towards CCL4, a CCR5 ligand (Dairaghi et al. 1998).

Positively selected DP thymocytes (DP CD69$^+$ thymocytes) downregulate CXCR4 expression (Suzuki et al. 1998). In contrast, CCR4 and CCR7 are upregulated (Suzuki et al. 1999; Davalos-Misslitz et al. 2007), whereas CCR9 expression in maintained (Uehara et al. 2002b). DP CD69$^+$ thymocytes continue responsive to CCL25 and CXCL12, although CXCR4 is downregulated, and start to respond to CCL19, CCL21, CCL17 and CCL22; the last two being CCR4 ligands (Campbell et al. 1999; Bleul and Boehm 2000; Cowan et al. 2014; Hu et al. 2015). The gain of responsiveness towards these molecules allows these cells to move towards the medulla, where these chemokines are strongly expressed (Chantry et al. 1999; Annunziato et al. 2000; Ueno et al. 2002; Cowan et al. 2014).

Apparently, CCR4 is important for the medullary entry of positively selected DP thymocytes. CCL17 and CCL22, CCR4 ligands, are expressed by thymic dendritic cells (DC) and medullary TEC. Thymic DCs are located in the corticomedullary junction and together with medullary TEC are important for negative selection. After this process, the remaining DP CD69$^+$ thymocytes loose CD4 or CD8 expression, becoming SP cells. Only part of immature CD4 SP thymocytes express CCR4; immature CD8 SP and mature CD4 and CD8 SP cells do not express this receptor, suggesting that it is no longer important for thymocyte migration after negative selection (Cowan et al. 2014; Hu et al. 2015). Nevertheless, it was found that positively selected DP thymocytes from CCR4 deficient-thymi have diminished interaction with DCs, impairing their ability to undergo negative selection to endogenous autoantigens. In fact, higher numbers of regulatory T cells and an increase in autoreactive naive CD4 T cells in secondary lymphoid organs were detected in these animals (Hu et al. 2015). These data demonstrate the importance of CCR4/CCL17 and CCR4/CCL22 interactions, not only for the entry of positively selected DP thymocytes in the medulla, but also in maintaining central tolerance.

CD4 and CD8 SP thymocytes also express CCR7 and migrate towards CCL19 and CCL21, with the percentage of migrating cells being higher than the one observed for positively selected DP CD69$^+$ thymocytes, since CCR7 expression is upregulated in these subpopulation (Kim et al. 1998; Campbell et al. 1999; Bleul and Boehm 2000; Misslitz et al. 2004). CCR7-deficient mice and *plt/plt* mice, which do not express CCL19 and has reduced CCL21 expression, showed poor accumulation of CD4 and CD8 SP thymocytes in the medulla, although these cells were found in cortical areas of the thymic lobules (Ueno et al. 2004; Nitta et al. 2009). The same effects were observed in CCL21-deficent mice (Kozai et al. 2017). By contrast, immature thymocytes from CCR7 transgenic mice are able to respond to CCL19 and CCL21 and DP thymocytes invade medullary areas. The effects of premature CCR7 expression were abrogated in *plt/plt* mice (Kwan and Killeen 2004). Conjointly, these results demonstrate that CCR7/CCL19 and CCR7/CCL21 interactions direct the migration of DP CD69+ and CD4 and CD8 SP thymocytes towards the medulla. However, no evidence of impaired T-cell development was detected in CCR7-deficient mice and in *plt/plt* mice; and regular frequencies of thymocytes and normal positive and negative selection events were observed, with depletion of autoreactive clones. Moreover, export of mature thymocytes from the thymus was normal (Ueno et al. 2004).

Nevertheless, lacrimal gland analysis indicated that CCR7-deficient and *plt/plt* mice exhibited autoimmune exocrinopathy resembling Sjogren's syndrome, suggesting that CCR7-induced migration to the thymic medulla is essential for self-tolerance maintenance (Kurobe et al. 2006). In the same vein, CCR7-deficient and *plt/plt* mice fail to efficiently deplete CD4 and CD8 SP thymocytes with transgenic TCR specific to tissue restricted antigens. Interestingly, it was shown that a CCR7-independent mechanism mediates depletion of thymocytes reactive for tissue restricted antigens, possible being accounted to thymic DCs, which are localized abundantly in the medulla and sparsely in the cortex (Nitta et al. 2009).

Corroborating all the data described above, CCL21 was directly implicated in the process of establishment of self-tolerance. As thymocytes from CCL21-deficent mice accumulate in thymic cortical areas, deletion of tissue restricted antigen reactive thymocytes was defective and these animals developed autoimmune dacryoadenitis (Kozai et al. 2017). These data indicate that CCR7/CCL19 and CCR7/CCL21 interactions not only control thymocyte traffic from thymic cortical regions to thymic medulla but also control negative selection process and the central tolerance maintenance.

The role of CCR9 in positively selected thymocyte migration to the medulla is still not clear, as CCL25 are mainly found in cortical areas. DP thymocytes express this receptor and DP and DP CD69+ thymocytes are able to respond, migrating towards CCL25. By contrast, no impact in SP subpopulations are observed in CCR9-deficent mice (Uehara et al. 2002a). It was suggested that this receptor can mediate migration away from the cortical region; however, as positively selected thymocyte and CD4 SP thymocytes migrate towards CCL25 *in vitro* (Bleul and Boehm 2000), this does not seem to be the case. Hence, whether CCR9/CCL25 interaction plays a role in thymocyte migration after positive selection needs further investigation.

In addition to the chemokines described above, others are found in thymic medulla. CXCL9, CXCL10 and CXCL11, which interact with CXCR3 receptor, are produced by different populations of medullary thymic epithelial cells. Small amounts of CXCL10 were also detected in thymic cortical areas. CXCR3 expression in the human thymus was mainly seen in the medullary region, in DP, CD8 SP and TCRγδ⁺ thymocytes as well as in natural killer cells. Only few positive cells were detected in the cortical region of the thymic lobules. All three chemokines were able to induce human CD8 SP thymocyte migration. Nevertheless, other evidence was provided sustaining that CD4 and CD8 SP murine thymocytes do not express CXCR3 and do not migrate towards their cognate ligands (Drennan et al. 2009).

Overall, these data indicate that CXCR4/CXCL12 and CCR9/CCL25 mediated interactions largely govern DP subpopulation migration through the inner cortex, as also orchestrate DN subpopulation migration. Furthermore, after positive selection these cells change their chemokine receptor expression pattern, which will allow them to migrate to the medulla, which in turn is mediated by CCR4/CCL17 and CCR4/CCL22 as well as CCR7/CCL19 and CCR7/CCL21 interactions, being also directly involved in central tolerance maintenance.

5.7 Thymocyte Emigration

After reaching the thymic medulla and having survived to negative selection, imma-ture CD4 and CD8 SP thymocytes finish their maturation process before leaving the organ. During this maturation period several molecules are regulated, including S1P1, mediating mature CD4 and CD8 SP thymocyte emigration.

The role of S1P1/S1P interactions upon thymocyte emigration was demonstrated since murine hematopoietic cells that do not express S1P1 do not appear in the peripheral blood and secondary lymphoid organs. Moreover, mature CD4 and CD8 SP thymocytes accumulate in the thymus of these animals (Matloubian et al. 2004). Interestingly, it was found that DP thymocytes from normal mice express less S1P1 than CD4 and CD8 SP thymocytes (Allende et al. 2004), which were able to migrate towards S1P higher concentrations. Human thymocytes also express S1P1, with DP cells expressing less S1P1 contents than SP thymocytes. Functionally, they are also able to migrate towards S1P and this migration is directly correlated with S1P1 expression levels in these cells (Mendes-da-Cruz et al. 2014).

S1P1 expression in thymocytes is controlled by the transcriptional factor Kruppel-like factor 2 (KLF2), which is expressed by mature thymocytes after posi-tive selection, although its deficiency does not impact thymocyte development. However, increased numbers of CD4 and CD8 SP thymocytes were found in the thymus of KLF2-deficient animals, while reduced numbers were found in the periphery (Carlson et al. 2006).

In a second vein, S1P1 expression is indirectly regulated by the interaction between CCR7 and CCL19, inducing ERK5 phosphorylation, which leads to increased expression of KLF2 (Shannon et al. 2012).

Indirect regulation of S1P1 expression is also mediated by the transcriptional factor forkhead box protein O1 (FoxO1) and CCR2 activation. FoxO1 control KLF2 expression and deficiency in FoxO1 impaired thymic emigration, as observed in KLF2, S1P1 (Fabre et al. 2008; Kerdiles et al. 2009) and CCR2 deficiency. CCR2 is expressed by mature thymocytes, whereas its ligand, CCL2, is produced by endo-thelial and by epithelial cells of the organ.

Their direct role upon thymocyte emigration was discarded since no gradient was found between the thymus and the blood vessels. However, CCR2 activation induces Stat3 phosphorylation, increase FoxO1, Klf2 and S1pr1 gene expression and enhance S1P1 membrane expression. In this respect, it has been defined that CCL2 can induce an increase in chemokinesis of mature thymocytes, which become more responsive to CCL2 and S1P than to S1P alone (Aili et al. 2018).

While FoxO1, KLF2 and CCR2 control the S1P1 transcriptional expression, CD69 acts directly on the receptor. Thymocyte maturation in the medulla is characterized by CD69 downregulation, a tissue retention signal, and by S1P1, CD62L and integrin β7 upregulation, which facilitate the recirculation of mature T cells (Carlson et al. 2006; Shiow et al. 2006).

S1P1 deficiency in lymphocytes induces an altered phenotype, with T cells presenting high levels of CD69 (Alfonso et al. 2006). Treatment of T lymphocytes with poly(I:C), a compound that mimics double-stranded RNA molecules, induces a decrease in S1P1 protein expression and inhibition of the S1P chemotactic function. These events seem to be dependent on CD69, since they are not detected in CD69-deficient mice. In addition, co-immunoprecipitation assays indicate that CD69 interacts directly with S1P1, forming a molecular complex with this receptor, which leads to its internalization and degradation. Thus, CD69 negatively regulates S1P1 function and promotes T cell retention in lymphoid organs, including the thymus (Shiow et al. 2006).

Pericytes and endothelial cells produce S1P near the blood vessels in the thymic medullary region and the transport of the S1P produced by the endothelial cells to the extracellular environment is mediated by the sphingolipid transporter 2 (Spns2) (Hisano et al. 2012). In the extracellular compartment, this lipid not only interacts with S1P1 present in SP thymocytes (as described above), but also with S1P1 and S1P2 expressed by endothelial cells. The balance between S1P1 and S1P2 activation in the thymic endothelium controls vascular permeability, which is inhibited by S1P1 (Garcia et al. 2001; Sanna et al. 2006) and increased by S1P2 (Sammani et al. 2010). Together the data demonstrate how complex is the regulation of mature thymocyte emigration, ultimately preventing that immature thymocytes that bypassed positive and negative selection reach the periphery.

Disturbances in S1P/S1P1 interaction in pathological conditions were shown to interfere with thymocyte development and emigration. As an example, experimental *Trypanosoma cruzi* acute infection is known to cause thymic atrophy in part due to premature release of immature thymocytes. Thymuses from infected animals presented smaller S1P amounts and altered levels of the enzymes involved in the metabolism of this lipid. Moreover, infected DN thymocytes presented increased expression of S1P1 and were chemotactic responsive to S1P. Infected mice treatment with FTY720, a specific inhibitor of S1P receptors, reversed the increased contents of DN T cells found in the periphery. These data indicate that the premature release of immature thymocytes is dependent of S1P/S1P1 and S1P/S1P3 axes and the presence of these cells in the periphery is possibly associated with the modulation of immune responses, since after activation they were able to induce pro-inflammatory cytokine genes (Lepletier et al. 2014). Premature escape of immature thymocytes may also be related to changes in CXCR4/CXCL12 and CCR5/CCL4 interactions, which are further accentuated in the presence of fibronectin (Mendes-da-Cruz et al. 2006). Therefore, different interactions may be acting together to modulate the premature release of immature thymocytes, which culminates with changes in shaping the T lymphocyte repertoire in the periphery.

Other molecules have been implicated in mature thymocyte emigration, however, so far none of them showed as much importance as S1P1. Although mature thymocytes express high levels of CCR7 and migrate towards CCL19 and CCL21, adult CCR7-deficient and *plt/plt* mice thymocytes show no impaired egress. In contrast, during neonatal period defective appearance of circulating T cells was detected in CCL19 neutralized mice and in CCR7-deficient animals (Ueno et al. 2002, 2004). Fugetaxis induced by CXCR4/CXCL12 interaction has been implicated in CD4 SP thymocyte emigration from fetal thymus (Vianello et al. 2005).

Thymocyte egress from the thymus is also controlled by fibronectin, since intrathymically administration of RGD synthetic peptides leads to the reduction of the recent thymic emigrants in the spleen (Savino et al. 2003). Such control was also seen in the non-obese diabetic (NOD) mouse model, which develops autoimmune type 1 diabetes dependent on T cells (Pearson et al. 2016). These animals present a natural defect in VLA-5 expression on thymocytes, impairing migration and adhesion to fibronectin (Cotta-de-Almeida et al. 2004). As a consequence, an accumulation of mature SP thymocytes, including Foxp3+ regulatory T cells (Treg) is observed in the thymus of NOD mice, forming giant perivascular spaces (PVS) in the medullary region (Savino et al. 1991; Cotta-de-Almeida et al. 2004; Mendes-da-Cruz et al. 2008). In parallel, reduced numbers of Tregs are observed in peripheral lymphoid organs, and such decrease can impact the ratio between effector and regulatory T cells and play a role in autoimmune events observed during the pathogenesis of type 1 diabetes. Moreover, in NOD mouse thymus, mature CD4 and CD8 SP thymocytes bearing the VLA-5 negative phenotype exhibit lower S1P1 expression and reduced S1P-driven migratory response when compared with controls. These results suggest that VLA-5 expression could influence thymocyte behavior under S1P stimulation, favoring or inhibiting the egress of mature thymocytes that may be involved in autoimmune events during type 1 diabetes (Lemos et al. 2018).

Lastly, hyaluronan also seems to be involved in this final step of thymocyte migration: blockade of CD44 leads to partial inhibition of the in vivo thymocyte egress in an arrangement of ECM molecules (Scollay and Godfrey 1995; Savino et al. 2000).

5.8 Ligand-Receptor Pairs Typically Expressed in the Nervous Systems Are Also Found in the Thymus and Control Thymocyte Migration and Development

Some hormones regulate the survival and proliferation of lymphoid and microenvironmental components of the thymus, acting via specific receptors (Savino et al. 2016). The intrathymic migration of developing T cells can be controlled by hormones produced not only by classic endocrine glands but also produced intra-thymically. As an example, the growth hormone (GH) enhances the deposition of proteins involved in cell migration, such as laminin and CXCL12. Accordingly, the migration of thymocytes derived from GH-transgenic mice or from mice injected with GH directly into the thymus is enhanced towards laminin and CXCL12. Importantly, the combined action of these two stimuli has a synergic effect further augmenting the migration of thymocytes. In both animal models, the numbers of cells derived from the thymus are also increased in peripheral lymphoid organs, suggesting that GH may affect the repertoire of both thymic and peripheral T cells (Savino et al. 2002b; Smaniotto et al. 2004, 2005).

As another example, the systemic treatment with T3, as well as the intrathymic injection of this hormone, enhances thymocyte adhesion and migration towards laminin and fibronectin. Such biological effects occur via the activation of T3 nuclear receptors, which are expressed in both developing thymocytes and TECs, and through the upregulation of the VLA-4, VLA-5 and VLA-6 integrins (Villa-Verde et al. 1993).

In addition to hormones, several typical neuron guiding molecules and their specific receptors have been reported in the thymus (Mendes-da-Cruz et al. 2009, 2012; Mendes-da-Cruz and Savino 2014). They can modulate thymocyte proliferation, survival and migration, and consequently modulate thymocyte development. This is the case of two well-known families of repulsive molecules originally reported in the nervous system, *ephrins* and *semaphorins*.

Ephrins are proteins highly expressed in the nervous system during development, being important for neuron growth, cone collapse and cell attachment (Henderson and Dalva 2018). The expression of ephrins and their corresponding Eph receptors have been reported in thymocytes and thymic microenvironmental cells, and disruption of Eph/ephrin signaling induces a marked hypocellularity, apparently associated with increased apoptosis (Muñoz et al. 2009).

Ephrins and Eph are both subdivided into two families, A and B, depending on their gene sequence similarities and ligand binding. As membrane-bound proteins, they can transmit cytoplasmic signals (forward and reverse) to both interacting cells (Mendes-da-Cruz et al. 2012). Considering specific interactions, EphB2-deficient bone marrow progenitors have reduced capacity to colonize wild-type fetal thymic lobes, as this apparently occurs due to the ability to block laminin-mediated migration of the precursors (Stimamiglio et al. 2010). EphB3 seems to be important in the DP to SP stage of thymocyte maturation (Montero-Herradón et al. 2017). Ephs A1, A2 and A3 receptors are involved in thymocyte survival, and Ephrin A1 attenuates thymocyte apoptosis induced by anti-CD3/CD28 stimulation, affecting mainly the DP thymocyte subpopulation (Freywald et al. 2006).

Some Eph ligands can be repulsive of thymocytes and also affect migration induced by other molecules. Ephrins A1, A3, and B2 (but not ephrin B3), are able to impair CXCL12-induced migration of human thymocytes by modifying CXCR4 receptor signaling and inducing actin polymerization. Ephrin B1/EphB2 signaling also inhibits chemokine (CXCL12, CCL21 and CCL25) and as well as laminin- and fibronectin-induced thymocyte migration in mice (Sharfe et al. 2002).

Another group of typical neuron guiding molecules found in the thymus comprises the class 3 semaphorins (Sema3A, Sema3F), glycoproteins that can bind neuropilin receptors and plexins (Mendes-da-Cruz et al. 2012).

We have demonstrated that developing thymocytes and thymic epithelial cells express Semaphorins 3A and 3F, as well as plexins, transmembrane proteins that are linked to neuropilins, allowing signal transduction into the cytoplasm (Lepelletier et al. 2007; Garcia et al. 2012; Mendes-da-Cruz et al. 2014). Both Semaphorin-3A and -3F can induce thymocyte repulsive migratory responses and affect migration induced by other molecules, being chemorepulsive for both human and mouse thymocytes. Semaphorin-3A inhibits laminin- and fibronectin-induced thymocyte migration, whereas Sema-3F does not. However, both class 3 semaphorins inhibit thymocyte chemotaxis induced by CXCL12, and this role is associated with a decrease in the membrane expression of corresponding receptor CXCR4; besides inhibiting the phosphorylation of kinases implicated in the signaling pathways induced by this chemokine (Lepelletier et al. 2007; Garcia et al. 2012). Additionally, Sema3F inhibits S1P-induced migration of human thymocytes (Mendes-da-Cruz et al. 2014). Conjointly, the data discussed above illustrate the importance of these interactions in thymocyte guidance alone or combined with different cell migration-related molecules.

5.9 The Concept of Multivectorial Migration for Developing Thymocytes

As discussed along this chapter, thymocyte migration can be influenced by a variety of soluble moieties and ECM molecules; some of them acting as attractors whereas others rather induce repulsion of developing thymocytes (Fig. 5.1a). Importantly,

---

Fig. 5.1 (continued) (DN) thymocytes. Thymocyte migration continues to the inner cortex in response mainly to the ECM molecules fibronectin, laminin and type IV collagen, while cells develop to double-positive (DP) thymocytes expressing both CD4 and CD8 (3). Cells migrating to the medulla loss CD4 or CD8 expression and become single-positive (SP) mature CD4 or CD8 thymocytes, as well as CD4 Foxp3+ regulatory T cells. This migration is coordinated by the chemokines CCL19/CCL21, CCL17/CCL22 and CXCL12 (4). Mature thymocytes then leave the thymus in response to sphingosine-1-phosphate (S1P), the chemokines CCL19/CCL21, fibronectin and hyaluronan (5), going to peripheral lymphoid organs to finish their proper maturation and form the peripheral T cell pool. During their journey into the thymus, thymocytes interact with microenvironmental cells represented by cortical thymic epithelial cells (cTECs), medullary thymic epithelial cells (mTECs), macrophages and dendritic cells that play an essential role in thymic positive and negative selection processes. (**b**) Representative scheme of the concept of multivectorial thymocyte migration. The migration of developing T cells can be described as a multivectorial process, in which the final vector is derived from the result of several simultaneous and/or sequential interactions of individual vectors. These individual vectors can promote thymocyte chemotaxis, in the case of chemokines and sphingosine-1-phosphate, haptotactic responses, in the case of extracellular matrix molecules, and fugetaxis, in the case of semaphorins and Eprins. This figure was adapted from Mendes-da-Cruz et al. (2008, 2012)

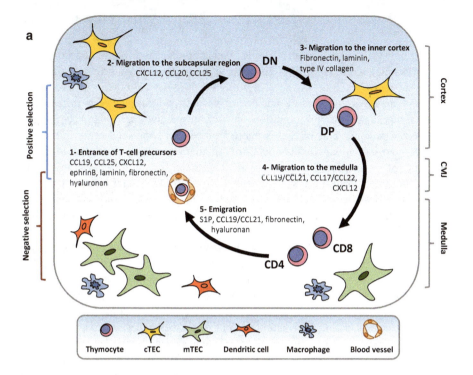

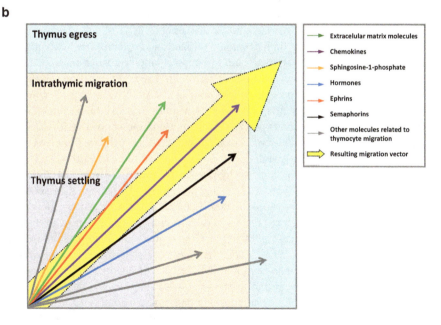

Fig. 5.1 Thymocyte differentiation and migration. (**a**) Bone marrow derived T-cell precursors enter the thymus in the corticomedullary junction (CMJ), in response to different migratory stimuli, including extracellular matrix (ECM), soluble and membrane-bound molecules such as CCL19, CCL25, CXCL12, ephrinB, laminin, fibronectin and hyaluronan (1). Cells next migrate to the subcapsular region following the CXCL12, CCL20 and/or CCL25 stimuli (2). At this stage, thymocytes lack the expression of the CD4 and CD8 co-receptors and are named double-negative

the combined action of these molecules can modulate tunely-oriented migratory responses and cell guidance for a given niche in the thymus. We proposed the multivectorial model of intrathymic T-cell migration (Mendes-da-Cruz et al. 2008), assuming that it results from a balance of several and simultaneous interactions between modulators of cell migration and their respective receptors. Accordingly, each ligand-receptor pair mediated interaction can be conceived as an individual vector. These individual components form together a resulting vector that leads the guidance of developing T-cells, from their entrance into the thymus, migration within the organ from the cortex toward the medulla, and final egress of mature thymocytes from the organ. Moreover, one given interaction may interfere with another one, with consequences upon the resulting vector driving oriented migration of a thymocyte.

Finally, the velocity and strength of these interactions, impact other thymocyte interactions with microenvironmental cells, including the selective processes, fate and repertoire of thymocytes. Accordingly, in some pathological conditions, the expression pattern of ligands and receptors involved in thymocyte migration can be altered. These alterations can impact cell-cell interactions and consequently control negative selection and the formation of a proper T-cell repertoire that will avoid autoimmune events in the periphery of the immune system. Examples were reported by our group in terms of glucocorticoid-induced thymic atrophy (Dalmau et al. 1999), *T. cruzi* acute infection (Mendes-da-Cruz et al. 2006; Lepletier et al. 2014) and type 1 diabetes, seen in the non-obese diabetic (NOD) mouse prior to the onset of diabetes (Mendes-da-Cruz et al. 2008; Lemos et al. 2018).

Intrathymic contents of both chemokine ligands and receptors can vary in pathological conditions. For example, acute infection by the protozoan parasite *T. cruzi* (the causative agent of Chagas disease) results in a severe thymic atrophy mainly due to the depletion and premature release of immature thymocytes (Savino et al. 1989; Leite-de-Morais et al. 1991; Leite-de-Moraes et al. 1992; Mendes-da-Cruz et al. 2003). Infected thymus presented an increase in the deposition CXCL12 and CCL4. Concomitantly, thymocytes from infected animals exhibited higher CXCR4 and CCR5 membrane levels and, as expected, migrated more towards CXCL12 and CCL4, as compared with uninfected thymocytes (Mendes-da-Cruz et al. 2006). It is important to note that T cells expressing prohibited Vβ segments in secondary lymphoid organs were described in experimental *T. cruzi* infection (Mendes-da-Cruz et al. 2003). Therefore, the increased migration of immature thymocytes induced by CXCR4/CXCL12 and CCR5/CCL4, which was accentuated in the presence of fibronectin, during *T. cruzi* infection may be related to the premature release of these cells to the periphery and may be the core of changes in T lymphocytes repertoire. These data reveal that disruptions in the interactions described can be related to the generation of autoimmunity in pathological conditions.

Lastly, it is noteworthy that, in a rather simplified way, using four ligand/receptor pairs, we developed a cellular automata-based mathematical model for thymocyte development, which predicts thymocyte expansion, migration, death and differentiation (Souza-e-Silva et al. 2009). Evolving this model will hopefully include many more cell migration-related individual vectors, and allow its application in both normal and pathological conditions.

5.10 Chemokines, Extracellular Matrix, Cell Migration and Central Tolerance

Besides the induction of thymocyte migration, chemokines and their receptors have been implicated in positive and negative selective processes in the thymus in different ways. Briefly, during negative selection peripheral tissue antigens (PTAs) are presented by medullary epithelial cells and DCs to developing thymocytes, which scan these cells to find their cognate peptide. Thymocytes whose T cell receptor (TCR) presents high affinity for a given PTA are eliminated by apoptosis, preventing the release of autoreactive T cells to the periphery (Passos et al. 2015). Chemokines can facilitate thymocytes/DC encounter by increasing cell motility. CCR4/CCL17 and CCR4/CCL22 interactions between CD4 SP thymocytes and DC, participate in the maintenance of central tolerance (Hu et al. 2015), whereas CCL21 increases the contact between peripheral T-lymphocytes and DC, accelerating CD69 up-regulation and enhancing antigen specific T cells proliferation in co-culture assays (Stachowiak et al. 2006).

Moreover, chemokine receptors are involved in DC recruitment to the thymus and central tolerance. CCR9/CCL25 interactions promote immature plasmocytoid DCs (pDCs) entry in the thymus and these cells are able to transport particle and soluble peripheral antigens to the organ. This transport was shown to be mediated by CCR9 and α_4 integrin (Hadeiba et al. 2012). Additionally, CCR2 is required to Sirpα^+ conventional DC (cDC) entry in the thymus, since CCR2-deficient mice present decreased numbers of these cells in the thymus and impairment of negative selection against blood-borne antigens (Baba et al. 2009).

Therefore, chemokine ligands and receptors, which participate in thymocyte migration, also favor cell motility and recruitment of cells from the thymic microenvironment that play a major role in the maintenance of central tolerance.

The role of fibronectin in intrathymic selection processes was provided by the fact that treatment with anti-CD3 monoclonal antibody in combination with fibronectin lead to the activation-induced cell death of DP thymocytes, an event mediated by VLA-5 (Takayama et al. 1998). Moreover, positively selected DP thymocytes upregulate VLA-5 expression and become able to migrate, responding to higher concentrations of fibronectin in the medulla (Crisa et al. 1996; Savino et al. 2002a).

A series set of evidence strongly indicate that laminin-mediated interaction also play a role (even if indirect) in generation of the T-cell repertoire. For example, the *dy/dy* mutant mouse model, with lack laminin-211 expression, presents an aberrant thymocyte development and an atrophic and disorganized thymus with a reduced cortical region (Magner et al. 2000). The lack of laminin-211 leads to an impairment in DN-DP progression, ascertained by increased apoptosis of immature cells in the subcapsular region, where DN thymocytes undergo β-selection.

The role of laminin and corresponding integrins in the thymus can be further extended to positive and negative selection, since these molecules can be present in the adhesive interactions in immunological synapses (Richie et al. 2002). As mentioned before, ITGA6 knock-down in human TECs leads to the modulation of synapse-related genes such as class I and class II MHC, as well as other molecules involved in cell adhesion and cell-cell interaction (Golbert et al. 2013, 2018). Conceptually, if we take into account that proper thymocyte selection depends on multiple interactions with microenvironmental cells, control of the various adhesion events may be critical for T cell maturation.

5.11 Concluding Remarks

The bulk of evidence summarized in the present chapter clearly demonstrates that intrathymic T-cell migration is a complex biological event, controlled by various numbers of interactions of developing thymocytes with the thymic microenvironment, simultaneously to the typical differential events including positive and negative selection of the T-cell repertoire. Accordingly, we cannot think of thymocyte differentiation without taking into account the cell migration molecular interactions that are occurring in each particular niche of the thymic lobules and vice-versa.

Acknowledgements Experiments developed in the context of the work presented in this chapter were funded with grants by The Oswaldo Cruz Foundation (Fiocruz), The National Institute of Science and Technology on Neuroimmunomodulation (INCT-NIM, CNPq), Capes, Faperj (Brazil) and FOCEM (Mercosur).

References

Aili A, Zhang J, Wu J et al (2018) CCR2 signal facilitates thymic egress by priming thymocyte responses to sphingosine-1-phosphate. Front Immunol 9:1263. https://doi.org/10.3389/fimmu.2018.01263

Aiuti A, Tavian M, Cipponi A et al (1999) Expression of CXCR4, the receptor for stromal cell-derived factor-1 on fetal and adult human lymphohematopoietic progenitors. Eur J Immunol 29:1823–1831. https://doi.org/10.1002/(SICI)1521-4141(199906)29:06<1823::AID-IMMU1823>3.0.CO;2-B

Alfonso C, McHeyzer-Williams MG, Rosen H (2006) CD69 down-modulation and inhibition of thymic egress by short- and long-term selective chemical agonism of sphingosine 1-phosphate receptors. Eur J Immunol 36:149–159. https://doi.org/10.1002/eji.200535127

Allende ML, Dreier JL, Mandala S, Proia RL (2004) Expression of the sphingosine 1-phosphate receptor, S1P1, on T-cells controls thymic emigration. J Biol Chem 279:15396–15401. https://doi.org/10.1074/jbc.M314291200

Allende ML, Bektas M, Lee BG et al (2011) Sphingosine-1-phosphate lyase deficiency produces a pro-inflammatory response while impairing neutrophil trafficking. J Biol Chem 286:7348–7358. https://doi.org/10.1074/jbc.M110.171819

Anderson G, Anderson KL, Jenkinson EJ (1997) Fibroblast dependency during early thymocyte development maps to the CD25+CD44+stage and involves interactions with fibroblast matrix molecules. Eur J Immunol 27:1200–1206

Annunziato F, Romagnani P, Cosmi L et al (2000) Macrophage-derived chemokine and EBI1-ligand chemokine attract human thymocytes in different stage of development and are produced by distinct subsets of medullary epithelial cells: possible implications for negative selection. J Immunol 165:238–246. https://doi.org/10.4049/jimmunol.165.1.238

Annunziato F, Romagnani P, Cosmi L et al (2001) Chemokines and lymphopoiesis in human thymus. Trends Immunol 22:277–281. https://doi.org/10.1016/S1471-4906(01)01889-0

Ayres-Martins S, Lannes-Vieira J, Farias-De-Oliveira DA et al (2004) Phagocytic cells of the thymic reticulum interact with thymocytes via extracellular matrix ligands and receptors. Cell Immunol 229:21–30. https://doi.org/10.1016/j.cellimm.2004.06.002

Baba T, Nakamoto Y, Mukaida N (2009) Crucial contribution of thymic sirpα+ conventional dendritic cells to central tolerance against blood-borne antigens in a CCR2-dependent manner. J Immunol 183:3053–3063. https://doi.org/10.4049/jimmunol.0900438

Barallobre MJ, Pascual M, Del Río JA, Soriano E (2005) The Netrin family of guidance factors: emphasis on Netrin-1 signalling. Brain Res Rev 49:22–47. https://doi.org/10.1016/j.brainresrev.2004.11.003

Barondes SH, Castronovo V, Cooper DNW et al (1994) Galectins: a family of animal β-galactoside-binding lectins. Cell 76:597–598. https://doi.org/10.1016/0092-8674(94)90498-7

Baxter JB, Blalock JE, Weigent DA (1991) Expression of immunoreactive growth hormone in leukocytes in vivo. J Neuroimmunol 33:43–54

Benz C, Heinzel K, Bleul CC (2004) Homing of immature thymocytes to the subcapsular microenvironment within the thymus is not an absolute requirement for T cell development. Eur J Immunol 34:3652–3663. https://doi.org/10.1002/eji.200425248

Berrih S, Savino W, Cohen S (1985) Extracellular matrix of the human thymus: immunofluorescence studies on frozen sections and cultured epithelial cells. J Histochem Cytochem 33:655–664. https://doi.org/10.1177/33.7.3891843

Bleul CC, Boehm T (2000) Chemokines define distinct microenvironments in the developing thymus. Eur J Immunol 30:3371–3379. https://doi.org/10.1002/1521-4141(2000012)30:12<3371::AID-IMMU3371>3.0.CO;2-L

Bofill M, Janossy G, Willcox N et al (1985) Microenvironments in the normal thymus and the thymus in myasthenia gravis. Am J Pathol 119:462–473

Bourdon MA, Ruoslahti E (1989) Tenascin mediates cell attachment through an RGD-dependent receptor. J Cell Biol 108:1149–1155. https://doi.org/10.1083/jcb.108.3.1149

Bunting MD, Comerford I, McColl SR (2011) Finding their niche: chemokines directing cell migration in the thymus. Immunol Cell Biol 89:185–196. https://doi.org/10.1038/icb.2010.142

Bunting MD, Comerford I, Kara EE et al (2014) CCR6 supports migration and differentiation of a subset of DN1 early thymocyte progenitors but is not required for thymic nTreg development. Immunol Cell Biol 92:489–498. https://doi.org/10.1038/icb.2014.14

Buono M, Facchini R, Matsuoka S et al (2016) A dynamic niche provides Kit ligand in a stage-specific manner to the earliest thymocyte progenitors. Nat Cell Biol 18:157–167. https://doi.org/10.1038/ncb3299

Campbell JJ, Pan J, Butcher EC (1999) Cutting edge: developmental switches in chemokine responses during T cell maturation. J Immunol 163:2353–2357. ji_v163n5p2353 [pii]

Carlson CM, Endrizzi BT, Wu J et al (2006) Kruppel-like factor 2 regulates thymocyte and T-cell migration. Nature 442:299–302. https://doi.org/10.1038/nature04882

Chaffin KE, Perlmutter RM (1991) A pertussis toxin-sensitive process controls thymocyte emigration. Eur J Immunol 21:2565–2573. https://doi.org/10.1002/eji.1830211038

Chang AC, Wadsworth S, Coligan JE (1993) Expression of merosin in the thymus and its interaction with thymocytes. J Immunol 151:1789–1801

Chantry D, Romagnani P, Raport CJ et al (1999) Macrophage-derived chemokine is localized to thymic medullary epithelial cells and is a chemoattractant for CD3(+), CD4(+), CD8(low) thymocytes. Blood 94:1890–1898

Corset V, Nguyen-Ba-Charvet KT, Forcet C et al (2000) Netrin-1-mediated axon outgrowth and cAMP production requires interaction with adenosine A2b receptor. Nature 407:747–750. https://doi.org/10.1038/35037600

Cotta-de-Almeida V, Villa-Verde DMS, Lepault F et al (2004) Impaired migration of NOD mouse thymocytes: a fibronectin receptor-related defect. Eur J Immunol 34:1578–1587. https://doi.org/10.1002/eji.200324765

Coulthard MG, Lickliter JD, Subanesan N et al (2001) Characterization of the Epha1 receptor tyrosine kinase: expression in epithelial tissues. Growth Factors 18:303–317

Cowan JE, Mccarthy NI, Parnell SM et al (2014) Differential requirement for CCR4 and CCR7 during the development of innate and adaptive abT cells in the adult thymus. J Immunol 193:1204–1212. https://doi.org/10.4049/jimmunol.1400993

Crisa L, Cirulli V, Ellisman MH et al (1996) Cell adhesion and migration are regulated at distinct stages of thymic T cell development: the roles of fibronectin, VLA4, and VLA5. J Exp Med 184:215–228. https://doi.org/10.1084/jem.184.1.215

Dairaghi DJ, Franz-bacon K, Callas E et al (1998) Macrophage inflammatory protein-1 β induces migration and activation of human thymocytes. Blood 91:2905–2913

Dalmau SR, Freitas CS, Savino W (1999) Upregulated expression of fibronectin receptors under-lines the adhesive capability of thymocytes to thymic epithelial cells during the early stages of differentiation: lessons from sublethally irradiated mice. Blood 93:974–990

Davalos-Misslitz ACM, Worbs T, Willenzon S et al (2007) Impaired responsiveness to T-cell receptor stimulation and defective negative selection of thymocytes in CCR7-deficient mice. Blood 110:4351–4359. https://doi.org/10.1182/blood-2007-01-070284

Drennan MB, Franki A, Dewint P et al (2009) Cutting edge: the chemokine receptor CXCR3 retains invariant NK T cells in the thymus. J Immunol 183:2213–2216. https://doi.org/10.4049/jimmunol.0901213

Fabre S, Carrette F, Chen J et al (2008) FOXO1 Regulates L-selectin and a network of human T cell homing molecules downstream of phosphatidylinositol 3-kinase. J Immunol 181:2980–2989. https://doi.org/10.4049/jimmunol.181.5.2980

Falk M, Ferletta M, Forsberg E, Ekblom P (1999) Restricted distribution of laminin alpha1 chain in normal adult mouse tissues. Matrix Biol 18:557–568. S0945-053X(99)00047-5 [pii]

Franz-bacon BK, Dairaghi DJ, Boehme SA et al (1999) Human thymocytes express CCR-3 and are activated by eotaxin. Blood 93:3233–3241

Freitas CS, Lyra JS, Dalmau SR, Savino W (1995) In vivo and in vitro expression of tenascin by human thymic microenvironmental cells. Dev Immunol 4:139–147

Freywald A, Sharfe N, Miller CD et al (2006) EphA receptors inhibit anti-CD3-induced apoptosis in thymocytes. J Immunol 176:4066–4074

Fyrst H, Saba JD (2010) An update on sphingosine-1-phosphate and other sphingolipid mediators. Nat Chem Biol 6:489–497. https://doi.org/10.1038/nchembio0910-689a

Garcia JG, Liu F, Verin AD et al (2001) Sphingosine 1-phosphate promotes endothelial cell barrier integrity by Edg-dependent cytoskeletal rearrangement. J Clin Invest 108:689–701. https://doi.org/10.1172/JCI12450

Garcia F, Lepelletier Y, Smaniotto S et al (2012) Inhibitory effect of semaphorin-3A, a known axon guidance molecule, in the human thymocyte migration induced by CXCL12. J Leukoc Biol 91:7–13. https://doi.org/10.1189/jlb.0111031

García-Ceca J, Alfaro D, Montero-Herradón S, Zapata AG (2013) Eph/ephrinB signalling is involved in the survival of thymic epithelial cells. Immunol Cell Biol 91:130–138. https://doi.org/10.1038/icb.2012.59

Gares SL, Pilarski LM (1999) Beta1-integrins control spontaneous adhesion and motility of human progenitor thymocytes and regulate differentiation-dependent expression of the receptor for hyaluronan-mediated motility. Scand J Immunol 50:626–634

Gares SL, Pilarski LM (2000) Balancing thymocyte adhesion and motility: a functional linkage between beta1 integrins and the motility receptor RHAMM. Dev Immunol 7:209–225

Golbert DC, Correa-de-Santana E, Ribeiro-Alves M et al (2013) ITGA6 gene silencing by RNA interference modulates the expression of a large number of cell migration-related genes in human thymic epithelial cells. BMC Genomics 14:1–9. https://doi.org/10.1186/1471-2164-14-S6-S3

Golbert DCF, Santana-van-Vliet E, Mundstein AS et al (2014) Laminin-database v.2.0: an update on laminins in health and neuromuscular disorders. Nucleic Acids Res 42:D426–D429. https://doi.org/10.1093/nar/gkt901

Golbert DCF, Santana-Van-Vliet E, Ribeiro-Alves M et al (2018) Small interference ITGA6 gene targeting in the human thymic epithelium differentially regulates the expression of immunological synapse-related genes. Cell Adh Migr 12:152–167. https://doi.org/10.1080/19336918.2017.1327513

Goldman R, Harvey J, Hogg N (1992) VLA???2 is the integrin used as a collagen receptor by leukocytes. Eur J Immunol 22:1109–1114. https://doi.org/10.1002/eji.1830220502

Gorski A, Kupiec-Weglinski JW (1995) Extracellular matrix proteins, regulators of T-cell functions in healthy and diseased individuals. Clin Diagn Lab Immunol 2:646–651

Gossens K, Naus S, Corbel SY et al (2009) Thymic progenitor homing and lymphocyte homeostasis are linked via S1P-controlled expression of thymic P-selectin/CCL25. J Exp Med 206:761–778. https://doi.org/10.1084/jem.20082502

Green JA, Suzuki K, Cho B et al (2011) The sphingosine 1-phosphate receptor $S1P_2$ maintains the homeostasis of germinal center B cells and promotes niche confinement. Nat Immunol 12:672–680. https://doi.org/10.1038/ni.2047

Guo XK, Liu YF, Zhou Y et al (2013) The expression of Netrin-1 in the thymus and its effects on thymocyte adhesion and migration. Clin Dev Immunol 2013:462152. https://doi.org/10.1155/2013/462152

Hadeiba H, Lahl K, Edalati A et al (2012) Plasmacytoid dendritic cells transport peripheral antigens to the thymus to promote central tolerance. Immunity 36:438–450. https://doi.org/10.1016/j.immuni.2012.01.017

Halkias J, Melichar HJ, Taylor KT et al (2014) Tracking migration during human T cell development. Cell Mol Life Sci 71:3101–3117. https://doi.org/10.1007/s00018-014-1607-2

Hemesath TJ, Stefansson K (1994) Expression of tenascin in thymus and thymic nonlymphoid cells. J Immunol (Baltimore, Md 1950) 152:422–428

Henderson NT, Dalva MB (2018) EphBs and ephrin-Bs: trans-synaptic organizers of synapse development and function. Mol Cell Neurosci 91:108. https://doi.org/10.1016/j.mcn.2018.07.002

Henry L (1967) Involution of the human thymus. J Pathol Bacteriol 93:661–671. https://doi.org/10.1002/path.1700930227

Hernandez-Lopez C, Varas A, Sacedon R et al (2002) Stromal cell-derived factor 1/CXCR4 signaling is critical for early human T-cell development. Blood 99:546–554. https://doi.org/10.1182/blood.V99.2.546

Hisano Y, Nishi T, Kawahara A (2012) The functional roles of S1P in immunity. J Biochem 152:305–311. https://doi.org/10.1093/jb/mvs090

Hong K, Hinck L, Nishiyama M et al (1999) A ligand-gated association between cytoplasmic domains of UNC5 and DCC family receptors converts netrin-induced growth cone attraction to repulsion. Cell 97:927–941. https://doi.org/10.1016/S0092-8674(00)80804-1

Hu Z, Lancaster JN, Sasiponganan C, Ehrlich LIR (2015) CCR4 promotes medullary entry and thymocyte – dendritic cell interactions required for central tolerance. J Exp Med 212:1947–1965. https://doi.org/10.1084/jem.20150178

Hughes CE, Nibbs RJB (2018) A guide to chemokines and their receptors. FEBS J 285:2944–2971. https://doi.org/10.1111/febs.14466

Ishii M, Kikuta J (2013) Sphingosine-1-phosphate signaling controlling osteoclasts and bone homeostasis. Biochim Biophys Acta 1831:223–227. https://doi.org/10.1016/j.bbalip.2012.06.002

Ishii M, Kikuta J, Shimazu Y et al (2010) Chemorepulsion by blood S1P regulates osteoclast precursor mobilization and bone remodeling in vivo. J Exp Med 207:2793–2798. https://doi.org/10.1084/jem.20101474

Iwao M, Fukada S, Harada T et al (2000) Interaction of merosin (laminin 2) with very late activation antigen-6 is necessary for the survival of CD4+CD8+immature thymocytes. Immunology 99:481–488. https://doi.org/10.1046/j.1365-2567.2000.00990.x

Jenne CN, Enders A, Rivera R et al (2009) T-bet-dependent S1P5 expression in NK cells promotes egress from lymph nodes and bone marrow. J Exp Med 206:2469–2481. https://doi.org/10.1084/jem.20090525

Kerdiles YM, Beisner DR, Tinoco R et al (2009) Foxo1 links homing and survival of naive T cells by regulating Lselectin, CCR7 and interleukin 7 receptor. Nat Immunol 10:176–184. https://doi.org/10.1038/ni.1689.Foxo1

Keul P, Lucke S, von Wnuck Lipinski K et al (2011) Sphingosine-1-phosphate receptor 3 promotes recruitment of monocyte/macrophages in inflammation and atherosclerosis. Circ Res 108:314–323. https://doi.org/10.1161/CIRCRESAHA.110.235028

Kim CH, Pelus LM, White JR, Broxmeyer HE (1998) Differential chemotactic behavior of developing T cells in response to thymic chemokines. Blood 91:4434–4443

Kim MG, Lee G, Lee SK et al (2000) Epithelial cell-specific laminin 5 is required for survival of early thymocytes. J Immunol 165:192–201. https://doi.org/10.4049/jimmunol.165.1.192

Kohno T, Matsuyuki H, Inagaki Y, Igarashi Y (2003) Sphingosine 1-phosphate promotes cell migration through the activation of Cdc42 in Edg-6/S1P4-expressing cells. Genes Cells 8:685–697

Kozai M, Kubo Y, Katakai T et al (2017) Essential role of CCL21 in establishment of central self-tolerance in T cells. J Exp Med 214:1925–1935. https://doi.org/10.1084/jem.20161864

Kremer L, Carramolino L, Goya Í et al (2001) The transient expression of C-C chemokine receptor 8 in thymus identifies a thymocyte subset committed to become CD4 + single-positive T cells. J Immunol 166:218–225. https://doi.org/10.4049/jimmunol.166.1.218

Krueger A, Willenzon S, Łyszkiewicz M et al (2010) CC chemokine receptor 7 and 9 double-deficient hematopoietic progenitors are severely impaired in seeding the adult thymus. Blood 115:1906–1912. https://doi.org/10.1182/blood-2009-07-235721.An

Kurobe H, Liu C, Ueno T et al (2006) CCR7-dependent cortex-to-medulla migration of positively selected thymocytes is essential for establishing central tolerance. Immunity 24:165–177. https://doi.org/10.1016/j.immuni.2005.12.011

Kutleša S, Siler U, Speiser A et al (2002) Developmentally regulated interactions of human thymocytes with different laminin isoforms. Immunology 105:407–418. https://doi.org/10.1046/j.1365-2567.2002.01384.x

Kwan J, Killeen N (2004) CCR7 directs the migration of thymocytes into the thymic medulla. J Immunol 172:3999–4007. https://doi.org/10.4049/jimmunol.172.7.3999

Lannes-Vieira J, Dardenne M, Savino W (1991) Extracellular matrix components of the mouse thymus microenvironment: ontogenetic studies and modulation by glucocorticoid hormones. J Histochem Cytochem 39:1539–1546. https://doi.org/10.1177/39.11.1918928

Lannes-Vieira J, Chammas R, Villa-Verde DMS et al (1993) Extracellular matrix components of the mouse thymic microenvironment. Iii. thymic epithelial cells express the vla6 complex that is involved in laminin-mediated interactions with thymocytes. Int Immunol 5:1421–1430. https://doi.org/10.1093/intimm/5.11.1421

Leite-de-Moraes MC, Hontebeyrie-Joskowicz M, Dardenne M, Savino W (1992) Modulation of thymocyte subsets during acute and chronic phases of experimental Trypanosoma cruzi infection. Immunology 77:95–98

Leite-de-Morais MC, Hontebeyrie-joskowicz M, Leboulenger F et al (1991) Studies on the thymus in Chagas' Disease II. thymocyte subset fluctuations in trypanosoma cruzi-infected mice: relationship to stress. Scand J Immunol 33:267–275. https://doi.org/10.1111/j.1365-3083.1991.tb01772.x

Lemos JP, Smaniotto S, Messias CV et al (2018) Sphingosine-1-phosphate receptor 1 is involved in non-obese diabetic mouse thymocyte migration disorders. Int J Mol Sci 19:1–14. https://doi.org/10.3390/ijms19051446

Lepelletier Y, Smaniotto S, Hadj-Slimane R et al (2007) Control of human thymocyte migration by Neuropilin-1/Semaphorin-3A-mediated interactions. Proc Natl Acad Sci U S A 104:5545–5550. https://doi.org/10.1073/pnas.0700705104

Lepletier A, de Almeida L, Santos L et al (2014) Early double-negative thymocyte export in Trypanosoma cruzi infection is restricted by sphingosine receptors and associated with human Chagas disease. PLoS Negl Trop Dis 8:e3203. https://doi.org/10.1371/journal.pntd.0003203

Lind EF, Prockop SE, Porritt HE, Petrie HT (2001) Mapping precursor movement through the postnatal thymus reveals specific microenvironments supporting defined stages of early lymphoid development. J Exp Med 194:127–134. https://doi.org/10.1084/jem.194.2.127

Linhares-Lacerda L, Ribeiro-Alves M, Nogueira ACMDA et al (2010) RNA interference-mediated knockdown of CD49e (α5 integrin chain) in human thymic epithelial cells modulates the expression of multiple genes and decreases thymocyte adhesion. BMC Genomics 11:S2. https://doi.org/10.1186/1471-2164-11-S5-S2

Lynch F, Ceredig R (1988) Ly-24 (Pgp-1) expression by thymocytes and peripheral T cells. Immunol Today 9:7–10. https://doi.org/10.1016/0167-5699(88)91347-3

Magner WJ, Chang AC, Owens J et al (2000) Aberrant development of thymocytes in mice lacking laminin-2. Dev Immunol 7:179–193. https://doi.org/10.1155/2000/90943

Matloubian M, Lo CG, Cinamon G et al (2004) Lymphocyte egress from thymus and peripheral lymphoid organs is dependent on S1P receptor 1. Nature 427:355–360. https://doi.org/10.1038/nature02284

Mayol K, Biajoux V, Marvel J et al (2011) Sequential desensitization of CXCR4 and S1P5 controls natural killer cell trafficking. Blood 118:4863–4871. https://doi.org/10.1182/blood-2011-06-362574

Meireles De Souza LR, Trajano V, Savino W (1993) Is there an interspecific diversity of the thymic microenvironment? Dev Immunol 3:123–135. https://doi.org/10.1155/1993/48056

Mendelson K, Evans T, Hla T (2014) Sphingosine 1-phosphate signalling. Development 141:5–9. https://doi.org/10.1242/dev.094805

Mendes-da-Cruz D, Savino W (2014) Typical neuron guiding molecules constitutively control thymus physiology. Adv Neuroimmune Biol 5:61–67. https://doi.org/10.3233/NIB-140087

Mendes-da-Cruz DA, De Meis J, Cotta-de-Almeida V, Savino W (2003) Experimental Trypanosoma cruzi infection alters the shaping of the central and peripheral T-cell repertoire. Microbes Infect 5:825–832. https://doi.org/10.1016/S1286-4579(03)00156-4

Mendes-da-Cruz DA, Silva JS, Cotta-de-Almeida V, Savino W (2006) Altered thymocyte migration during experimental acute Trypanosoma cruzi infection: combined role of fibronectin and the chemokines CXCL12 and CCL4. Eur J Immunol 36:1486–1493. https://doi.org/10.1002/eji.200535629

Mendes-da-Cruz DA, Smaniotto S, Keller AC et al (2008) Multivectorial abnormal cell migration in the NOD mouse thymus. J Immunol 180:4639–4647. 180/7/4639 [pii]

Mendes-da-Cruz DA, Lepelletier Y, Brignier AC et al (2009) Neuropilins, semaphorins, and their role in thymocyte development. Ann N Y Acad Sci 1153:20–28. https://doi.org/10.1111/j.1749-6632.2008.03980.x

Mendes-da-Cruz DA, Stimamiglio MA, Muñoz JJ et al (2012) Developing T-cell migration: role of semaphorins and ephrins. FASEB J 26:4390–4399. https://doi.org/10.1096/fj.11-202952

Mendes-da-Cruz DA, Brignier AC, Asnafi V et al (2014) Semaphorin 3F and neuropilin-2 control the migration of human T-cell precursors. PLoS One 9:e103405. https://doi.org/10.1371/journal.pone.0103405

Mertani HC, Delehaye-Zervas MC, Martini JF et al (1995) Localization of growth hormone receptor messenger RNA in human tissues. Endocrine 3:135–142. https://doi.org/10.1007/BF02990065

Messias CV, Santana-Van-Vliet E, Lemos JP et al (2016) Sphingosine-1-phosphate induces dose-dependent chemotaxis or fugetaxis of T-ALL blasts through S1P1 activation. PLoS One 11:e0148137. https://doi.org/10.1371/journal.pone.0148137

Misslitz A, Pabst O, Hintzen G et al (2004) Thymic T cell development and progenitor localization depend on CCR7. J Exp Med 200:481–491. https://doi.org/10.1084/jem.20040383

Mojcik CF, Salomon DR, Chang a C, Shevach EM (1995) Differential expression of integrins on human thymocyte subpopulations. Blood 86:4206–4217

Montero-Herradón S, García-Ceca J, Zapata AG (2017) EphB receptors, mainly EphB3, contribute to the proper development of cortical thymic epithelial cells. Organogenesis 13:192–211. https://doi.org/10.1080/15476278.2017.1389368

Mori K, Itoi M, Tsukamoto N et al (2007) The perivascular space as a path of hematopoietic progenitor cells and mature T cells between the blood circulation and the thymic parenchyma. Int Immunol 19:745–753. https://doi.org/10.1093/intimm/dxm041

Muñoz JJ, Alonso-C LM, Sacedón R et al (2002) Expression and function of the Eph A receptors and their ligands ephrins A in the rat thymus. J Immunol 169:177–184

Muñoz JJ, García-Ceca J, Alfaro D et al (2009) Organizing the thymus gland. Ann N Y Acad Sci 1153:14–19. https://doi.org/10.1111/j.1749-6632.2008.03965.x

Murata N, Sato K, Kon J et al (2000) Quantitative measurement of sphingosine 1-phosphate by radioreceptor-binding assay. Anal Biochem 282:115–120. https://doi.org/10.1006/abio.2000.4580

Murray P, Edgar D (2000) Regulation of programmed cell death by basement membranes in embryonic development. J Cell Biol 150:1215–1221. https://doi.org/10.1083/jcb.150.5.1215

Nishiuchi R, Takagi J, Hayashi M et al (2006) Ligand-binding specificities of laminin-binding integrins: a comprehensive survey of laminin-integrin interactions using recombinant alpha-3beta1, alpha6beta1, alpha7beta1 and alpha6beta4 integrins. Matrix Biol 25:189–197. https://doi.org/10.1016/j.matbio.2005.12.001

Nitta T, Nitta S, Lei Y et al (2009) CCR7-mediated migration of developing thymocytes to the medulla is essential for negative selection to tissue-restricted antigens. Proc Natl Acad Sci U S A 106:17129–17133

Nussbaum C, Bannenberg S, Keul P et al (2015) Sphingosine-1-phosphate receptor 3 promotes leukocyte rolling by mobilizing endothelial P-selectin. Nat Commun 6:6416. https://doi.org/10.1038/ncomms7416

Ocampo JSP, de BJM, Corrêa-de-Santana E et al (2008) Laminin-211 controls thymocyte-thymic epithelial cell interactions. Cell Immunol 254:1–9. https://doi.org/10.1016/j.cellimm.2008.06.005

Oliveira-dos-Santos AJ, Rieker-Geley T, Recheis H, Wick G (1997) Murine thymic nurse cells and rosettes: analysis of adhesion molecule expression using confocal microscopy and a simplified enrichment method. J Histochem Cytochem 45:1293–1297. https://doi.org/10.1177/002215549704500912

Passos GA, Mendes-da-Cruz DA, Oliveira EH (2015) The thymic orchestration involving aire, miRNAs, and cell–cell interactions during the induction of central tolerance. Front Immunol 6:1–7. https://doi.org/10.3389/fimmu.2015.00352

Patel DD, Hale LP, Whichard LP et al (1995) Expression of CD44 molecules and CD44 ligands during human thymic fetal development: expression of CD44 isoforms is developmentally regulated. Int Immunol 7:277–286. https://doi.org/10.1093/intimm/7.2.277

Pearson JA, Wong FS, Wen L (2016) The importance of the non obese diabetic (NOD) mouse model in autoimmune diabetes. J Autoimmun 66:76–88. https://doi.org/10.1016/j.jaut.2015.08.019

Person PL, Korngold R, Teuscher C (1992) Pertussis toxin-induced lymphocytosis is associated with alterations in thymocyte. J Immunol 148:1506–1511

Petrie HT, Zúñiga-Pflücker JC (2007) Zoned out: functional mapping of stromal signaling microenvironments in the thymus. Annu Rev Immunol 25:649–679. https://doi.org/10.1146/annurev.immunol.23.021704.115715

Plotkin J, Prockop SE, Lepique A, Petrie HT (2003) Critical role for CXCR4 signaling in progenitor localization and T cell differentiation in the postnatal thymus. J Immunol 171:4521–4527. https://doi.org/10.4049/jimmunol.171.9.4521

Poznansky MC, Olszak IT, Foxall R et al (2000) Active movement of T cells away from a chemokine. Nat Med 6:543–548. https://doi.org/10.1038/75022

Richie LI, Ebert PJR, Wu LC et al (2002) Imaging synapse formation during thymocyte selection: inability of CD3ζ to form a stable central accumulation during negative selection. Immunity 16:595–606. https://doi.org/10.1016/S1074-7613(02)00299-6

Robertson P, Means TK, Luster AD, Scadden DT (2006) CXCR4 and CCR5 mediate homing of primitive bone marrow-derived hematopoietic cells to the postnatal thymus. Exp Hematol 34:308–319. https://doi.org/10.1016/j.exphem.2005.11.017

Romagnani P, Annunziato F, Lazzeri E et al (2001) Interferon-inducible protein 10, monokine induced by interferon gamma, and interferon-inducible T-cell alpha chemoattractant are produced by thymic epithelial cells and attract T-cell receptor (TCR) $\alpha\beta^+$ CD8$^+$ single-positive T cells, TCR$\gamma\delta^+$ T cell. Blood 97:601–608

Ruiz P, Wiles MV, Imhof BA (1995) 6 integrins participate in pro???T cell homing to the thymus. Eur J Immunol 25:2034–2041. https://doi.org/10.1002/eji.1830250735

Saga Y, Tsukamoto T, Jing N et al (1991) Murine tenascin: cDNA cloning, structure and temporal expression of isoforms. Gene 104:177–185

Salomon DR, Mojcik CF, Chang AC et al (1994) Constitutive activation of integrin alpha 4 beta 1 defines a unique stage of human thymocyte development. J Exp Med 179:1573–1584. https://doi.org/10.1084/jem.179.5.1573

Sammani S, Moreno-Vinasco L, Mirzapoiazova T et al (2010) Differential effects of sphingosine 1-phosphate receptors on airway and vascular barrier function in the murine lung. Am J Respir Cell Mol Biol 43:394–402. https://doi.org/10.1165/rcmb.2009-0223OC

Sanna MG, Wang S-K, Gonzalez-Cabrera PJ et al (2006) Enhancement of capillary leakage and restoration of lymphocyte egress by a chiral S1P1 antagonist in vivo. Nat Chem Biol 2:434–441. https://doi.org/10.1038/nchembio804

Savagner P, Imhof BA, Yamada KM, Thiery JP (1986) Homing of hemopoietic precursor cells to the embryonic thymus: characterization of an invasive mechanism induced by chemotactic peptides. J Cell Biol 103:2715–2727. https://doi.org/10.1083/jcb.103.6.2715

Savino W, Leite-de-Moraes MC, Hontebeyrie-Joskowicz M, Dardenne M (1989) Studies on the thymus in Chagas' disease. I. Changes in the thymic microenvironment in mice acutely infected with Trypanosoma cruzi. Eur J Immunol 19:1727–1733. https://doi.org/10.1002/eji.1830190930

Savino W, Boitard C, Bach J-F, Dardenne M (1991) Studies on the thymus in nonobese diabetic mouse. I. Changes in the microenvironmental compartments. Lab Invest 64:405

Savino W, Dalmau SR, Dealmeida VC (2000) Role of extracellular matrix-mediated interactions in thymocyte migration. Dev Immunol 7:279–291. https://doi.org/10.1155/2000/60247

Savino W, Mendes-da-Cruz DA, Silva JS et al (2002a) Intrathymic T-cell migration: a combinatorial interplay of extracellular matrix and chemokines? Trends Immunol 23:305–313

Savino W, Postel-Vinay M, Smaniotto S, Dardenne M (2002b) The thymus gland: a target organ for growth hormone. Scand J Immunol 55:442–452

Savino W, Ayres Martins S, Neves-dos-Santos S et al (2003) Thymocyte migration: an affair of multiple cellular interactions? Braz J Med Biol Res 36:1015–1025. https://doi.org/10.1590/S0100-879X2003000800007

Savino W, Mendes-da-Cruz DA, Ferreira Golbert DC et al (2015) Laminin-mediated interactions in thymocyte migration and development. Front Immunol 6:579. https://doi.org/10.3389/fimmu.2015.00579

Savino W, Mendes-da-Cruz DA, Lepletier A, Dardenne M (2016) Hormonal control of T-cell development in health and disease. Nat Rev Endocrinol 12:77–89. https://doi.org/10.1038/nrendo.2015.168

Sawada M, Nagamine J, Takeda K et al (1992) Expression of VLA-4 on thymocytes. Maturation stage-associated transition and its correlation with their capacity to adhere to thymic stromal cells. J Immunol 149:3517–3524

Schabath R, Müller G, Schubel A et al (1999) The murine chemokine receptor CXCR4 is tightly regulated during T cell development and activation. J Leukoc Biol 66:996–1004. https://doi.org/10.1002/jlb.66.6.996

Schmeissner PJ, Xie H, Smilenov LB et al (2001) Integrin functions play a key role in the differentiation of thymocytes in vivo. J Immunol 167:3715–3724. https://doi.org/10.4049/jimmunol.167.7.3715

Schwab SR, Pereira JP, Matloubian M et al (2005) Lymphocyte sequestration through S1P lyase inhibition and disruption of S1P gradients. Science 309:1735–1739. https://doi.org/10.1126/science.1113640

Schwarz BA, Sambandam A, Maillard I et al (2007) Selective thymus settling regulated by cytokine and chemokine receptors. J Immunol 178:2008–2017. https://doi.org/10.4049/jimmunol.178.4.2008

Scollay R, Godfrey DI (1995) Thymic emigration: conveyor belts or lucky dips? Immunol Today 16:268–273. https://doi.org/10.1016/0167-5699(95)80179-0

Serafini T, Colamarino SA, Leonardo ED et al (1996) Netrin-1 is required for commissural axon guidance in the developing vertebrate nervous system. Cell 87:1001–1014. https://doi.org/10.1016/S0092-8674(00)81795-X

Shannon LA, McBurney TM, Wells MA et al (2012) CCR7/CCL19 controls expression of EDG-1 in T cells. J Biol Chem 287:11656–11664. https://doi.org/10.1074/jbc.M111.310045

Sharfe N, Freywald A, Toro A et al (2002) Ephrin stimulation modulates T‐cell chemotaxis. Eur J Immunol 32:3745–3755. https://doi.org/10.1002/1521-4141(200212)32:12<3745::AID-IMMU3745>3.0.CO;2-M

Shimizu Y, Shaw S (1991) Lymphocyte interactions with extracellular matrix. FASEB J 5:2292

Shiow LR, Rosen DB, Brdicková N et al (2006) CD69 acts downstream of interferon-alpha/beta to inhibit S1P1 and lymphocyte egress from lymphoid organs. Nature 440:540–544. https://doi.org/10.1038/nature04606

Simon MM, Kramer MD, Prester M, Gay S (1991) Mouse T-cell associated serine proteinase 1 degrades collagen type IV: a structural basis for the migration of lymphocytes through vascular basement membranes. Immunology 73:117–119

Smaniotto S, Ribeiro-Carvalho MM, Dardenne M et al (2004) Growth hormone stimulates the selective trafficking of thymic CD4+CD8- emigrants to peripheral lymphoid organs. Neuroimmunomodulation 11:299–306. https://doi.org/10.1159/000079410

Smaniotto S, De Mello-Coelho V, Villa-Verde DMS et al (2005) Growth hormone modulates thymocyte development in vivo through a combined action of laminin and CXC chemokine ligand 12. Endocrinology 146:3005–3017. https://doi.org/10.1210/en.2004-0709

Smyth N, Vatansever HS, Murray P et al (1999) Absence of basement membranes after targeting the LAMC1 gene results in embryonic lethality due to failure of endoderm differentiation. J Cell Biol 144:151–160. https://doi.org/10.1083/jcb.144.1.151

Souza-e-Silva H, Savino W, Feijóo RA, Vasconcelos ATR (2009) A cellular automata-based mathematical model for thymocyte development. PLoS One 4:e8233. https://doi.org/10.1371/journal.pone.0008233

Spiegel S, Milstien S (2011) The outs and the ins of sphingosine-1-phosphate in immunity. Nat Rev Immunol 11:403–415. https://doi.org/10.1038/nri2974

Stachowiak AN, Wang Y, Huang Y-C, Irvine DJ (2006) Homeostatic lymphoid chemokines synergize with adhesion ligands to trigger T and B lymphocyte chemokinesis. J Immunol 177:2340–2348. https://doi.org/10.4049/jimmunol.177.4.2340

Stimamiglio MA, Jiménez E, Silva-Barbosa SD et al (2010) EphB2-mediated interactions are essential for proper migration of T cell progenitors during fetal thymus colonization. J Leukoc Biol 88:483–494. https://doi.org/10.1189/jlb.0210079

Suniara BRK, Jenkinson EJ, Owen JJT (2000) An essential role for thymic mesenchyme in early T cell development. J Exp Med 191:1051–1056

Suzuki G, Nakata Y, Dan Y et al (1998) Loss of SDF-1 receptor expression during positive selection in the thymus. Int Immunol 10:1049–1056. https://doi.org/10.1093/intimm/10.8.1049

Suzuki G, Sawa H, Kobayashi Y et al (1999) Pertussis toxin-sensitive signal controls the trafficking of thymocytes across the corticomedullary junction in the thymus. J Immunol 162:5981–5985. https://doi.org/10.4049/jimmunol.181.7.4723

Takabe K, Paugh SW, Milstien S, Spiegel S (2008) "Inside-Out" signaling of sphingosine-1-phosphate: : therapeutic targets. Pharmacol Rev 60:181–195. https://doi.org/10.1124/pr.107.07113

Takahashi K, Ishida M, Hirokawa K, Takahashi H (2008) Expression of the semaphorins *Sema 3D* and *Sema 3F* in the developing parathyroid and thymus. Dev Dyn 237:1699–1708. https://doi.org/10.1002/dvdy.21556

Takayama E, Kina T, Katsura Y, Tadakuma T (1998) Enhancement of activation-induced cell death by fibronectin in murine CD4+ CD8+ thymocytes. Immunology 95:553–558

Thuy AV, Reimann C-M, Hemdan NY a, Gräler MH (2014) Sphingosine 1-phosphate in blood: function, metabolism, and fate. Cell Physiol Biochem 34:158–171. https://doi.org/10.1159/000362992

Trampont PC, Tosello-Trampont AC, Shen Y et al (2010) CXCR4 acts as a costimulator during thymic B-selection. Nat Immunol 11:162–170. https://doi.org/10.1038/ni.1830

Uehara S, Grinberg A, Farber JM, Love PE (2002a) A role for CCR9 in T lymphocyte development and migration. J Immunol 168:2811–2819. https://doi.org/10.4049/jimmunol.168.6.2811

Uehara S, Song K, Farber JM, Love PE (2002b) Characterization of CCR9 expression and CCL25/thymus-expressed chemokine responsiveness during T cell development: CD3highCD69+ thymocytes and TCR+ thymocytes preferentially respond to CCL25. J Immunol 168:134–142. https://doi.org/10.4049/jimmunol.168.1.134

Ueno T, Hara K, Willis MS et al (2002) Role for CCR7 ligands in the emigration of newly generated T lymphocytes from the neonatal thymus. Immunity 16:205–218. https://doi.org/10.1016/S1074-7613(02)00267-4

Ueno T, Saito F, Gray DHD et al (2004) CCR7 signals are essential for cortex – medulla migration of developing thymocytes. J Exp Med 200:493–505. https://doi.org/10.1084/jem.20040643

Utsumi K, Sawada M, Narumiya S et al (1991) Adhesion of immature thymocytes to thymic stromal cells through fibronectin molecules and its significance for the induction of thymocyte differentiation. Proc Natl Acad Sci U S A 88:5685–5689. https://doi.org/10.1073/pnas.88.13.5685

Vianello F, Kraft P, Mok YT et al (2005) A CXCR4-dependent chemorepellent signal contributes to the emigration of mature single-positive CD4 cells from the fetal thymus. J Immunol 175:5115–5125. 175/8/5115 [pii]

Villa-Verde DM, de Mello-Coelho V, Farias-de-Oliveira DA et al (1993) Pleiotropic influence of triiodothyronine on thymus physiology. Endocrinology 133:867–875. https://doi.org/10.1210/endo.133.2.8344222

Villa-Verde DMS, Lagrota-Candido MJ et al (1994) Extracellular matrix components of the mouse thymus microenvironment. IV. Modulation of thymic nurse cells by extracellular matrix ligands and receptors. Eur J Immunol 24:659–664. https://doi.org/10.1002/eji.1830240326

Virtanen I, Lohi J, Tani T et al (1996) Laminin chains in the basement membranes of human thymus. Histochem J 28:643–650. https://doi.org/10.1007/BF02331385

Virtanen I, Gullberg D, Rissanen J et al (2000) Laminin α1-chain shows a restricted distribution in epithelial basement membranes of fetal and adult human tissues. Exp Cell Res 257:298–309. https://doi.org/10.1006/excr.2000.4883

Vivinus-Nebot M, Ticchioni M, Mary F et al (1999) Laminin 5 in the human thymus: control of T cell proliferation via alpha6beta4 integrins. J Cell Biol 144:563–574

Wadsworth S, Halvorson MJ, Coligan JE (1992) Developmentally regulated expression of the beta 4 integrin on immature mouse thymocytes. J Immunol 149:421–428

Wadsworth S, Halvorson MJ, Chang AC, Coligan JE (1993) Multiple changes in VLA protein glycosylation, expression, and function occur during mouse T cell ontogeny. J Immunol 150:847–857

Walzer T, Chiossone L, Chaix J et al (2007) Natural killer cell trafficking in vivo requires a dedicated sphingosine 1-phosphate receptor. Nat Immunol 8:1337–1344. https://doi.org/10.1038/ni1523

Watterson K, Sankala H, Milstien S, Spiegel S (2003) Pleiotropic actions of sphingosine-1-phosphate. Prog Lipid Res 42:344–357. https://doi.org/10.1016/S0163-7827(03)00015-8

Werneck CC, Cruz MS, Silva LCF et al (2000) Is there a glycosaminoglycan-related heterogeneity of the thymic epithelium? J Cell Physiol 185:68–79. https://doi.org/10.1002/1097-4652(200010)185:1<68::AID-JCP6>3.0.CO;2-D

Wurbel MA, Malissen B, Campbell JJ (2006) Complex regulation of CCR9 at multiple discrete stages of T cell development. Eur J Immunol 36:73–81. https://doi.org/10.1002/eji.200535203

Yebra M, Montgomery AMP, Diaferia GR et al (2003) Recognition of the neural chemoattractant netrin-1 by integrins α6β4 and α3β1 regulates epithelial cell adhesion and migration. Dev Cell 5:695–707. https://doi.org/10.1016/S1534-5807(03)00330-7

Zingoni A, Soto H, Hedrick JA et al (1998) Cutting edge: the chemokine receptor CCR8 is preferentially expressed in Th2 but not Th1 cells. J Immunol 161:547–551

Zlotnik A, Yoshie O (2000) Chemokines: a new classification review system and their role in immunity. Immunity 12:121–127

Zlotnik A, Yoshie O (2012) The chemokine superfamily revisited. Immunity 36:705–712. https://doi.org/10.1016/j.immuni.2012.05.008

Zlotoff DA, Sambandam A, Logan TD et al (2010) CCR7 and CCR9 together recruit hematopoietic progenitors to the adult thymus. Blood 115:1897–1905. https://doi.org/10.1182/blood-2009-08-237784

Chapter 6
Thymic Crosstalk: An Overview of the Complex Cellular Interactions That Control the Establishment of T-Cell Tolerance

Magali Irla

Abstract The thymus ensures the generation of a self-tolerant T-cell repertoire capable of recognizing foreign antigens. The selection of the T-cell repertoire is dictated by the thymic microenvironment. Among stromal cells, medullary thymic epithelial cells (mTECs) play a pivotal role in this process through their unique ability to express thousands of tissue-restricted self-antigens. In turn, developing T cells control the pool and maturation of mTECs. This phenomenon of bidirectional interactions between TECs and thymocytes is referred to as thymic crosstalk. In this chapter, I discuss the discovery of thymic crosstalk and our current understanding of bidirectional interactions between mTECs and thymocytes. Finally, I summarize recent advances indicating that thymic crosstalk is not restricted to TECs and thymocytes but also occurs between TECs and dendritic cells, as well as B cells and thymocytes. This complex cellular interplay is essential for efficient T-cell selection.

Abbreviations

Aire	Autoimmune Regulator
cDCs	Conventional dendritic cells
cTECs	Cortical thymic epithelial cells
DN	Double negative cells
DP	Double positive cells
Fezf2	Fez family zinc-finger 2
LTα	Lymphotoxin α

M. Irla (✉)
Centre d'Immunologie de Marseille-Luminy (CIML), INSERM U1104, CNRS UMR7280,
Aix-Marseille Université UM2, Marseille Cedex 09, France
e-mail: Magali.Irla@inserm.fr

© Springer Nature Switzerland AG 2019   149
G. A. Passos (ed.), *Thymus Transcriptome and Cell Biology*,
https://doi.org/10.1007/978-3-030-12040-5_6

LTβR   Lymphotoxin beta receptor
mTECs   Medullary thymic epithelial cells
nTregs   Natural regulatory T cells
OVA   Ovalbumin
pDCs   Plasmacytoid dendritic cells
RANKL   Receptor Activator of Nuclear factor Kappa-B Ligand
SP   Single positive cells
TECs   Thymic epithelial cells
TCR   T-cell receptor
TRAs   Tissue-restricted self-antigens
WT   Wild-type

6.1   Introduction

The thymus is composed of two main anatomical regions: the cortex and the medulla. Cortical and medullary compartments support early and late stages of T-cell development, respectively (Takahama 2006). T-cell progenitors derived from the bone marrow colonize the thymus *via* venules localized at the cortico-medullary junction (Petrie and Zuniga-Pflucker 2007). At this early stage, these cells do not express the CD4 and CD8 co-receptors and are commonly described as CD4/CD8 double-negative (DN) cells. DN cells can be phenotypically distinguished into four major developmental stages during their migration outward to the subcapsule (Lind et al. 2001; Porritt et al. 2003). These cells are termed DN1 ($CD44^+CD25^-$), DN2 ($CD44^+CD25^+$), DN3 ($CD44^-CD25^+$) and DN4 ($CD44^-CD25^-$) (Godfrey et al. 1993). Further differentiation results in the simultaneous acquisition of CD4 and CD8 co-receptors and thus in the generation of $CD4^+CD8^+$double-positive (DP) cells that migrate back inward through the cortex. Only a small fraction of DP thymocytes (3–5%) showing low affinity engagement of their T-cell receptors (TCRs) with peptide/MHC complexes survive at this stage (Takahama et al. 2010). In contrast, DP thymocytes that do not receive a survival signal due to absent or insufficient TCR engagement die by neglect whereas cells that show a high-affinity for self-peptide/MHC complexes are also eliminated. Positively-selected DP thymocytes loose either the CD4 or CD8 coreceptor and are designated as single-positive (SP) cells. This process of selection is referred to as positive selection. SP thymocytes relocate into the medulla where they undergo a second step of selection, called negative selection (also known as clonal deletion) that consists in the deletion of SP cells that harbour a TCR with high affinity for self-peptide/MHC complexes (Palmer 2003; Klein et al. 2009). Furthermore, TCR/MHCII interactions of intermediate affinity preferentially leads to the generation of natural regulatory T cells (nTregs) that express the transcription factor forkhead box P3 (Foxp3). nTregs are important actors of peripheral tolerance through their ability to control potentially hazardous autoreactive T cells that have escaped the negative selection process (Josefowicz et al. 2012). Thymic selection processes occurring in the cortex and the medulla lead to the generation of a self-tolerant and diversified peripheral T-cell repertoire capable of responding to foreign pathogens and malignant transformation of self.

The survival, proliferation and differentiation of developing T cells are under the control of thymic stromal cells. Cortical thymic epithelial cells (cTECs) induce the positive selection of DP thymocytes whereas medullary TECs (mTECs) and dendritic cells (DCs) mediate the negative selection of SP thymocytes and the induction of nTreg cells (Lopes et al. 2015; Klein et al. 2014). mTECs represent less than 1% of the total thymic cells but are essential actors involved in the selection of the T-cell repertoire by their unique ability to express thousands of tissue-restricted self-antigens (TRAs) (Derbinski et al. 2001; Sansom et al. 2014; Passos et al. 2015). It is estimated that mTECs express in total more than 19,000 protein-coding genes and thus approximately 85% of the mouse genome, which represents the highest proportion of genes expressed in any cell type. The Autoimmune Regulator (Aire) was the first transcription factor identified for driving the expression of numerous TRAs (Anderson et al. 2002). Subsequent studies based on high-throughput RNA-sequencing on purified mTECs found that Aire regulates between 3,000 and 4,000 TRAs (Sansom et al. 2014; Meredith et al. 2015). More recently, a second transcription factor called Fezf2 (Fez family zinc-finger 2) has been shown to induce the expression of several Aire-independent TRAs (Takaba et al. 2015). Nevertheless, the precise number of Fezf2-regulated genes remains to be established. Importantly, both *Aire*$^{-/-}$ and *Fezf2*$^{-/-}$ mice exhibit signs of autoimmunity characterized by inflammatory cell infiltration and autoantibodies directed against several peripheral tissues (Takaba et al. 2015; Anderson et al. 2002), highlighting the importance of these two transcriptional regulators in the induction of central tolerance.

mTECs can be identified by histology by the expression of several markers such as the cytokeratin-5, cytokeratin 14, ER-TR5 and MTS10 (Klug et al. 1998; Van Vliet et al. 1984; Godfrey et al. 1990). mTECs also show reactivity with Ulex Europaeus Agglutinin 1 (UEA-1) and Tetragonolobus Purpureas Agglutinin (TPA) lectins on thymic sections (Farr and Anderson 1985). Progress in methods of TEC isolation and the generation of monoclonal antibodies have greatly aided the study of mTEC biology. In particular, clones such as G8.8, BP-1 and 5H12 respectively recognize the pan-epithelial determinant EpCAM (Epithelial Cellular Adhesion Molecule), the glycoprotein cell-surface marker Ly51 and the transcriptional regulator Aire. The identification of mTECs based on enzymatically disaggregated thymic preparations revealed that they constitute a heterogeneous cell population. Total mTECs are classically defined as CD45^{-}EpCAM^{+}Ly51$^{lo/-}$ by flow cytometry. mTEC subsets can be further identified based on cell-surface levels of MHC class II (MHCII) and the CD80 costimulatory molecule. MHCII^{lo}CD80^{lo} and MHCII^{hi}CD80^{hi} cells are commonly referred to as mTEC^{lo} and mTEC^{hi}, respectively (Gray et al. 2006). mTEC differentiation proceeds along distinct maturational stages with MHCII$^{-/lo}$CD80$^{-/lo}$Aire^{-} immature cells that give rise to MHCII^{hi}CD80^{hi}Aire^{+} mature cells through a MHCII^{hi}CD80^{hi}Aire^{-} intermediate stage (Gray et al. 2007; Gabler et al. 2007; Rossi et al. 2007). Cell fate mapping studies have identified that a fraction of MHCII[hi]CD80[hi]Aire[+] mature mTECs further progress to a post-Aire stage expressing reduced levels of MHCII and CD80 molecules that is thus defined as MHCII[lo]CD80[lo]Aire[−]mTECs (Yano et al. 2008; Nishikawa et al. 2010; Metzger et al. 2013). A small subset of these Aire[−]mTECs expresses involucrin, a component of the cross-linked cornified envelope, which is a marker of keratinocyte terminal differentiation and Hassall's Corpuscles. These latter are complex anatomical structures located in the medulla that have been proposed to be involved in the removal of apoptotic thymocytes and maturation of SP thymocytes (Watanabe et al. 2005; Senelar et al. 1976). Importantly, mature mTECs show a turnover period of around 2–3 weeks at the post-natal stage (Gabler et al. 2007; Gray et al. 2007), indicating that this population is continuously replenished from an mTEC progenitor.

6.1.1 The Discovery of Thymic Crosstalk

The biological relevance of the thymus remained enigmatic for centuries. The role of this organ in the establishment of a normal immune system was recognized in 1961. Pioneer experiments of thymectomy in mice after birth performed by Jacques Miller have resulted in a marked lymphocyte deficiency observed in blood circulation and lymphoid tissues (Miller 1961). Thymectomised mice were consequently highly susceptible to infections and died prematurely. These experiments are the first demonstration of the crucial role of the thymus in the lymphocyte production (Miller 2002). Historically, the development of T cells was initially described as dependent on interactions with stromal cells (Fink and Bevan 1978; Kyewski et al. 1989) whereas the impact of developing T cells on stromal cell biology was unknown at that time. The hypothesis that developing T cells could regulate the thymic epithelium comes from experiments based on the disruption of the hematopoietic compartment by treating mice with irradiation or cyclosporine A (Adkins et al. 1988; Kanariou et al. 1989). These treatments were shown to result in a marked reduction of the medullary regions that partially recovered after treatment cessation. Nevertheless, these experiments could not conclude whether the loss of the medullary regions and mTECs was due to an altered T-cell compartment or to toxic effects related to these treatments. Subsequent experiments found that severe combined immunodeficient (SCID) mice that have an early block in T-cell development at the DN3 stage show an organized cortex but a disorganized medulla containing few scattered ER-TR5[+] mTECs (Shores et al. 1991). Interestingly, in 1991, Shores *et al.* demonstrated that the introduction of TCR bearing thymocytes into these mice leads to the development and maturation of the medullary epithelium (Shores et al. 1991). Indeed, the infusion of wild-type (WT) bone marrow cells into SCID recipients results in the differentiation of TCR[+] thymocytes and subsequent restoration of the medullary compartment. These results are the first compelling demonstration that lympho-stromal interactions are bidirectional. The development of thymocytes and TECs is thus interdependent. Furthermore, in 1992, Surh *et al.* demonstrated that the intravenous administration of peripheral T cells from WT mice into SCID recipients also leads to mTEC growth, further highlighting the importance of TCR$^+$ T cells in the development of the medulla (Surh et al. 1992). Conversely, the depletion of SP thymocytes in mice treated with antibodies against the αβ TCR leads to reduced medullary regions containing few mTECs (van Ewijk et al. 1994). Subsequent studies have reported that medullary size, composition and organization are severely perturbed in both *Tcra*$^{-/-}$ and *Zap70*$^{-/-}$ mice lacking SP thymocytes due to a block in thymocyte development at the DP stage (Negishi et al. 1995; Palmer et al. 1993). Similarly, *Rag-1*$^{-/-}$ and *Rag-2*$^{-/-}$ mice that show an arrest in T-cell development at the DN3 stage have a well-developed cortex but an almost absent medulla (Hollander et al. 1995; van Ewijk et al. 2000). Interestingly, since *Tcra*$^{-/-}$ mice produce normal numbers of γδ T cells, it was initially suggested that the development of the medulla is exclusively regulated by αβ T cells (Palmer et al. 1993). This hypothesis was confirmed later with *Tcrd*$^{-/-}$ mice, which definitely showed that γδ T cells are dispensable for mTEC differentiation and medulla formation at the post-natal stage (Hikosaka et al. 2008). Therefore, these observations based on different mouse strains with altered T-cell development further underline that αβ TCR-bearing SP thymocytes are indispensable for the development of the medullary epithelium. This phenomenon of bidirectional interactions between mTECs and SP thymocytes has been designated as "thymic crosstalk" (see for review van Ewijk et al. 1994; Lopes et al. 2015). In 1995, Hollander *et al.* discovered that thymocyte crosstalk not only controls the development of the medullary epithelium but also that of the cortical epithelium (Hollander et al. 1995). Transgenic mice expressing high copies of the human CD3ε chain (tgε26) that have an early block in T-cell development at the DN1 stage show a highly disturbed cortical and medullary epithelium with poor cortico-medullary demarcation (Hollander et al. 1995; van Ewijk et al. 2000; Wang et al. 1994). Interestingly, the transplantation of bone marrow cells from *Rag-2*$^{-/-}$ mice, which have a subsequent block at the DN3 stage, into tgε26 recipients or the transplantation of tgε26 thymi under the kidney capsule of *Rag-1*$^{-/-}$ mice restores the organization of the cortical epithelium (van Ewijk et al. 2000; Klug et al. 1998). These pioneer observations revealed that the transition from the DN1 to DN3 stage controls the development of the cortical epithelium. Importantly, this defect can be corrected during foetal development and up to 6 days post-natal by the transplantation of hematopoietic cells from either foetal or adult origin (Hollander et al. 1995). Interestingly, a subsequent transplantation of WT bone marrow cells into tgε26 recipients, initially transplanted with *Rag-2*$^{-/-}$ bone marrow cells, restores the organization of the medullary microenvironment (van Ewijk et al. 2000). Distinct thymocyte subsets thus control the development of the thymic epithelium in a stepwise manner.

In sum, these pioneer studies have established that the thymic crosstalk represents a symbiotic relationship, whereby TECs regulate the different stages of T-cell development while distinct thymocytes reciprocally control the formation and integrity of the cortical and medullary epithelium.

6.1.2   Crosstalk Between mTECs and SP Thymocytes

6.1.2.1   CD4⁺ SP Thymocytes Control Medulla Expansion and Composition

Although pioneer studies identified that positively-selected SP thymocytes are essential for the development of the medulla (Shores et al. 1991, 1994; Surh et al. 1992; van Ewijk et al. 1994), the respective contribution of CD4$^+$ and CD8$^+$ SP thymocytes in this process remained to be determined. This question was solved by analysing thymic sections from mice deficient in MHCII (*H2-Aa$^{-/-}$* mice) and MHCI (*β2m$^{-/-}$* mice) expression, which consequently lack CD4$^+$ and CD8$^+$ SP thymocytes, respectively. Antibody staining against the cTEC marker cytokeratin 8 and the mTEC marker cytokeratin 14 showed a clear cortico-medullary demarcation in both knockout mice. Nevertheless, in contrast to *β2m$^{-/-}$* mice that have well-developed medullary regions closely resembling to those observed in WT mice, *H2-Aa$^{-/-}$* mice have markedly under-developed medulla (Irla et al. 2012, 2013). The importance of CD4$^+$ SP thymocytes in medulla development was confirmed using *Ciita$^{IV-/IV-}$* mice that lack the promoter IV of the gene encoding the MHCII transactivator (CIITA), which regulates MHCII expression in cTECs (Waldburger et al. 2001). Consequently, *Ciita$^{IV-/IV-}$* mice have a block in the positive selection of CD4$^+$ SP thymocytes. Our laboratory further investigated how CD4$^+$ SP thymocytes influence the 3D topology of the medulla. For this, we developed an automated process called Full Organ Reconstruction in 3D (For3D) that allows the visualization and quantification of thymic lobes from a collection of serial immunolabeled sections for cTEC- and mTEC-specific markers (Serge et al. 2015). Using For3D, we found that under physiological conditions, the medulla exhibits a complex topology, composed of ~200 small medullae that are disconnected to a major medullary compartment (Irla et al. 2013). Since each individual medullary islet derives from a single progenitor (Rodewald et al. 2001), mTEC islets likely grow and fuse together, which leads to larger islets and ultimately to a major central compartment. Interestingly, the medulla exhibits fractal properties, which results in a considerable cortico-medullary area, which is particularly enriched in Aire$^+$mTECs (Irla et al. 2013). This distribution in Aire$^+$ mTECs is more pronounced in neonates compared to adults, which is consistent with the fact that Aire is essential to prevent autoimmune disorders during the perinatal period (Guerau-de-Arellano et al. 2009; Yang et al. 2015). Furthermore, the fractal geometry of the cortico-medullary region ensures short distances between the cortex and medulla, and thus efficient migration of positively-selected SP thymocytes into the medulla as assessed by mathematical models (Irla et al. 2013). Interestingly, 3D reconstructions of thymic lobes from *H2-Aa$^{-/-}$* mice revealed that CD4$^+$ SP thymocytes control the complex topology of the medulla (Irla et al. 2013; Nasreen et al. 2003). These mice lack the usual large medullary compartment and instead have small individual islets, which leads to a globally reduced medullary volume. Nevertheless, the development of the medulla in *H2-Aa$^{-/-}$* mice is less severely affected than in mice exhibiting a complete absence of SP thymocytes such as $Tcra^{-/-}$ or $Zap70^{-/-}$ mice. These observations suggest that other cell type(s) are implicated in the patterning of the medulla. 3D reconstructions of thymic lobes from $\beta2m^{-/-}$ mice indicate that CD8$^+$ SP thymocytes have a minor role compared with CD4$^+$ SP thymocytes since these mice do not display any obvious defect in the 3D medullary organization. The contribution of other cell type(s) that dependent on the expression of $Tcra$ or $Zap70$ genes for their development, such as invariant NKT cells, remains to be determined. Importantly, CD4$^+$ SP thymocytes not only control the size and 3D topology of the medulla but also its composition in mTECs. In contrast to $\beta2m^{-/-}$ mice, both $H2\text{-}Aa^{-/-}$ and $Ciita^{IV-/IV-}$ mice have strongly reduced numbers of UEA-1$^+$ and CD80^{hi}Aire$^+$ mature mTECs (Irla et al. 2008). In accordance with the critical role of Aire in the establishment of T-cell tolerance through clonal deletion and nTreg induction (see for review Perniola 2018), these mice exhibit an abnormal infiltration of activated CD69^{hi} and CD44^{hi} CD8$^+$ T cells in several peripheral tissues, indicative of impaired tolerance. The capacity of CD4$^+$ SP thymocytes to promote the differentiation of mature mTECs was also recapitulated $in\ vitro$ using reaggregated thymic organ cultures in which foetal thymic stromal cells were reaggregated with either purified DP or CD4$^+$ SP thymocytes. Whereas DP thymocytes have no effect, CD4$^+$ SP thymocytes induce a significant increase in the number of CD80^{hi} mature mTECs (Irla et al. 2008), further highlighting the capacity of CD4$^+$ SP thymocytes to promote mTEC differentiation.

Although CD4$^+$ SP thymocytes are important, it remained to be determined whether they control medulla organization and the cellularity of mature mTECs through the secretion of soluble mediators or by direct cell-cell contact interactions with mTECs. The generation of an elegant transgenic mouse model, in which antigen presentation to CD4$^+$ T cells is selectively abrogated in mTECs, has demonstrated that TCR-MHCII mediated contacts between CD4$^+$ SP thymocytes and mTECs control the pool of CD80^{hi}Aire$^+$ mature mTECs (Irla et al. 2008; Zhu and Fu 2008). These observations are reinforced by the fact that several MHCII-restricted TCR transgenic mice that do not express the cognate antigen, such as OTII-$Rag2^{-/-}$, female Marilyn-$Rag2^{-/-}$ and B3K508-$Rag1^{-/-}$ mice, have a strongly impaired medulla development, despite the development of CD4$^+$ SP thymocytes (Irla et al. 2012). Interestingly, when OTII-$Rag2^{-/-}$ mice, which express an MHCII-restricted TCR specific for ovalbumin (OVA) are backcrossed with RIP-mOVA transgenic mice in which the rat insulin promoter (RIP) drives the synthesis of membrane-bound OVA in mTECs (Kurts et al. 1996), the differentiation of CD80^{hi}Aire$^+$mature mTECs and the expansion of medullary regions are observed (Irla et al. 2012, 2008). Similar observations were made with Marilyn-$Rag2^{-/-}$ females administered with the exogenous cognate H-Y peptide (Irla et al. 2012). Thus antigen-specific interactions with autoreactive CD4$^+$ SP thymocytes play a pivotal role in both medulla patterning and the differentiation of CD80^{hi}Aire$^+$ mature mTECs. In sum, the interactions between CD4$^+$ SP thymocytes and mTECs act as a feedback loop in which mTECs expand and differentiate to accommodate to the pool of self-reactive T cells.

6.1.2.2 Molecular Signals Involved in Medulla Expansion and Composition During mTEC-Thymocyte Crosstalk

Instructive signals provided by interactions with SP thymocytes to mTECs mainly involve three ligands/receptors of the tumor necrosis factor (TNF) superfamily: RANKL (Receptor Activator of Nuclear factor Kappa-B Ligand)/RANK, CD40L (CD40 Ligand)/CD40 and LTα1β2 (Lymphotoxin α1β2)/LTβR (Lymphotoxin β receptor). These TNF ligands and receptors are expressed by SP thymocytes and mTECs, respectively. RANKL/RANK axis plays a major role in the differentiation of Aire⁺ mTECs. Both *Tnfsf11* (RANKL) and *Tnfrsf11a* (RANK)-deficient mice have dramatically reduced numbers of Aire⁺ mTECs in their thymi at the postnatal stage (Hikosaka et al. 2008; Akiyama et al. 2008; Saade et al. 2010; Rossi et al. 2007). Conversely, mice deficient for osteoprotegerin, a decoy receptor for RANKL, exhibit large medulla with increased numbers of Aire⁺ mTECs (Hikosaka et al. 2008; Khan et al. 2014). In contrast to the embryonic thymus that shows a complete absence of Aire⁺ cells (Akiyama et al. 2008), the postnatal thymus contains some Aire⁺cells in *Tnfrsf11a*⁻/⁻ or *Tnfsf11*⁻/⁻ mice, suggesting that after birth other signal(s) are implicated in mTEC differentiation. In line with these observations, CD40L/CD40 axis was identified to contribute to this process. Although *Cd40lg*⁻/⁻ and *Cd40*⁻/⁻ mice have a subtler defect in Aire⁺ mTEC cellularity compared to *Tnfsf11*⁻/⁻ and *Tnfrsf11a*⁻/⁻ mice (Irla et al. 2008; Akiyama et al. 2008), mice deficient for both *Tnfrsf11a* and *Cd40* have a more severely reduced number of Aire⁺ mTECs than mice deficient for *Tnfrsf11a* or *Cd40* alone (Akiyama et al. 2008). Further evidence indicating that RANKL and CD40L signals are both required for mTEC differentiation was provided by experiments based on the *in vitro* stimulation of 2-deoxyguanosine-treated fetal thymic lobes with recombinant RANKL and/or CD40L proteins. Whereas RANKL and CD40L each induces a modest increase in mature mTECs (Mouri et al. 2011; Irla et al. 2010, 2012), RANKL and CD40L added together substantial increase these cells, similarly to control levels induced by thymocytes. These findings are fully consistent with the observation that RANKL is expressed at higher levels in CD4⁺ SP thymocytes than in CD8⁺ SP thymocytes (Hikosaka et al. 2008; Desanti et al. 2012; Irla et al. 2008) and that CD40L is exclusively expressed in CD4⁺ SP thymocytes (Irla et al. 2008; Hikosaka et al. 2008). Thus, RANKL and CD40L signals provided by CD4⁺ SP thymocytes cooperate for driving the differentiation of mTECs in the post-natal thymus (Irla et al. 2010).

In contrast to *Tnfrsf11a*⁻/⁻, *Tnfsf11*⁻/⁻, *Cd40lg*⁻/⁻ and *Cd40*⁻/⁻ mice, mice deficient for *Ltbr* (LTβR) or its ligands *Lta* (LTα), *Ltb* (LTβ) or *Tnfsf14* (LIGHT) have normal differentiation of Aire⁺ mTECs (Venanzi et al. 2007; Lopes et al. 2017; Boehm et al. 2003; Seach et al. 2008). Nevertheless, *Ltbr*⁻/⁻ mice have highly disorganized medullary regions with a substantially reduction in different mTEC subsets such as UEA-1⁺ mTECs, terminally differentiated involucrin⁺ mTECs and CCL21⁺ mTEC^lo cells (Venanzi et al. 2007; Boehm et al. 2003; Lkhagvasuren et al. 2013; White et al. 2010). Interestingly, although the treatment of

2-deoxyguanosine-treated fetal thymic lobes with agonistic anti-LTβR antibodies does not induce the differentiation of CD80^{hi} mature mTECs, it triggers a significant increase in numbers of total mTECs (Irla et al. 2012). In contrast, the treatment with RANKL and CD40L, alone or in combination, has a little effect on the cellularity of total mTECs. However, when agonistic anti-LTβR antibodies were combined with RANKL, CD40L or both, a synergistic increase was observed, similarly to that observed induced by thymocytes. These observations indicate that the LTβR signalling has a dominant role in determining the total mTEC cellularity, which is line with the fact that $Ltbr^{-/-}$ mice have reduced numbers of mTEC^{lo} and mTEC^{hi} cells and that the lymphotoxin axis is required for the regeneration of mTEC subsets (Lopes et al. 2017; Venanzi et al. 2007). Moreover, the generation of $Ltbr/Cd40$-double deficient mice indicated no obvious cooperation between these two TNF receptors since these mice do not show any additional defects in medullary organization or mTEC differentiation than those observed in $Ltbr^{-/-}$ mice (Mouri et al. 2011). In contrast, mice deficient for both $Ltbr$ and $Tnfsf11$ show a greater reduction in mTEC^{lo} and mTEC^{hi} cells including UEA-1^{+} and Aire^{+} mTECs compared to single knockout mice. These observations indicate that LTβR and RANKL cooperate for optimal mTEC differentiation. In contrast to $Ltbr^{-/-}$ mice, mice deficient in LTβR ligands Lta, Ltb and $Tnfsf14$ (expressed as LTα1β2 heterotrimer and LIGHT3 homotrimer) have a less pronounced phenotype than $Ltbr^{-/-}$ mice (Irla et al. 2013; Boehm et al. 2003; Venanzi et al. 2007). Nevertheless, the 3D reconstructions of thymic lobes of $Lta^{-/-}$ mice revealed a medullary topology that resembles to that observed in $H2$-$Aa^{-/-}$ mice, which is characterized by isolated islets without any large central compartment (Irla et al. 2013). The lymphotoxin pathway thus plays an important role in controlling the medullary architecture.

In the context of antigen-specific interactions between mTECs and CD4^{+} SP thymocytes, the CD28-CD80/86 and CD40L-CD40 costimulatory axes play an important role in the patterning of the medulla. Similarly to $Lta^{-/-}$ mice, the 3D reconstruction of thymic lobes from $Cd80/86^{-/-}$ mice has shown that the medulla of these mice is mostly devoid of any major compartment (Irla et al. 2013). Furthermore, the combined absence of CD40L-CD40 and CD28-CD80/86 axes results in profound alterations in the mTEC compartment that are comparable to those observed in the absence of SP thymocytes (Williams et al. 2014). Interestingly, CD4^{+} SP thymocytes purified from CD80/86$^{-/-}$ mice express reduced levels of LTα, indicating that crosstalk involving the CD28-CD80/86 axis controls the delivery of LTβR ligands to mTECs (Irla et al. 2012, 2013; Williams et al. 2014).

In sum, studies during the last decade revealed a division of labour between three TNF members mainly delivered by CD4^{+} SP thymocytes, with RANKL/RANK and CD40L/CD40 preferentially involved in the differentiation of mature mTECs and LTα1β2/LTβR implicated in mTEC cellularity and medulla patterning.

6.1.2.3 Implication of mTEC-CD4⁺ Thymocyte Crosstalk in the Thymic Entry of Peripheral Antigen-Presenting Cells Involved in T-Cell Tolerance Induction

Our laboratory recently found that the crosstalk between mTECs and CD4$^+$ SP thymocytes not only regulates the size and composition of the mTEC compartment but also that of DCs (Lopes et al. 2018). DCs represent around 0.5% of the total thymic cells and constitute a heterogeneous cell population containing three distinct subsets: CD11c^{int}PDCA-1^{hi} plasmacytoid DCs (pDCs), CD11c^{hi}CD8α^{hi}Sirpα$^-$ resident and CD11c^{hi}CD8α^{lo}Sirpα$^+$ migratory conventional DCs (cDCs) (Wu and Shortman 2005; Lopes et al. 2015). Besides mTECs, DCs play an important role in the negative selection of autoreactive T cells and the generation of Foxp3$^+$nTregs (Bonasio et al. 2006; Proietto et al. 2008; Baba et al. 2009; Koble and Kyewski 2009). A constitutive ablation of DCs leads to defective clonal deletion characterized by an enlargement of the CD4$^+$ SP cell compartment and consequently to fatal autoimmunity (Ohnmacht et al. 2009). In particular, migratory cDCs and pDCs continuously circulate into the thymus where they contribute to the deletion of autoreactive thymocytes by displaying innocuous peripheral self-antigens that would be otherwise not presented to thymocytes (Bonasio et al. 2006; Baba et al. 2009; Hadeiba et al. 2012; Derbinski and Kyewski 2010). In contrast to CD8α^{hi}Sirpα$^-$ resident cDCs, the pool of CD8α^{lo}Sirpα$^+$ migratory cDCs and CD11c^{int}PDCA-1^{hi} pDCs as well as of F4/80$^+$CD11b$^+$macrophages is regulated by high affinity interactions between mTECs and CD4$^+$ thymocytes (Lopes et al. 2018). Compared to OTII-$Rag2^{-/-}$ mice, numbers of these three cell types were indeed increased in the thymus of RipmOVAxOTII-$Rag2^{-/-}$ mice, in which interactions between OVA-expressing mTECs and OTII CD4$^+$ SP thymocytes occur. In accordance with these observations, the thymus of NOD.$Tcra^{-/-}$mice, lacking crosstalk between mTECs and SP thymocytes, contains reduced numbers of cDCs with diminished levels of CD40 and CD86 co-stimulatory molecules (Spidale et al. 2014). Adoptive transfer experiments of peripheral DC and macrophage-enriched cells in OTII-$Rag2^{-/-}$ and RipmOVAxOTII-$Rag2^{-/-}$ recipients revealed that this enhanced cellularity was due to an increase thymic entry (Lopes et al. 2018). Thus, antigen-specific interactions between mTECs and CD4$^+$ SP thymocytes not only regulate medulla organization and mature mTEC cellularity but also the thymic entry of peripheral antigen-presenting cells that are implicated in the induction of T-cell tolerance. This regulatory circuit is negatively regulated by LTα, which is induced in CD4$^+$ SP thymocytes upon crosstalk (Lopes et al. 2018; Irla et al. 2012). LTα in turn represses the expression of CCL2, CCL8 and CCL12 chemokines (CCR2 ligands) in mTECs. Consequently, $Lta^{-/-}$ mice have an increased recruitment of CD8α^{lo}Sirpα$^+$ migratory cDCs, CD11c^{int}PDCA-1^{hi} pDCs and F4/80$^+$CD11b$^+$macrophages in their thymi due to an enhanced expression of CCL2, CCL8 and CCL12 chemokines. This chemokine upregulation is likely due to an augmented expression of the classical NF-κB subunit p65, known to regulate CCL2 and CCL8 (Ping et al. 1996; Ueda et al. 1994; Deng et al. 2013; Lopes et al. 2018). Therefore, the thymic entry of CD8α^{lo}Sirpα^+ migratory cDCs, CD11c^{int}PDCA-1^{hi} pDCs and F4/80$^+$CD11b$^+$ macrophages that relies on the expression of the chemokine receptor CCR2 is attenuated by LTα induced upon mTEC-thymocyte crosstalk. This tight regulation of peripheral antigen-presenting cell entry into the thymus ultimately regulates clonal deletion of autoreactive T cells (Lopes et al. 2018). This is illustrated by the fact that (1) autoreactive thymocytes are highly deleted at the DP, CD4^{lo}CD8^{lo} and CD4$^+$ SP stages in $Lta^{-/-}$ mice and (2) the adoptive transfer of OVA$_{323-339}$-loaded DC and macrophage-enriched cells in OTII-$Rag2^{-/-}$x$Lta^{-/-}$ recipients leads to increased deletion of OTII DP and CD4$^+$ SP cells compared OTII-$Rag2^{-/-}$ recipients. Thus, crosstalk between mTECs and CD4$^+$ SP thymocytes regulates the pool of peripheral DCs and macrophages in the thymus, which fine-tunes the selection of the T-cell repertoire.

6.2   Identification of Other Types of Crosstalk in the Thymus

6.2.1   Thymic Crosstalk Between mTECs and DCs

The notion of a possible crosstalk between mTECs and DCs emerged from studies demonstrating that CD8α^{hi}Sirpα^- resident cDCs efficiently cross-present mTEC-derived self-antigens (Gallegos and Bevan 2004; Koble and Kyewski 2009). Considering that a given particular self-antigen is expressed by only 1–3% of mTECs at any given time (Derbinski et al. 2008), this unidirectional transfer of self-antigens from mTECs to DCs is suspected to reinforce negative selection of autoreactive CD4$^+$ and CD8$^+$ SP thymocytes. Furthermore, mice that bear a natural mutation in the NF-KB-inducing kinase (NIK), which is an upstream kinase acting primarily in the non-canonical NF-KB pathway, show defective mTEC and medulla organization (Kajiura et al. 2004). Interestingly, the generation of bone marrow chimeras indicate that mice lacking NIK specifically in the thymic stroma have a reduced number of pDCs, CD8α^{hi}Sirpα^- and CD8α^{lo}Sirpα^+ cDCs (Mouri et al. 2014). In these mice, CD8α^{hi}Sirpα^- cDCs express diminished levels of the CD80 co-stimulatory molecule. These observations thus indicate that NIK expression in the stroma controls the cellularity and maturation of DCs. Moreover, the expression of Aire in mTECs controls the production of the chemokine XCL1, which regulates the positioning in medullary regions of CD8α^{hi}Sirpα^- cDCs, which express the chemokine receptor XCR1 (Lei et al. 2011). Thus, mTECs regulate the cellularity, maturation and medullary localization of DCs. In turn, CD8α^{hi}Sirpα^- resident cDCs by cross-presenting mTEC-derived self-antigens, as well as CD8α^{lo}Sirpα^+ migratory cDCs and pDCs, whose thymic entry is controlled by mTEC-derived chemokines, by transporting peripheral self-antigens participate in the elimination of autoreactive thymocytes.

6.2.2   Thymic Crosstalk Between B Cells and Thymocytes

B cells represent around 0.3% of the thymic cellularity and thus a non-negligible cell type in the mouse thymus. Similarly to mTECs and DCs, they are mainly located in the medulla and at the cortico-medullary junction. Their functional role remained enigmatic for a long time. Thymic B cells exhibit distinct phenotypic features compared to resting B cells observed in the periphery such as in the spleen or lymph nodes. They express higher levels of MHCII as well as CD40, CD80 and CD86 co-stimulatory molecules than their peripheral counterparts (Perera et al. 2013). The majority of thymic B cells develop intrathymically from *Rag*-expressing progenitors (Perera et al. 2013). Furthermore, bone marrow-derived B cells have been also shown to contribute to the thymic B cell pool (Yamano et al. 2015). Since thymic B cells exhibit an activated phenotype with typical features of antigen-presenting cells and because they are located in the medulla where negative selection occurs, they have been suspected to play a role in the selection of developing T cells. In accordance with this assumption, the size of the $CD4^+$ SP cell compartment was increased (Xing et al. 2015; Yamano et al. 2015) whereas that of nTreg cells (Walters et al. 2014) was decreased in mice lacking B cells. Several works based on the use of transgenic mice have demonstrated that B cells act as *bona fide* antigen-presenting cells implicated in the negative selection of autoreactive T cells (Perera et al. 2013; Kleindienst et al. 2000; Frommer and Waisman 2010). In agreement with their role in T-cell selection, Aire-reporter transgenic mice have revealed that around 3% of thymic B cells express Aire but at a lesser extent compared to mTECs (Yamano et al. 2015). Similarly, around 5% of B cells express Aire in the human thymus (Gies et al. 2017). The comparison of gene expression between WT and $Aire^{-/-}$ B cells revealed that Aire regulates the expression of hundred genes in mice, including some self-antigens. RNA-seq transcriptomic analysis also showed that human thymic B cells express more self-antigens than their peripheral counterpart. Thus thymic B cells play a non-negligible role in the selection of developing T cells.

   Given the implication of B cells in the establishment of T-cell tolerance, the identification of the cellular and molecular mechanisms that sustain their maturation and homeostasis raised a special interest. The CD40L/CD40 costimulatory axis was found to play an important role in thymic B cell biology (Kleindienst et al. 2000; Frommer and Waisman 2010; Ferrero et al. 1999). *Cd40*-deficient mice show reduced numbers of thymic B cells and express low levels of CD69 and CD86 (Ferrero et al. 1999). *Cd40lg*-deficient mice also show reduced numbers of B cells in their thymus (Akirav et al. 2011). Given that CD40L is exclusively expressed by $CD4^+SP$ thymocytes (Irla et al. 2008), mice lacking SP thymocytes such as $Tcra^{-/-}$ mice have a reduced thymic B cell cellularity (Fujihara et al. 2014). A recent work further implicates that the licensing of B cells implicates the CD40L/CD40 costimulatory signal with that CD40 stimulation on B cells induced by autoreactive $CD4^+$ SP thymocyte expressing CD40L (Yamano et al. 2015). Thus, developing $CD4^+$ T cells through CD40L-CD40 axis maintain the pool and maturation of thymic B cells, which are implicated in the selection of the T-cell repertoire.

6.3 Conclusion

Pioneer studies have identified the thymic crosstalk as an interdependent cellular relationship, whereby the selection of SP thymocytes depends on the medullary epithelium and reciprocally the development and integrity of the medullary epithelium relies on the differentiation of SP thymocytes. Studies conducted over the last decade have revealed the precise molecular and cellular basis of this crosstalk. Furthermore, these studies have shown that the thymic crosstalk not only involves mTECs and SP thymocytes but also other cellular actors such as B cells, cDCs and pDCs (Fig. 6.1). This complex cellular interplay constitutes a fine-tuning mechanism that allows the thymus to adapt its capacity to efficiently delete autoreactive T cells, which critically prevents the development of autoimmune disorders. Additional studies are needed to define how the respective cellular interactions affect the distinct cell types involved in thymic crosstalk. Moreover, future investigations based at molecular and epigenetic levels are expected to shed new light on the

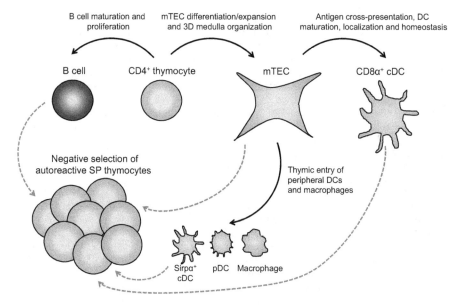

Fig. 6.1 Medullary thymic crosstalk involves different cell types that collaborate for efficient negative selection. Thymic crosstalk has been initially identified as a dialogue between SP thymocytes and mTECs during which mTECs control the selection of SP thymocytes while reciprocally SP thymocytes regulate the size and composition of mTECs. Of note, CD4+ SP thymocytes play a dominant role in process. These reciprocal interactions between mTECs and CD4+ SP thymocytes also control the thymic entry of DCs and macrophages that display innocuous self-antigens captured in the periphery and thus participate in the deletion of autoreactive thymocytes. Moreover, interactions between mTECs and cDCs regulate the maturation and medullary positioning of CD8α+ cDCs, which are involved in the negative selection of thymocytes by cross-presenting self-antigens expressed by mTECs. Finally, CD4+ SP thymocytes control the pool and maturation of B cells, which in turn also control the selection of SP thymocytes. Dashed grey lines indicate the negative selection of autoreactive SP thymocytes

mechanisms by which the thymic crosstalk regulates the establishment of T-cell tolerance.

Acknowledgments I gratefully thank Prof. Arnauld Sergé (Centre de Recherche en Cancérologie de Marseille) and Teshika Jayewickreme (Harvard University) for constructive comments. This work was supported by the Marie Curie Actions (Career Integration Grants, CIG_SIGnEPI4Tol_618541 to MI), institutional grants from Institut National de la Santé et de la Recherche Médicale, Centre National de la Recherche Scientifique and Aix-Marseille Université.

References

Adkins B, Gandour D, Strober S, Weissman I (1988) Total lymphoid irradiation leads to transient depletion of the mouse thymic medulla and persistent abnormalities among medullary stromal cells. J Immunol 140(10):3373–3379

Akirav EM, Xu Y, Ruddle NH (2011) Resident B cells regulate thymic expression of myelin oligodendrocyte glycoprotein. J Neuroimmunol 235(1-2):33–39. https://doi.org/10.1016/j.jneuroim.2011.03.013

Akiyama T, Shimo Y, Yanai H, Qin J, Ohshima D, Maruyama Y, Asaumi Y, Kitazawa J, Takayanagi H, Penninger JM, Matsumoto M, Nitta T, Takahama Y, Inoue J (2008) The tumor necrosis factor family receptors RANK and CD40 cooperatively establish the thymic medullary microenvironment and self-tolerance. Immunity 29(3):423–437

Anderson MS, Venanzi ES, Klein L, Chen Z, Berzins SP, Turley SJ, von Boehmer H, Bronson R, Dierich A, Benoist C, Mathis D (2002) Projection of an immunological self shadow within the thymus by the aire protein. Science 298(5597):1395–1401

Baba T, Nakamoto Y, Mukaida N (2009) Crucial contribution of thymic Sirp alpha+ conventional dendritic cells to central tolerance against blood-borne antigens in a CCR2-dependent manner. J Immunol 183(5):3053–3063. https://doi.org/10.4049/jimmunol.0900438

Boehm T, Scheu S, Pfeffer K, Bleul CC (2003) Thymic medullary epithelial cell differentiation, thymocyte emigration, and the control of autoimmunity require lympho-epithelial cross talk via LTbetaR. J Exp Med 198(5):757–769

Bonasio R, Scimone ML, Schaerli P, Grabie N, Lichtman AH, von Andrian UH (2006) Clonal deletion of thymocytes by circulating dendritic cells homing to the thymus. Nat Immunol 7(10):1092–1100. https://doi.org/10.1038/ni1385

Deng X, Xu M, Yuan C, Yin L, Chen X, Zhou X, Li G, Fu Y, Feghali-Bostwick CA, Pang L (2013) Transcriptional regulation of increased CCL2 expression in pulmonary fibrosis involves nuclear factor-kappaB and activator protein-1. Int J Biochem Cell Biol 45(7):1366–1376. https://doi.org/10.1016/j.biocel.2013.04.003

Derbinski J, Kyewski B (2010) How thymic antigen presenting cells sample the body's self-antigens. Curr Opin Immunol 22(5):592–600. https://doi.org/10.1016/j.coi.2010.08.003

Derbinski J, Schulte A, Kyewski B, Klein L (2001) Promiscuous gene expression in medullary thymic epithelial cells mirrors the peripheral self. Nat Immunol 2(11):1032–1039

Derbinski J, Pinto S, Rosch S, Hexel K, Kyewski B (2008) Promiscuous gene expression patterns in single medullary thymic epithelial cells argue for a stochastic mechanism. Proc Natl Acad Sci U S A 105(2):657–662. https://doi.org/10.1073/pnas.0707486105

Desanti GE, Cowan JE, Baik S, Parnell SM, White AJ, Penninger JM, Lane PJ, Jenkinson EJ, Jenkinson WE, Anderson G (2012) Developmentally regulated availability of RANKL and CD40 ligand reveals distinct mechanisms of fetal and adult cross-talk in the thymus medulla. J Immunol 189(12):5519–5526. https://doi.org/10.4049/jimmunol.1201815

van Ewijk W, Shores EW, Singer A (1994) Crosstalk in the mouse thymus. Immunol Today 15(5):214–217 van Ewijk W, Hollander G, Terhorst C, Wang B (2000) Stepwise development of thymic microenvironments in vivo is regulated by thymocyte subsets. Development 127(8):1583–1591

Farr AG, Anderson SK (1985) Epithelial heterogeneity in the murine thymus: fucose-specific lectins bind medullary epithelial cells. J Immunol 134(5):2971–2977

Ferrero I, Anjuere F, Martin P, Martinez del Hoyo G, Fraga ML, Wright N, Varona R, Marquez G, Ardavin C (1999) Functional and phenotypic analysis of thymic B cells: role in the induction of T cell negative selection. Eur J Immunol 29(5):1598–1609. https://doi.org/10.1002/(SICI)1521-4141(199905)29:05<1598::AID-IMMU1598>3.0.CO;2-O

Fink PJ, Bevan MJ (1978) H-2 antigens of the thymus determine lymphocyte specificity. J Exp Med 148(3):766–775

Frommer F, Waisman A (2010) B cells participate in thymic negative selection of murine autoreactive CD4+ T cells. PLoS One 5(10):e15372. https://doi.org/10.1371/journal.pone.0015372

Fujihara C, Williams JA, Watanabe M, Jeon H, Sharrow SO, Hodes RJ (2014) T cell-B cell thymic cross-talk: maintenance and function of thymic B cells requires cognate CD40-CD40 ligand interaction. J Immunol 193(11):5534–5544. https://doi.org/10.4049/jimmunol.1401655

Gabler J, Arnold J, Kyewski B (2007) Promiscuous gene expression and the developmental dynamics of medullary thymic epithelial cells. Eur J Immunol 37(12):3363–3372

Gallegos AM, Bevan MJ (2004) Central tolerance to tissue-specific antigens mediated by direct and indirect antigen presentation. J Exp Med 200(8):1039–1049

Gies V, Guffroy A, Danion F, Billaud P, Keime C, Fauny JD, Susini S, Soley A, Martin T, Pasquali JL, Gros F, Andre-Schmutz I, Soulas-Sprauel P, Korganow AS (2017) B cells differentiate in human thymus and express AIRE. J Allergy Clin Immunol 139(3):1049–1052 e1012. https://doi.org/10.1016/j.jaci.2016.09.044

Godfrey DI, Izon DJ, Tucek CL, Wilson TJ, Boyd RL (1990) The phenotypic heterogeneity of mouse thymic stromal cells. Immunology 70(1):66–74

Godfrey DI, Kennedy J, Suda T, Zlotnik A (1993) A developmental pathway involving four phenotypically and functionally distinct subsets of CD3-CD4-CD8- triple-negative adult mouse thymocytes defined by CD44 and CD25 expression. J Immunol 150(10):4244–4252

Gray DH, Seach N, Ueno T, Milton MK, Liston A, Lew AM, Goodnow CC, Boyd RL (2006) Developmental kinetics, turnover, and stimulatory capacity of thymic epithelial cells. Blood 108(12):3777–3785

Gray D, Abramson J, Benoist C, Mathis D (2007) Proliferative arrest and rapid turnover of thymic epithelial cells expressing aire. J Exp Med 204(11):2521–2528

Guerau-de-Arellano M, Martinic M, Benoist C, Mathis D (2009) Neonatal tolerance revisited: a perinatal window for aire control of autoimmunity. J Exp Med 206(6):1245–1252. https://doi.org/10.1084/jem.20090300

Hadeiba H, Lahl K, Edalati A, Oderup C, Habtezion A, Pachynski R, Nguyen L, Ghodsi A, Adler S, Butcher EC (2012) Plasmacytoid dendritic cells transport peripheral antigens to the thymus to promote central tolerance. Immunity 36(3):438–450. https://doi.org/10.1016/j.immuni.2012.01.017

Hikosaka Y, Nitta T, Ohigashi I, Yano K, Ishimaru N, Hayashi Y, Matsumoto M, Matsuo K, Penninger JM, Takayanagi H, Yokota Y, Yamada H, Yoshikai Y, Inoue J, Akiyama T, Takahama Y (2008) The cytokine RANKL produced by positively selected thymocytes fosters medullary thymic epithelial cells that express autoimmune regulator. Immunity 29(3):438–450

Hollander GA, Wang B, Nichogiannopoulou A, Platenburg PP, van Ewijk W, Burakoff SJ, Gutierrez-Ramos JC, Terhorst C (1995) Developmental control point in induction of thymic cortex regulated by a subpopulation of prothymocytes. Nature 373(6512):350–353

Irla M, Hugues S, Gill J, Nitta T, Hikosaka Y, Williams IR, Hubert FX, Scott HS, Takahama Y, Hollander GA, Reith W (2008) Autoantigen-specific interactions with CD4+ thymocytes control mature medullary thymic epithelial cell cellularity. Immunity 29(3):451–463

Irla M, Hollander G, Reith W (2010) Control of central self-tolerance induction by autoreactive CD4+ thymocytes. Trends Immunol 31(2):71–79. https://doi.org/10.1016/j.it.2009.11.002

Irla M, Guerri L, Guenot J, Serge A, Lantz O, Liston A, Imhof BA, Palmer E, Reith W (2012) Antigen recognition by autoreactive cd4(+) thymocytes drives homeostasis of the thymic medulla. PLoS One 7(12):e52591. https://doi.org/10.1371/journal.pone.0052591

Irla M, Guenot J, Sealy G, Reith W, Imhof BA, Serge A (2013) Three-dimensional visualization of the mouse thymus organization in health and immunodeficiency. J Immunol 190(2):586–596. https://doi.org/10.4049/jimmunol.1200119

Josefowicz SZ, Lu LF, Rudensky AY (2012) Regulatory T cells: mechanisms of differentiation and function. Annu Rev Immunol 30:531–564. https://doi.org/10.1146/annurev.immunol.25.022106.141623

Kajiura F, Sun S, Nomura T, Izumi K, Ueno T, Bando Y, Kuroda N, Han H, Li Y, Matsushima A, Takahama Y, Sakaguchi S, Mitani T, Matsumoto M (2004) NF-kappa B-inducing kinase establishes self-tolerance in a thymic stroma-dependent manner. J Immunol 172(4):2067–2075

Kanariou M, Huby R, Ladyman H, Colic M, Sivolapenko G, Lampert I, Ritter M (1989) Immunosuppression with cyclosporin A alters the thymic microenvironment. Clin Exp Immunol 78(2):263–270

Khan IS, Mouchess ML, Zhu ML, Conley B, Fasano KJ, Hou Y, Fong L, Su MA, Anderson MS (2014) Enhancement of an anti-tumor immune response by transient blockade of central T cell tolerance. J Exp Med 211(5):761–768. https://doi.org/10.1084/jem.20131889

Klein L, Hinterberger M, Wirnsberger G, Kyewski B (2009) Antigen presentation in the thymus for positive selection and central tolerance induction. Nat Rev Immunol 9(12):833–844. https://doi.org/10.1038/nri2669

Klein L, Kyewski B, Allen PM, Hogquist KA (2014) Positive and negative selection of the T cell repertoire: what thymocytes see (and don't see). Nat Rev Immunol 14(6):377–391. https://doi.org/10.1038/nri3667

Kleindienst P, Chretien I, Winkler T, Brocker T (2000) Functional comparison of thymic B cells and dendritic cells in vivo. Blood 95(8):2610–2616

Klug DB, Carter C, Crouch E, Roop D, Conti CJ, Richie ER (1998) Interdependence of cortical thymic epithelial cell differentiation and T-lineage commitment. Proc Natl Acad Sci U S A 95(20):11822–11827

Koble C, Kyewski B (2009) The thymic medulla: a unique microenvironment for intercellular self-antigen transfer. J Exp Med 206(7):1505–1513

Kurts C, Heath WR, Carbone FR, Allison J, Miller JF, Kosaka H (1996) Constitutive class I-restricted exogenous presentation of self antigens in vivo. J Exp Med 184(3):923–930

Kyewski BA, Schirrmacher V, Allison JP (1989) Antibodies against the T cell receptor/CD3 complex interfere with distinct intra-thymic cell-cell interactions in vivo: correlation with arrest of T cell differentiation. Eur J Immunol 19(5):857–863. https://doi.org/10.1002/eji.1830190512

Lei Y, Ripen AM, Ishimaru N, Ohigashi I, Nagasawa T, Jeker LT, Bosl MR, Hollander GA, Hayashi Y, Malefyt Rde W, Nitta T, Takahama Y (2011) Aire-dependent production of XCL1 mediates medullary accumulation of thymic dendritic cells and contributes to regulatory T cell development. J Exp Med 208(2):383–394. https://doi.org/10.1084/jem.20102327

Lind EF, Prockop SE, Porritt HE, Petrie HT (2001) Mapping precursor movement through the postnatal thymus reveals specific microenvironments supporting defined stages of early lymphoid development. J Exp Med 194(2):127–134

Lkhagvasuren E, Sakata M, Ohigashi I, Takahama Y (2013) Lymphotoxin beta receptor regulates the development of CCL21-expressing subset of postnatal medullary thymic epithelial cells. J Immunol 190(10):5110–5117. https://doi.org/10.4049/jimmunol.1203203

Lopes N, Serge A, Ferrier P, Irla M (2015) Thymic crosstalk coordinates medulla organization and T-cell tolerance induction. Front Immunol 6:365. https://doi.org/10.3389/fimmu.2015.00365

Lopes N, Vachon H, Marie J, Irla M (2017) Administration of RANKL boosts thymic regeneration upon bone marrow transplantation. EMBO Mol Med 9:835. https://doi.org/10.15252/emmm.201607176

Lopes N, Charaix J, Cedile O, Serge A, Irla M (2018) Lymphotoxin alpha fine-tunes T cell clonal deletion by regulating thymic entry of antigen-presenting cells. Nat Commun 9(1):1262. https://doi.org/10.1038/s41467-018-03619-9

Meredith M, Zemmour D, Mathis D, Benoist C (2015) Aire controls gene expression in the thymic epithelium with ordered stochasticity. Nat Immunol 16(9):942–949. https://doi.org/10.1038/ni.3247

Metzger TC, Khan IS, Gardner JM, Mouchess ML, Johannes KP, Krawisz AK, Skrzypczynska KM, Anderson MS (2013) Lineage tracing and cell ablation identify a post-aire-expressing thymic epithelial cell population. Cell Rep 5(1):166–179

Miller JF (1961) Immunological function of the thymus. Lancet 2(7205):748–749

Miller JF (2002) The discovery of thymus function and of thymus-derived lymphocytes. Immunol Rev 185:7–14

Mouri Y, Yano M, Shinzawa M, Shimo Y, Hirota F, Nishikawa Y, Nii T, Kiyonari H, Abe T, Uehara H, Izumi K, Tamada K, Chen L, Penninger JM, Inoue JI, Akiyama T, Matsumoto M (2011) Lymphotoxin signal promotes thymic organogenesis by eliciting RANK expression in the embryonic thymic stroma. J Immunol 186:5047

Mouri Y, Nishijima H, Kawano H, Hirota F, Sakaguchi N, Morimoto J, Matsumoto M (2014) NF-kappaB-inducing kinase in thymic stroma establishes central tolerance by orchestrating cross-talk with not only thymocytes but also dendritic cells. J Immunol 193(9):4356–4367. https://doi.org/10.4049/jimmunol.1400389

Nasreen M, Ueno T, Saito F, Takahama Y (2003) In vivo treatment of class II MHC-deficient mice with anti-TCR antibody restores the generation of circulating CD4 T cells and optimal architecture of thymic medulla. J Immunol 171(7):3394–3400

Negishi I, Motoyama N, Nakayama K, Nakayama K, Senju S, Hatakeyama S, Zhang Q, Chan AC, Loh DY (1995) Essential role for ZAP-70 in both positive and negative selection of thymocytes. Nature 376(6539):435–438

Nishikawa Y, Hirota F, Yano M, Kitajima H, Miyazaki J, Kawamoto H, Mouri Y, Matsumoto M (2010) Biphasic aire expression in early embryos and in medullary thymic epithelial cells before end-stage terminal differentiation. J Exp Med 207(5):963–971. https://doi.org/10.1084/jem.20092144

Ohnmacht C, Pullner A, King SB, Drexler I, Meier S, Brocker T, Voehringer D (2009) Constitutive ablation of dendritic cells breaks self-tolerance of CD4 T cells and results in spontaneous fatal autoimmunity. J Exp Med 206(3):549–559. https://doi.org/10.1084/jem.20082394

Palmer E (2003) Negative selection--clearing out the bad apples from the T-cell repertoire. Nat Rev Immunol 3(5):383–391. https://doi.org/10.1038/nri1085

Palmer DB, Viney JL, Ritter MA, Hayday AC, Owen MJ (1993) Expression of the alpha beta T-cell receptor is necessary for the generation of the thymic medulla. Dev Immunol 3(3):175–179

Passos GA, Mendes-da-Cruz DA, Oliveira EH (2015) The thymic orchestration involving aire, miRNAs, and cell-cell interactions during the induction of central tolerance. Front Immunol 6:352. https://doi.org/10.3389/fimmu.2015.00352

Perera J, Meng L, Meng F, Huang H (2013) Autoreactive thymic B cells are efficient antigen-presenting cells of cognate self-antigens for T cell negative selection. Proc Natl Acad Sci U S A 110(42):17011–17016. https://doi.org/10.1073/pnas.1313001110

Perniola R (2018) Twenty years of AIRE. Front Immunol 9:98. https://doi.org/10.3389/fimmu.2018.00098

Petrie HT, Zuniga-Pflucker JC (2007) Zoned out: functional mapping of stromal signaling microenvironments in the thymus. Annu Rev Immunol 25:649–679

Ping D, Jones PL, Boss JM (1996) TNF regulates the in vivo occupancy of both distal and proximal regulatory regions of the MCP-1/JE gene. Immunity 4(5):455–469

Porritt HE, Gordon K, Petrie HT (2003) Kinetics of steady-state differentiation and mapping of intrathymic-signaling environments by stem cell transplantation in nonirradiated mice. J Exp Med 198(6):957–962. https://doi.org/10.1084/jem.20030837

Proietto AI, van Dommelen S, Zhou P, Rizzitelli A, D'Amico A, Steptoe RJ, Naik SH, Lahoud MH, Liu Y, Zheng P, Shortman K, Wu L (2008) Dendritic cells in the thymus contribute to T-regulatory cell induction. Proc Natl Acad Sci U S A 105(50):19869–19874

Rodewald HR, Paul S, Haller C, Bluethmann H, Blum C (2001) Thymus medulla consisting of epithelial islets each derived from a single progenitor. Nature 414(6865):763–768

Rossi SW, Kim MY, Leibbrandt A, Parnell SM, Jenkinson WE, Glanville SH, McConnell FM, Scott HS, Penninger JM, Jenkinson EJ, Lane PJ, Anderson G (2007) RANK signals from CD4(+)3(-) inducer cells regulate development of aire-expressing epithelial cells in the thymic medulla. J Exp Med 204(6):1267–1272

Saade M, Irla M, Yammine M, Boulanger N, Victorero G, Vincentelli R, Penninger JM, Hollander GA, Chauvet S, Nguyen C (2010) Spatial (Tbata) expression in mature medullary thymic epithelial cells. Eur J Immunol 40(2):530–538. https://doi.org/10.1002/eji.200939605

Sansom SN, Shikama-Dorn N, Zhanybekova S, Nusspaumer G, Macaulay IC, Deadman ME, Heger A, Ponting CP, Hollander GA (2014) Population and single-cell genomics reveal the aire dependency, relief from Polycomb silencing, and distribution of self-antigen expression in thymic epithelia. Genome Res 24:1918. https://doi.org/10.1101/gr.171645.113

Seach N, Ueno T, Fletcher AL, Lowen T, Mattesich M, Engwerda CR, Scott HS, Ware CF, Chidgey AP, Gray DH, Boyd RL (2008) The lymphotoxin pathway regulates aire-independent expression of ectopic genes and chemokines in thymic stromal cells. J Immunol 180(8):5384–5392

Senelar R, Escola MJ, Escola R, Serrou B, Serre A (1976) Relationship between Hassall's corpuscles and thymocytes fate in guinea-pig foetus. Biomedicine 24(2):112–122

Serge A, Bailly AL, Aurrand-Lions M, Imhof BA, Irla M (2015) For3D: full organ reconstruction in 3D, an automatized tool for deciphering the complexity of lymphoid organs. J Immunol Methods 424:32. https://doi.org/10.1016/j.jim.2015.04.019

Shores EW, Van Ewijk W, Singer A (1991) Disorganization and restoration of thymic medullary epithelial cells in T cell receptor-negative scid mice: evidence that receptor-bearing lymphocytes influence maturation of the thymic microenvironment. Eur J Immunol 21(7):1657–1661

Shores EW, Van Ewijk W, Singer A (1994) Maturation of medullary thymic epithelium requires thymocytes expressing fully assembled CD3-TCR complexes. Int Immunol 6(9):1393–1402

Spidale NA, Wang B, Tisch R (2014) Cutting edge: antigen-specific thymocyte feedback regulates homeostatic thymic conventional dendritic cell maturation. J Immunol 193(1):21–25. https://doi.org/10.4049/jimmunol.1400321

Surh CD, Ernst B, Sprent J (1992) Growth of epithelial cells in the thymic medulla is under the control of mature T cells. J Exp Med 176(2):611–616

Takaba H, Morishita Y, Tomofuji Y, Danks L, Nitta T, Komatsu N, Kodama T, Takayanagi H (2015) Fezf2 orchestrates a thymic program of self-antigen expression for immune tolerance. Cell 163(4):975–987. https://doi.org/10.1016/j.cell.2015.10.013

Takahama Y (2006) Journey through the thymus: stromal guides for T-cell development and selection. Nat Rev Immunol 6(2):127–135

Takahama Y, Nitta T, Mat Ripen A, Nitta S, Murata S, Tanaka K (2010) Role of thymic cortex-specific self-peptides in positive selection of T cells. Semin Immunol 22(5):287–293. https://doi.org/10.1016/j.smim.2010.04.012

Ueda A, Okuda K, Ohno S, Shirai A, Igarashi T, Matsunaga K, Fukushima J, Kawamoto S, Ishigatsubo Y, Okubo T (1994) NF-kappa B and Sp1 regulate transcription of the human monocyte chemoattractant protein-1 gene. J Immunol 153(5):2052–2063

Van Vliet E, Melis M, Van Ewijk W (1984) Monoclonal antibodies to stromal cell types of the mouse thymus. Eur J Immunol 14(6):524–529. https://doi.org/10.1002/eji.1830140608

Venanzi ES, Gray DH, Benoist C, Mathis D (2007) Lymphotoxin pathway and aire influences on thymic medullary epithelial cells are unconnected. J Immunol 179(9):5693–5700

Waldburger JM, Suter T, Fontana A, Acha-Orbea H, Reith W (2001) Selective abrogation of major histocompatibility complex class II expression on extrahematopoietic cells in mice lacking promoter IV of the class II transactivator gene. J Exp Med 194(4):393–406

Walters SN, Webster KE, Daley S, Grey ST (2014) A role for intrathymic B cells in the generation of natural regulatory T cells. J Immunol 193(1):170–176. https://doi.org/10.4049/jimmunol.1302519

Wang B, Biron C, She J, Higgins K, Sunshine MJ, Lacy E, Lonberg N, Terhorst C (1994) A block in both early T lymphocyte and natural killer cell development in transgenic mice with high-copy numbers of the human CD3E gene. Proc Natl Acad Sci U S A 91(20):9402–9406

Watanabe N, Wang YH, Lee HK, Ito T, Cao W, Liu YJ (2005) Hassall's corpuscles instruct dendritic cells to induce CD4+CD25+ regulatory T cells in human thymus. Nature 436(7054):1181–1185. https://doi.org/10.1038/nature03886

White AJ, Nakamura K, Jenkinson WE, Saini M, Sinclair C, Seddon B, Narendran P, Pfeffer K, Nitta T, Takahama Y, Caamano JH, Lane PJ, Jenkinson EJ, Anderson G (2010) Lymphotoxin signals from positively selected thymocytes regulate the terminal differentiation of medullary thymic epithelial cells. J Immunol 185(8):4769–4776. https://doi.org/10.4049/jimmunol.1002151

Williams JA, Zhang J, Jeon H, Nitta T, Ohigashi I, Klug D, Kruhlak MJ, Choudhury B, Sharrow SO, Granger L, Adams A, Eckhaus MA, Jenkinson SR, Richie ER, Gress RE, Takahama Y, Hodes RJ (2014) Thymic medullary epithelium and thymocyte self-tolerance require cooperation between CD28-CD80/86 and CD40-CD40L costimulatory pathways. J Immunol 192(2):630–640. https://doi.org/10.4049/jimmunol.1302550

Wu L, Shortman K (2005) Heterogeneity of thymic dendritic cells. Semin Immunol 17(4):304–312. https://doi.org/10.1016/j.smim.2005.05.001

Xing C, Ma N, Xiao H, Wang X, Zheng M, Han G, Chen G, Hou C, Shen B, Li Y, Wang R (2015) Critical role for thymic CD19+CD5+CD1dhiIL-10+ regulatory B cells in immune homeostasis. J Leukoc Biol 97(3):547–556. https://doi.org/10.1189/jlb.3A0414-213RR

Yamano T, Nedjic J, Hinterberger M, Steinert M, Koser S, Pinto S, Gerdes N, Lutgens E, Ishimaru N, Busslinger M, Brors B, Kyewski B, Klein L (2015) Thymic B cells are licensed to present self antigens for central T cell tolerance induction. Immunity 42:1048. https://doi.org/10.1016/j.immuni.2015.05.013

Yang S, Fujikado N, Kolodin D, Benoist C, Mathis D (2015) Immune tolerance. Regulatory T cells generated early in life play a distinct role in maintaining self-tolerance. Science 348(6234):589–594. https://doi.org/10.1126/science.aaa7017

Yano M, Kuroda N, Han H, Meguro-Horike M, Nishikawa Y, Kiyonari H, Maemura K, Yanagawa Y, Obata K, Takahashi S, Ikawa T, Satoh R, Kawamoto H, Mouri Y, Matsumoto M (2008) Aire controls the differentiation program of thymic epithelial cells in the medulla for the establishment of self-tolerance. J Exp Med 205(12):2827–2838. https://doi.org/10.1084/jem.20080046

Zhu M, Fu YX (2008) Coordinating development of medullary thymic epithelial cells. Immunity 29(3):386–388

Chapter 7
The Autoimmune Regulator (*AIRE*) Gene, the Master Activator of Self-Antigen Expression in the Thymus

Matthieu Giraud and Pärt Peterson

Abstract It has been 20 years that the *AIRE* gene was discovered. It is the causing gene of a rare and life-threatening autoimmune disease with severe manifestations against a variety of organs. Since *AIRE*'s identification and positional cloning, thorough investigations have revealed key insights into the understanding of the role of AIRE and into its mode of action. It has appeared clear that AIRE uniquely induces the expression of thousands of tissue-restricted self-antigens in the thymus. These self-antigens are presented to developing T cells, resulting in the deletion of the self-reactive ones and the generation of regulatory T cells, in order to establish and maintain immunological tolerance.

7.1 Introduction

7.1.1 AIRE, the Causing Gene of the Multiorgan Autoimmune Disease APECED

Autoimmune polyendocrinopathy candidiasis ectodermal dystrophy (APECED) is a rare and life-threatening autoimmune disease with a prevalence of 1–9:1,000,000 (Orphanet, http://www.orpha.net). Patients have severe autoimmune manifestations against a variety of organs. The prevalence of APECED is increased in certain populations such as the Finnish, Norwegian, Sardinian and Iranian Jewish, due to consanguinity or founder effects (Myhre et al. 2001). Most commonly, patients with

M. Giraud (✉)
Centre de Recherche en Transplantation et Immunologie, UMR 1064, Institut National de la Santé et de la Recherche Médicale (INSERM), Université de Nantes, Nantes, France
e-mail: matthieu.giraud@inserm.fr

P. Peterson (✉)
Molecular Pathology, Institute of Biomedicine and Translational Medicine,
University of Tartu, Tartu, Estonia
e-mail: part.peterson@ut.ee

© Springer Nature Switzerland AG 2019   169
G. A. Passos (ed.), *Thymus Transcriptome and Cell Biology*,
https://doi.org/10.1007/978-3-030-12040-5_7

APECED present two of the three major symptoms: hypoparathyroidism, adrenal insufficiency (Addison disease), and chronic mucocutaneous candidiasis (Ahonen et al. 1990). In addition to these symptoms, APECED is associated with a range of clinical features including several endocrine autoimmune disorders, such as, premature gonadal failure, hypothyroidism, pernicious anaemia, type 1 diabetes, autoimmune hepatitis and gastritis, as well as autoimmune skin diseases, such as alopecia and vitiligo (Perheentupa 2006). These autoimmune manifestations are associated with the presence of autoantibodies in the serum of the patients directed against multiple self-antigens whose encoding genes are normally expressed in specific peripheral tissues, and with lymphocyte infiltration in various endocrine glands (Peterson and Peltonen 2005).

In contrast to most autoimmune diseases, the mode of familial inheritance of APECED revealed that it is an autosomal recessive monogenic disorder. Classic genetic linkage analyses were performed and the APECED-causing gene, *AIRE*, was identified by positional cloning (Finnish-German APECED Consortium 1997; Nagamine et al. 1997). Since then, the autoimmune regulator (AIRE) has turned out to be a key molecule controlling the induction of immunological tolerance in the thymus. More than 100 different pathogenic mutations in the *AIRE* gene have been described so far. In Finnish and European patients, the most common mutation is an early stop mutation: R257X; in Persian Jews, it is the nonsense mutation Y85C and in Sardinians, the early stop codon mutation: R139X. A frame shift mutation due to a 13bp deletion is also commonly found among the *AIRE* mutations detected in Europeans and North Americans, reviewed in (Peterson et al. 2008; Abramson and Goldfarb 2016) (Fig. 7.1).

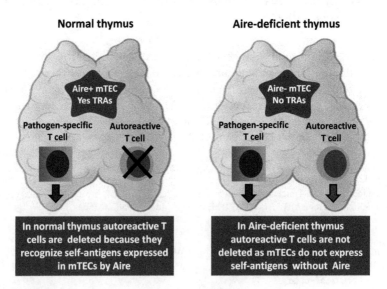

Fig. 7.1 AIRE controls thymic expression of self-antigens

Following *AIRE* identification, mouse *AIRE*-KO mouse models were generated with the aim to decipher the mechanisms linking AIRE to breakdown of immunological tolerance (Ramsey et al. 2002b; Anderson 2002). Although, the phenotype of *AIRE*-KO mice is much milder than in APECED, it is characterized by autoantibodies and lymphocyte infiltrations in several organs, therefore resembling the clinical features of APECED patients (Mathis and Benoist 2009).

7.1.2   Structural and Functional Properties of AIRE

AIRE is a 545-amino-acid protein with a molecular weight of 57.5 kDa. It is encoded by a gene composed of 14 exons, located in the region q22.3 of chromosome 21 (Finnish-German APECED Consortium 1997; Nagamine et al. 1997).

7.1.2.1   AIRE Cellular Localization

AIRE is a nuclear protein that localizes into discrete small nuclear speckles uniformly distributed in the nucleus (Heino et al. 1999). These speckles resemble, but are distinct from, promyelocytic leukemia (PML) nuclear bodies. The AIRE-positive speckles are dependent on the cell-cycle. They disaggregate before mitosis and restructure after cell division, varying in size and number from one cell to another (Björses et al. 1999). The nuclear speckles are tightly associated with the nuclear matrix, as well as AIRE, as shown by its exclusive location within the protein fraction of the nuclear matrix obtained by sequential extraction of nuclear proteins. Association of AIRE with the nuclear matrix and the chromatin is thought to form structures where distant portions of chromatin containing AIRE-responsive loci, super-enhancer DNA regions and regulatory factors recruited by AIRE are brought together to unlock the transcriptional machinery at AIRE-dependent genes (Akiyoshi et al. 2004; Tao et al. 2006; Abramson et al. 2010). Although AIRE is mainly located in the nucleus, it is also detectable in the cytoplasm where it forms fibrillar structures reminiscent of intermediate filaments or microtubules (Rinderle et al. 1999; Pitkänen et al. 2001).

7.1.2.2   AIRE Functional Domains

AIRE is composed of several domains that are characteristic of proteins specialized in the regulation of transcription and in chromatin binding (Perniola and Musco 2014). It has been shown that AIRE is structurally closely related to members of the speckled-protein SP100 family, which are chromatin binding proteins playing a role in regulation of the transcription (Bloch et al. 2000; Pitkänen and Peterson 2003).

- The N-terminal region of AIRE (aa 1–100) has revealed a six-helix structure and an evolutionary conserved caspase-recruitment domain (CARD) that shares similarity to Sp100 proteins (Pitkänen et al. 2001; Ferguson et al. 2008). Deletion of the CARD domain in *AIRE* expression constructs, as well as insertion of pathogenic mutations in the CARD coding sequence, revealed that this domain plays a role in AIRE's oligomerization and consequently in proper AIRE transactivation function (Pitkänen et al. 2001; Ramsey et al. 2002a). The CARD is also required for the localization of AIRE within nuclear speckles and the association of AIRE with nuclear matrix. AIRE, harboring mutations in the CARD domain, has been shown to block the formation of nuclear speckles and exerts no effect on the expression of AIRE-dependent genes (Pitkänen et al. 2001; Halonen et al. 2004). Interestingly, the CARD domain can also mediate the dimerization of proteins involved in inflammation or apoptosis (Park et al. 2007).

- Three conserved putative nuclear localization signals (NLS) (aa 110–114, 131–133 and 159–167) have been involved in the nuclear import of AIRE (Ilmarinen et al. 2006; Saltis et al. 2008), whereas the N-terminal region is involved in nuclear export (Pitkänen et al. 2001).

- The SAND domain (aa 180–280) is a conserved 80-amino acid sequence. Like the CARD domain, it plays a role in the nuclear speckle localization of AIRE. The SAND domain has been thought to function as a DNA-binding module, since it mediates anchor of some SAND-containing proteins to the genome. However, AIRE's SAND lacks canonical KDKW motifs that are required for DNA binding (Bottomley et al. 2001; Surdo et al. 2003). Lack of evidence that AIRE would bind DNA and act as a classical transcription factor, tipped the scale in favor of a scenario in which AIRE functions as an activator of gene expression impacting higher order structure than DNA, the chromatin.

- The C-terminal region of AIRE is composed of two plant-homeodomain (PHD) zinc fingers, related to PHD1 (aa 299–340) and PHD2 (aa 434–475). PHD fingers are enriched in cysteine domains and include two zinc ions (Aasland et al. 1995). These domains are known to interact, in general, with the chromatin and more specifically with histone H3 depending on its degree of methylation. Particularly, PHD1 has been described to interact with unmethylated histone H3 lysine 4 (H3K4me0) (Org et al. 2008; Koh et al. 2008). In contrast to PHD1, PHD2's surface is positively charged precluding interaction with histone H3 but permitting interaction with partners involved in chromatin binding (Gaetani et al. 2012; Yang et al. 2013). Regarding the impact of the PHD domains on the transactivation function of AIRE, mutations in each of these domains have been studied revealing deleterious effects on the induction of Aire-sensitive gene expression by AIRE.

- In addition to these features, AIRE has four nuclear-receptor-binding LXXLL motifs that are scattered throughout the protein sequence. They are known to bind to nuclear receptors resulting in transcriptional activation or repression. Notably, AIRE's LXXLL motif lying in the C-terminus is critical for AIRE transactivation activity (Meloni et al. 2008).

7.1.3   AIRE Expression in the Thymus

AIRE is expressed almost exclusively in the thymus, in a small and very specific cell population: the medullary thymic epithelial cells (mTECs) that express high levels of major histocompatibility complex (MHC) class II molecules and CD80 co-stimulatory marker (Heino et al. 1999; Heino et al. 2000; Derbinski et al. 2001). These MHCII^hi CD80^hi AIRE+ mTECs are called mTEC^hi and have the unique capacity to: (i) express a wide array of tissue-restricted self-antigens (TRAs) whose expression is normally restricted to particular peripheral tissues/organs; (ii) present TRA-derived antigenic peptides by MHCII molecules to developing thymocytes in the thymus. The expression of an important part of these TRAs, corresponding to thousands of gene-products, is activated by AIRE in mTEC^hi (Gray et al. 2007a; Rossi et al. 2007). Majority of mTEC^hi are thought to be post-mitotic and a fraction of them undergo apoptosis through the action of AIRE, probably through the action of its CARD domain which exhibits pro-apoptotic features (Gray et al. 2007b). mTEC^hi derive from immature MHCII^lo CD80^lo AIRE- mTECs, so-called mTEC^lo. mTEC^lo have been described as highly proliferating and already committed to the mTEC lineage (Gray et al. 2007b). Differentiation of mTEC^lo gives first rise to intermediate mTECs (MHCII^hi CD80^hi AIRE- mTECs) that express and present AIRE-independent TRAs. Following the path of differentiation, AIRE turns on in intermediate mTECs that differentiate into mature mTEC^hi able to induce the expression of thousands of AIRE-dependent TRAs (Gäbler et al. 2007; Gray et al. 2007b).

Although mTEC^hi had been thought to be end-product of differentiation, convincing evidence showed that a fraction of these cells, which do not undergo apoptosis, gives rise to post-AIRE mTECs (Nishikawa et al. 2010; Wang et al. 2012). Post-AIRE mTECs are characterized by a loss of expression of AIRE, of most AIRE-dependent TRAs and of MHCII and CD80 molecules. Loss of AIRE-dependent, as well as -independent TRAs, had been described in these cells (Wang et al. 2012; Metzger et al. 2013; Michel et al. 2017). Recently the lectin Tetragonolobus purpureas agglutinin (TPA) has been identified as a cell surface marker specific to post-AIRE cells (Michel et al. 2017). Further differentiation of these cells leads to the loss of their nuclei and their merging to form Hassall's Corpuscles (Yano et al. 2008; Wang et al. 2012). Although the precise role of Hassall's Corpuscles is still elusive, it has been proposed to favor Treg development through expression of TSLP (Watanabe et al. 2005). It has also been proposed that post-AIRE mTECs could reconstitute the immature mTEC^lo reservoir and differentiate into mTEC^hi again (Metzger et al. 2013).

7.2 Thymic "Crosstalk" and Signals Driving Aire and mTEC Differentiation

During mTEC differentiation, AIRE is expressed in MHCII[hi] CD80[hi] mTECs. However, isolated AIRE+ mTEC[hi] loose AIRE expression when they are cultured ex vivo. They rapidly de-differentiate and die. Identification of the signals triggering AIRE expression and maintaining mTEC[hi] in their state of differentiation has then been a subject of thorough investigations. It has notably been observed that a 3D medullary microarchitecture and a proper signaling were necessary for the expression of AIRE (Kont et al. 2008; Rossi et al. 2007). Indeed, disruption of the 3D medullary microarchitecture leads to the decrease of AIRE expression (Akiyama et al. 2005). In addition to the spatial 3D organization of mTEC[hi] in the medulla, several studies identified a member of the tumor necrosis factor (TNF) superfamily, namely, the receptor activator of nuclear factor kappa-B (RANK), and its corresponding ligand, RANKL, as factors controlling the emergence of AIRE+ mTEC[hi] in the embryonic thymus. RANK is expressed by mTEC[hi] and RANKL by lymphoid tissue inducer (LTi) cells before the development of single-positive (SP) thymocytes (Rossi et al. 2007). Convincingly, mice deficient for RANK or RANKL show a total absence of AIRE+ mTEC[hi] in the embryonic thymus, indicating that RANK can regulate the emergence of AIRE+ mTEC[hi] in the embryo (Rossi et al. 2007; Akiyama et al. 2008). In the post-natal thymus, deletion of RANK or RANKL also leads to a strong reduction of AIRE+ mTEC[hi] and TRA expression (Hikosaka et al. 2008). However, in contrast to the embryonic thymus, the AIRE+ mTEC[hi] reduction is not complete, leaving the possibility of other factors to control the expression of AIRE in post-natal MHCII[hi] CD80[hi] mTECs. Further investigations have led to the identification of CD40, a second member of the TNF superfamily that recognizes its ligand CD40L, and plays a role in the expression of AIRE by mTEC[hi]. However, the effect of CD40 on turning AIRE on is not as strong as the one exerted by RANK (Akiyama et al. 2008; Irla et al. 2008) when the two factors are taken individually. Together, both effects synergize in the post-natal thymus, strongly suggesting a marked cooperation between the CD40 and RANK signals (Akiyama et al. 2008).

While CD40 and RANK are known to be expressed by mTEC[hi], sources of CD40L and RANKL have been more difficult to identify. CD40L was found to be specifically expressed by CD4+ thymocytes (Irla et al. 2008; Hikosaka et al. 2008) whereas RANKL is expressed by CD4+ and CD8+ thymocytes with a preferential expression for CD4+ thymocytes (Desanti et al. 2012). Then, CD4+ thymocytes, expressing both RANKL and CD40L, may play an important role in the development of mature AIRE+ mTEC[hi] in adult mice. RANKL production has been also associated with thymic γδ T cell progenitors (Roberts et al. 2012).

7.3   AIRE-Induced Expression of Thousands of TRA Genes in mTEC^hi

Following AIRE's identification as the causing gene for APECED, *AIRE*-KO mouse models were generated with the aim to decipher the mechanisms linking AIRE to breakdown of self-tolerance (Ramsey et al. 2002b; Anderson 2002). Notably, the role of AIRE as a major transcriptional activator of TRA gene expression was conclusively demonstrated using the *AIRE*-KO mouse model generated in the Benoist and Mathis' lab (Anderson 2002). In this model, analysis of the differentially expressed genes between WT and *AIRE*-KO mTEC^hi revealed a significant excess of genes over-represented in WT mTEC^hi. It has also been found that the AIRE-dependent genes are not specific to mTEC^hi and are normally expressed in peripheral tissues or organs. These observations showed that AIRE is able to activate the expression of hundreds of TRA genes, directly or indirectly, and therefore functions as a main activator of promiscuous gene expression (PGE) in mTEC^hi. Initiation of PGE by AIRE in mTEC^hi was further demonstrated in fetal thymic organ cultures and adult thymic cultures (Sousa Cardoso et al. 2006; Chen et al. 2008). In addition to activate the expression of an important number of TRA genes, PGE outweighs the sex-, cell type- and differentiation state-regulation of gene expression (Anderson 2002; Gotter et al. 2004).

More recently, the range of TRA genes induced by AIRE in mTEC^hi has been re-evaluated by RNAseq (Sansom et al. 2014; St-Pierre et al. 2015; Danan-Gotthold et al. 2016). These studies showed that AIRE induces the expression of a wide array of thousands of TRA genes, providing convincing evidence that AIRE is the main factor contributing to PGE in mTEC^hi. Beyond the role of AIRE, it seemed reasonable to hypothesize that other factors could control the remaining thousands of AIRE-independent TRAs in mTEC^hi. In this regard, FEZF2 has recently been proposed to activate TRA genes but to a much smaller extent than AIRE does (Takaba et al. 2015)

7.4   AIRE-Dependent Expression of miRNAs

Following the identification of the effect of AIRE on the induction of thousands of TRA genes, the hypothesis that AIRE would also control the expression of miRNAs or would be regulated by specific miRNAs has emerged. miRNAs are small non-coding RNA molecules (21–25 nucleotides) that trigger RNA degradation after hybridization to specific DNA motifs in the 3'UTR ends of the transcripts. miRNAs could also repress translation of their targeted transcripts (Bartel 2009). The hypothesis of a link between AIRE and post-transcriptional control of gene expression has been strengthened by the observation that mTEC-specific deficiency of molecules involved in miRNA pathway resulted in impaired PGE and thymocyte development (Zuklys et al. 2012; Khan et al. 2014). Importantly variation of miRNA patterns was described throughout mTEC differentiation and a subset of miRNAs was identified as specifically repressing translation of AIRE mRNA (Papadopoulou et al. 2011; Ucar et al. 2013). Moreover, AIRE has been reported to directly control the expression of miRNAs in a murine mTEC line, with an effect on the miRNA composition and the level of expression of a set of miRNAs (Macedo et al. 2013). Interestingly, computational inferring of the impact of miRNAs on the downregulated expression of AIRE-dependent TRA genes in adult mouse mTEC^{hi}, in comparison to newborns, showed that almost none of miRNAs expressed in adult mice target the 3′UTR ends of several downregulated AIRE-dependent TRA genes (Macedo et al. 2015; Oliveira et al. 2016). This suggests that some AIRE-dependent TRA genes may be refractoriness to miRNA-mediated post-transcription regulation in mTEC^{hi}. Thus, these studies showed that, in addition to an action on transcription, AIRE is also involved in the post-transcriptional control of PGE notably through regulation of miRNA expression.

7.5 Link Between AIRE-Induced TRA Expression and Thymocyte Selection

7.5.1 Negative Selection of Self-Reactive Developing Thymocytes

Accumulating evidence revealed that impaired TRA activation due to AIRE deficiency resulted in impaired negative selection of developing self-reactive thymocytes in the thymus. Indeed, Liston and al. demonstrated that AIRE was controlling the effectiveness of eliminating self-reactive thymocytes in transgenic mice having thymocytes harboring TCRs specific to the neo-self-antigen hen egg lysosome (HEL) and HEL expression in mTEC^{hi} under the control of the insulin rat promoter (RIP) (Liston et al. 2003, 2004). The involvement of AIRE in negative selection of self-reactive thymocytes was confirmed by an independent study using transgenic mice having the OVA peptide expression in mTEC^{hi} under the control of the mouse RIP promoter, and thymocytes harboring OVA-specific TCRs (Anderson et al. 2005). It has thus become clear that mTEC^{hi} present AIRE-activated TRAs through MHCII molecules to developing thymocytes, resulting in negative selection of the self-reactive ones. However, AIRE-dependent TRA activation may not be the only cause to efficient negative selection. Indeed, AIRE-KO mice develop a Sjögren's syndrome-like autoimmune reaction against alpha-fodrin, which is an AIRE-independent TRA (Kuroda et al. 2005). Another example came from the study of the AIRE-KO NOD mouse model, by Mitsuru Matsumoto's lab, that develop autoimmune pancreatitis to isomerase A2, which is also an AIRE-independent TRA (Niki et al. 2006).

7.5.2   Generation of Regulatory T Cells

The role of the thymus in shaping immunological tolerance relies on negative selection of self-reactive thymocytes and on generation of thymic regulatory T cells (Treg) that are selected for their specificity to self-antigens and exported in the periphery to control autoimmunity. In this regard, it has been shown that mTEC[hi], not only control negative selection of self-reactive developing thymocytes, but also contribute to immunological tolerance by inducing naturally occurring Tregs (Aschenbrenner et al. 2007; Hinterberger et al. 2010). In addition to a role in negative selection, a role of AIRE in the generation of Tregs was investigated. Initial studies did not reveal major changes in the numbers and function of Tregs generated in transgenic (RIP-HEL or RIP-OVA) mouse models, in which the expression of HEL or OVA by mTEC[hi] is dependent on AIRE (Liston et al. 2003, 2004; Anderson et al. 2005). The study of a potential impact of AIRE on the diversity of the Treg TCR repertoire also failed to show any difference and therefore strengthened the idea that AIRE has no effect on the generation of Tregs (Daniely et al. 2010).

Nevertheless, recent studies have seriously challenged this idea. Indeed, Malchow et al. showed that the generation of tumor-specific Tregs in a mouse model of oncogen-driven prostate cancer was clearly dependent on AIRE expression (Malchow et al. 2013). Yang et al. also demonstrated that AIRE is able to promote, in the perinatal age, the generation of a specific compartment of Tregs that persist in adult mice (Yang et al. 2015). It has been also shown that AIRE deficiency impairs the mechanisms that enable Tregs to suppress autoimmune reactions in the periphery (Teh et al. 2010; Aricha et al. 2011). Together, these studies showed that the role of AIRE on immunological tolerance is not restricted to an effect on negative selection of self-reactive thymocytes but also on the generation of subsets of natural Tregs that escape the thymus and contribute to immunological tolerance in the periphery.

7.6   Co-localization of AIRE-Induced Genes in Genomic Clusters

AIRE activates the expression of a wide array of TRA genes in mTEC[hi]. Key understanding into the mode of action of AIRE was notably revealed by the analysis of the expression of AIRE-dependent genes. It has been shown that the AIRE-dependent genes, contributing to PGE, co-localize in genomic clusters (Gotter et al. 2004; Johnnidis et al. 2005). These observations were consistent with the idea that AIRE could function in a large complex composed of transcriptional regulators, RNA- and chromatin-binding proteins that can access specific chromatin regions. This hypothesis was sustained by the findings that (i) AIRE transactivation function is not controlled by the action of an individual domain of AIRE, suggesting that AIRE may be a co-activator of a multimolecular complex. The involvement of each domain of AIRE in its function has notably been shown in AIRE-negative cells after transfection of an AIRE expression plasmid deleted for each structural domain of AIRE (CARD, SAND, PHD1, PHD2) (Giraud et al. 2014), and that (ii) large AIRE-containing complexes are located in chromatin stretches enclosing transcription start sites (TSS) of AIRE-dependent genes (Bansal et al. 2017). These chromatin stretches are known as super-enhancers and are anchored by large cell-specific transcriptional structures that control, under the action of AIRE, the expression of AIRE-dependent genes clustered at these locations.

7.7 Coordinated AIRE-Induced Expression of Clusters of Co-expressed Genes Chosen Stochastically at the Single Cell Level

Although great advances had been made in the understanding of the coordination of the induced-expression of AIRE-dependent genes, one central question remained unanswered: does each single mTEC^{hi} express, under the action of AIRE, all the AIRE-dependent genes or just a fraction of them? And if so, does the choice of AIRE-dependent genes expressed in single mTEC^{hi} follow a stochastic or coordinated model?

Using a single-cell RT-qPCR approach, Villasenor and al. showed that genes of the S100 family, whose expression is induced by AIRE at the mTEC^{hi} population level, are not expressed in all single mTEC^{hi} (Villaseñor et al. 2008). At the same time, Derbinski et al. showed that the genes of the casein locus, which is a regulated by AIRE, are expressed in random combinations in single mTEC^{hi}. In contrast the genes of this locus are uniformly co-expressed in single cells of the mammary epithelium (Derbinski et al. 2008). These findings revealed that the AIRE-dependent genes are not expressed in each single mTEC^{hi}.

Later, pools of co-expressed AIRE-dependent genes were identified through selection of small sets of mTEC^{hi} expressing particular AIRE-dependent TRAs. Thus it appeared that the genes, whose expression is induced by AIRE at the mTEC^{hi} population level, are not expressed independently of each other, but with a certain degree of cooperation, as clusters of co-expressed overlapping and complementary genes (Pinto et al. 2013).

Invaluable insights into the understanding of the mode of expression associated with the induction by AIRE in mTEC^{hi} came from two recent reports deciphering the pattern of expression of the AIRE-dependent genes at the single-cell level (Meredith et al. 2015; Brennecke et al. 2015). These studies revealed that the co-expressed genes clustered in the genome in a distribution that is very likely to be set up at the DNA or epigenetic level by a stochastic determinism. These clusters are not shared between one mouse and another, and appear to be stable through mTEC divisions. In addition, expression of these clusters appeared to be coordinated within single mTEC^{hi}, each cluster of co-expressed genes been activated in a set of single mTEC^hi. This mode of expression ensures that every AIRE-dependent TRA is expressed, at least, in a small set of mTEC^hi, which is sufficient to mediate negative selection of developing self-reactive thymocytes. The coordinated expression of clusters of co-expressed genes induced by AIRE has potentially two advantages: (i) it may provide a more economic means for the cells than a model in which each cell chooses individually the genes that it expresses; (ii) the level of expression of the AIRE-dependent genes, activated in discrete sets of single mTEC^hi, is higher in comparison to what it would have been if each cell was uniformly expressing the AIRE-dependent genes. Higher expression at the single-cell level results in higher protein levels and consequently in a more efficient presentation of AIRE-dependent self-antigens by a single mTEC^hi to developing thymocytes in the thymic medulla.

7.8   Factors Driving AIRE Expression and Activation in mTEC^hi

Although the RANK and CD40 signaling pathways are known to play key roles in the induction of AIRE expression in post-natal mTEC^hi, other pathways are likely to be involved. Identification of these pathways has been the focus of thorough investigations.

The demethylation of a CpG island in the promoter region seems to be a prerequisite for the activation of the AIRE gene (Murumägi et al. 2003). AIRE is critically regulated by NF-kB signaling as the deletion of the enhancer region containing NF-kB responsive elements upstream of the transcription start site completely abolishes AIRE expression in mTECs (Haljasorg et al. 2015; LaFlam et al. 2015). In a recent report, Herzig et al. showed that the AIRE locus is insulated by the global chromatin organizer: CCCTC-binding factor (CTCF) in cells and tissues in which AIRE is not expressed (Herzig et al. 2016). CTCF is removed in mTEC^hi, leaving the AIRE locus accessible for demethylation and binding of transcription factors. Remarkably, a coordinated action of several factors including IRF4, IRF8, TBX21, TCF7 and CTCFL, has been shown to induce the expression of AIRE. Interestingly, IRF4 has been described as a major factor controlling antigen presentation by mTEC^hi and mTEC^hi-driven thymocyte development (Haljasorg et al. 2017). Furthermore, IRF8 has been shown in mTEC^hi to activate, with AIRE, the expression of the pathogenic self-antigen of the auto-immune myasthenia gravis, the alpha-subunit of the muscle acetylcholine receptor (Giraud et al. 2007). Identification of the factors controlling the activation of the AIRE locus in mTEC^hi has been a great step forward. Now it is of key importance to identify or specify the signals that trigger the expression of these factors to better understand the sequential events leading to mTEC^hi-maturation and AIRE expression.

Although the expression of AIRE is a prerequisite for the induction of an important number of TRAs in mTEC^hi, post-translational modifications of AIRE have been shown to impact its activation. Indeed, deacetylation of AIRE by the deacetylase sirtuin-1 (SIRT1) that targets lysine residues located between AIRE's NLS and SAND domains, leads to improved expression of AIRE-dependent genes in mTEC[hi] (Chuprin et al. 2015). SIRT1 is not specific to mTEC[hi] but is expressed at very high levels in these cells. AIRE's lysine residues deacetylated by SIRT1 have also been shown to be targeted by the CREB-binding protein (CBP) for acetylation. CBP is the first protein that was found to directly interact with AIRE and to co-localize with the latter in nuclear speckles (Pitkanen 2000). CBP had been shown to induce the expression of AIRE-sensitive reporter constructs in transfected cells (Pitkanen et al. 2005) and of several AIRE-endogenous targets in HEK293 cells (Ferguson et al. 2008). However, acetylation mediated by CBP at AIRE's lysine residues appears to have inhibitory rather than activating properties on AIRE itself (Saare et al. 2012; Chuprin et al. 2015). Therefore, the effect of CBP on the induction of the expression of AIRE-dependent genes most likely result from the effect of CBP on acetylation of histone lysine residues at AIRE-dependent loci rather than of AIRE itself. However, how the effect of the CBP-mediated acetylation reconciles with the upregulation of particular AIRE-target genes remains to be identified. Alternatively, CBP-mediated acetylation may affect the correct subcellular localization and stability of the AIRE protein. Thus, the disruption of the acetylation sites of CBP and its homologue protein p300 at AIRE NLS and SAND regions affected the size and number of AIRE nuclear bodies and its nuclear stability, which could be the result of changed hydrophobic interactions between proteins within nuclear bodies as the positive charge of the acetylated lysine is removed (Saare et al. 2012; Incani et al. 2014). Together, these observations suggest that SIRT1 is able to deacetylate the AIRE's lysine residues that are specifically acetylated by CBP, thereby keeping AIRE in an active deacetylated form while the CBP-mediated acetylation of the AIRE protein is maintained in an inactive form in nuclear bodies (Chuprin et al. 2015).

Acetylation of AIRE is not always associated with inhibition of its function. Indeed, it has been shown that the transfected AIRE in human thymic epithelial cells (4D6) is acetylated on its CARD domain and that this modification is likely to improve the recruitment of the BRD4 protein which plays a role in activation of poised transcription (Yoshida et al. 2015).

7.9 AIRE's Mode of Action

7.9.1 Involvement of RNAP-II Pause Release

To get insights into the mode of action of AIRE, a genome-scale RNA interference (RNAi) screen based on lentiviral shRNA delivery in 4D6 cells transfected with an *AIRE* expression construct was notably performed (Giraud et al. 2014). Almost 3000 chromatin-binding factors, transcription factors and RNA-binding proteins were screened for an effect on the induction of the expression of an AIRE-sensitive reporter construct and of two endogenous AIRE-target genes. This functional shRNA screen identified a number of factors with potential effect on AIRE-triggered gene expression. The identified factors were clustered together depending on the interactions that they have with one another, as reported in public protein databases. Remarkably, a cluster composed of candidate factors involved in RNA elongation appeared to be central in the reconstructed network of AIRE's functional partners. Conversely, none of the identified factors was related to RNA initiation. Thus, those findings determined RNA elongation as a main mechanism underlying AIRE-triggered gene expression. Among the factors involved in RNA elongation, members of the pTEFb complex were identified, namely cyclin T2 (CCNT2), HEXIM1 as well as other proteins, in particular the heterogeneous nuclear ribonucleoprotein L (HNRNPL) whose involvement in the recruitment of the pTEFb complex by AIRE has been demonstrated in *AIRE* transfected HEK293 cells (Giraud et al. 2014). In these cells, AIRE was shown to interact with the pTEFb complex composed of the cyclin-dependent kinase 9 (CDK9), CCNT2, HEXIM1 and the small 7SK-RNA. pTEFb-containing HEXIM1 and 7SK-RNA corresponds to the inactive form of the complex. Once recruited by AIRE at the promoters of AIRE-dependent genes, the inactive pTEFb gets activated by the release of 7SK-RNA-HEXIM1, resulting in release of the paused RNAP-II and productive elongation. It has been proposed that HNRNPL could help to recycle the active pTEFb from the site of transcription termination to the site of initiation through complexing with 7SK-RNA-HEXIM1. In addition to shed a new light on the sequential events leading to pTEFb recruitment at AIRE-dependent genes, these findings confirmed previous reports showing interactions of AIRE with the pTEFb complex (Oven et al. 2007; Zumer et al. 2011). This is also consistent with the observation that AIRE-activated gene expression is controlled by the release of the paused RNAP-II (Oven et al. 2007; Giraud et al. 2012). Later, Yoshida et al. found that pTEFb was also recruited by AIRE though the bridging action of the bromodomain-containing protein 4 (BRD4), which interacts with the acetylated lysine residues of AIRE's CARD domain (Yoshida et al. 2015). Importantly BRD4 has already been shown to be involved in some transcriptional mechanisms leading to the release of the paused RNAP-II, notably through its binding to acetylated histones H3/H4 (Kanno et al. 2014). If AIRE-mediated recruitment of CBP to target gene loci is able to acetylate histones H3 and H4, the BRD4 binding to acetylated histones could thereby allow RNAP-II to remain active and enable transcriptional elongation at AIRE-induced genes.

7.9.2   Involvement of DNA Damage Response

The identity of the molecular partners physically associated with AIRE was investigated in *AIRE* transfected HEK293 cells by co-immunoprecipitation and mass spectrometry analyses (Abramson et al. 2010; Gaetani et al. 2012). Abramson et al. studied the effect of the AIRE partners on its transactivation function by RNAi-mediated knockdown and quantification of the remaining expression of an AIRE endogenous target in HEK293 cells and of an AIRE-sensitive reporter construct in

4D6 cells. A number of AIRE's partners with functions including nuclear transport, chromatin modification, mRNA processing and transcription, were identified. Both reports identified DNA-activated protein kinase (DNA-PK), as one of the prominent AIRE interacting partners; a result that has also been reported in other studies (Liiv et al. 2008; Zumer et al. 2012). Along with DNA-PK, known to play a role in DNA damage repair, other AIRE's partners were found to take part in a molecular complex involved in DNA damage and repair. Among these factors, DNA-topoisomerase IIa (TOP2A) and poly-(ADP-ribose) polymerase 1 (PARP1) were identified as co-immunoprecipitating with AIRE and DNA-PK. Importantly knockdown of these proteins impaired AIRE-induced expression of an AIRE-sensitive endogenous target in *AIRE*-transfected HEK293 cells. Later, studies revealed AIRE interactions with other topoisomerase family members, TOP1 and TOP2B (Bansal et al. 2017). Notably, TOP1 knockdown reduced the association of Aire with TOP2B and TOP2A, but not vice versa, suggesting their role in more downstream effects. Furthermore, AIRE-mediated transcription is enhanced by etoposide and camptothecin that specifically crosslink TOP2 and TOP1, respectively, to chromatin and, by blocking their religation activity, induce double-strand breaks in DNA (Guha et al. 2017). As topoisomerases control the topology of the DNA during transcription through transient breaking of the DNA molecule and rejoining of the broken strands, they can, thereby, cause relaxation of the chromatin and ease superhelical tensions generated by the elongating RNAP-II (Mondal and Parvin 2001). DNA breaks may help AIRE recruit DNA-PK, PARP1 and other partners, that are known to have an effect on chromatin remodeling, such as the chromatin transcriptional elongation factor (FACT), which promotes the exchange of the H2AX histone variant in response to DNA damage (Heo et al. 2008), or the tripartite motif containing 28 (TRIM28), which is associated with condensed chromatin and promotes chromatin relaxation in response to DNA damage (Ziv et al. 2006). Although the precise sequential events linking the action of AIRE and topoisomerases with the final induction of AIRE-dependent gene expression still requires further investigations, it is acknowledged that double-stranded DNA breaks created by topoisomerases trigger the recruitment and activation of DNA damage-response factors. Thus, the AIRE-mediated DNA-damage and response at the promoters of AIRE-dependent genes is expected to result in chromatin relaxation and improved transcriptional elongation, following the release of the paused RNAP-II.

7.10   Target Gene Recognition by AIRE

Although important insights into the understanding of the mode of action of AIRE have been provided by a number of recent studies, signals and molecules controlling the specificity of AIRE for its target genes remain to be fully identified. In this regard, it has been shown that AIRE, through its PHD1 domain, directly binds to unmethylated histone H3 lysine 4 (H3K4me0) (Koh et al. 2008; Org et al. 2008). Recognition by AIRE of its target genes has then been thought to be driven by the binding of AIRE to H3K4me0 and by the specificity of AIRE target genes for this repressive mark. AIRE would specifically bind to its target genes epigenetically silenced by H3K4me0 and activate their transcription. However, this reasoning has been put into a new perspective by recent studies showing that (i) H3K4me0 is relatively frequent in the genome and encompasses the promoter of many AIRE-independent genes, (ii) disruption of AIRE's PHD1 domain impairs the induction of only part of the AIRE-dependent genes (Org et al. 2009), (iii) demethylation of H3K4 residues by a specific demethylase does not increase the number of genes induced by AIRE (Koh et al. 2010). Thus, it has appeared very unlikely that the binding of AIRE to H3K4me0 would be the sole signal triggering specific recognition of AIRE's target genes.

In addition to H3K4me0, the ATF7IP-MBD1 complex has been proposed to contribute to the specificity of AIRE for its target genes (Waterfield et al. 2014). Activating-transcription-factor-7-interacting protein (ATF7IP) is known to regulate positively or negatively gene transcription, and methyl-CpG-binding-domain protein 1 (MBD1) to bind to unmethylated CpG dinucleotides that characterize promoters of silenced genes. AIRE was shown to interact with ATF7IP and knockdown of the latter resulted in the decreased induction of several reporter genes sensitive to AIRE in HEK293 cells. Furthermore, the study of *MBD1*-KO mice revealed that MBD1 deficiency resulted in decreased activation of AIRE-induced genes in vivo. AIRE would then target its sensitive genes through interaction with the ATF7IP-MBD1 complex that is bound to unmethylated CpG dinucleotides characteristic of silenced genes. Similarly to H3K4me0, ATF7IP-MBD1 may also contribute to the specificity of AIRE for its target genes, but very unlikely explain the full specificity.

Other factors, contributing to the specificity of AIRE have been identified, notably marks of polycomb silencing, such as H3K27me3 and absent or low levels of H3K4me3 (Sansom et al. 2014). AIRE does not interact directly with H3K27me3 (Org et al. 2009) but probably indirectly through bridging by the chromodomain helicase DNA (CHD) 4 and 6 that have been reported to interact with AIRE (Gaetani et al. 2012; Yang et al. 2013). Thus, AIRE would identify the repressive chromatin marks H3K27me3 and lack of H3K4me3 through interaction with CHD molecules. This would result in the recruitment of AIRE at its target genes and would unlock transcription by overriding the repressive chromatin state (Sansom et al. 2014).

7.11   Conclusion

AIRE has appeared to be a key regulator of immunological tolerance through induced expression of thousands of TRA genes in mTEC^{hi} and their presentation to developing thymocytes. Huge progress has been made in the fine understanding of the infringing *modus operandi* of AIRE, revealing that AIRE mediates gene induction through cooperation with many other molecular factors, notably histones. Although many aspects of the mode of action of AIRE have been revealed, many questions remained unanswered, such as the fine events or signals involved in the regulation of AIRE expression or in targeting of AIRE-dependent genes by AIRE.

References

Aasland R, Gibson TJ, Stewart AF (1995) The PHD finger: implications for chromatin-mediated transcriptional regulation. Trends Biochem Sci 20:56–59

Abramson J, Goldfarb Y (2016) AIRE: from promiscuous molecular partnerships to promiscuous gene expression. Eur J Immunol 46:22–33. https://doi.org/10.1002/eji.201545792

Abramson J, Giraud M, Benoist C, Mathis D (2010) Aire's partners in the molecular control of immunological tolerance. Cell 140:123–135. https://doi.org/10.1016/j.cell.2009.12.030

Ahonen P, Myllärniemi S, Sipilä I, Perheentupa J (1990) Clinical variation of autoimmune polyendocrinopathy-candidiasis-ectodermal dystrophy (APECED) in a series of 68 patients. N Engl J Med 322:1829–1836. https://doi.org/10.1056/NEJM199006283222601

Akiyama T, Maeda S, Yamane S et al (2005) Dependence of self-tolerance on TRAF6-directed development of thymic stroma. Science 308:248–251. https://doi.org/10.1126/science.1105677

Akiyama T, Shimo Y, Yanai H et al (2008) The tumor necrosis factor family receptors RANK and CD40 cooperatively establish the thymic medullary microenvironment and self-tolerance. Immunity 29:423–437. https://doi.org/10.1016/j.immuni.2008.06.015

Akiyoshi H, Hatakeyama S, Pitkanen J et al (2004) Subcellular Expression of Autoimmune Regulator Is Organized in a Spatiotemporal Manner. Journal of Biological Chemistry 279:33984–33991. https://doi.org/10.1074/jbc.M400702200

Anderson MS (2002) Projection of an immunological self shadow within the thymus by the aire protein. Science 298:1395–1401. https://doi.org/10.1126/science.1075958

Anderson MS, Venanzi ES, Chen Z et al (2005) The cellular mechanism of aire control of T cell tolerance. Immunity 23:227–239. https://doi.org/10.1016/j.immuni.2005.07.005

Aricha R, Feferman T, Scott HS et al (2011) The susceptibility of aire(-/-) mice to experimental myasthenia gravis involves alterations in regulatory T cells. J Autoimmun 36:16–24. https://doi.org/10.1016/j.jaut.2010.09.007

Aschenbrenner K, D'Cruz LM, Vollmann EH et al (2007) Selection of Foxp3+ regulatory T cells specific for self antigen expressed and presented by Aire+ medullary thymic epithelial cells. Nat Immunol 8:351–358. https://doi.org/10.1038/ni1444

Bansal K, Yoshida H, Benoist C, Mathis D (2017) The transcriptional regulator aire binds to and activates super-enhancers. Nat Immunol 18:263–273. https://doi.org/10.1038/ni.3675

Bartel DP (2009) MicroRNAs: target recognition and regulatory functions. Cell 136:215–233. https://doi.org/10.1016/j.cell.2009.01.002

Björses P, Pelto-Huikko M, Kaukonen J et al (1999) Localization of the APECED protein in distinct nuclear structures. Hum Mol Genet 8:259–266

Bloch DB, Nakajima A, Gulick T et al (2000) Sp110 localizes to the PML-Sp100 nuclear body and may function as a nuclear hormone receptor transcriptional coactivator. Mol Cell Biol 20:6138–6146

Bottomley MJ, Collard MW, Huggenvik JI et al (2001) The SAND domain structure defines a novel DNA-binding fold in transcriptional regulation. Nat Struct Biol 8:626–633. https://doi.org/10.1038/89675

Brennecke P, Reyes A, Pinto S et al (2015) Single-cell transcriptome analysis reveals coordinated ectopic gene-expression patterns in medullary thymic epithelial cells. Nat Immunol 16:933–941. https://doi.org/10.1038/ni.3246

Chen J, Yang W, Yu C, Li Y (2008) Autoimmune regulator initiates the expression of promiscuous genes in thymic epithelial cells. Immunol Invest 37:203–214. https://doi.org/10.1080/08820130801967841

Chuprin A, Avin A, Goldfarb Y et al (2015) The deacetylase Sirt1 is an essential regulator of aire-mediated induction of central immunological tolerance. Nat Immunol 16:737–745. https://doi.org/10.1038/ni.3194

Danan-Gotthold M, Guyon C, Giraud M et al (2016) Extensive RNA editing and splicing increase immune self-representation diversity in medullary thymic epithelial cells. Genome Biol 17:219. https://doi.org/10.1186/s13059-016-1079-9

Daniely D, Kern J, Cebula A, Ignatowicz L (2010) Diversity of TCRs on natural Foxp3+ T cells in mice lacking aire expression. J Immunol 184:6865–6873. https://doi.org/10.4049/jimmunol.0903609

Derbinski J, Schulte A, Kyewski B, Klein L (2001) Promiscuous gene expression in medullary thymic epithelial cells mirrors the peripheral self. Nat Immunol 2:1032–1039. https://doi.org/10.1038/ni723

Derbinski J, Pinto S, Rösch S et al (2008) Promiscuous gene expression patterns in single medullary thymic epithelial cells argue for a stochastic mechanism. Proc Natl Acad Sci U S A 105:657–662. https://doi.org/10.1073/pnas.0707486105

Desanti GE, Cowan JE, Baik S et al (2012) Developmentally regulated availability of RANKL and CD40 ligand reveals distinct mechanisms of fetal and adult cross-talk in the thymus medulla. J Immunol 189:5519–5526. https://doi.org/10.4049/jimmunol.1201815

Ferguson BJ, Alexander C, Rossi SW et al (2008) AIRE's CARD revealed, a new structure for central tolerance provokes transcriptional plasticity. J Biol Chem 283:1723–1731. https://doi.org/10.1074/jbc.M707211200

Finnish-German APECED Consortium (1997) An autoimmune disease, APECED, caused by mutations in a novel gene featuring two PHD-type zinc-finger domains. Nat Genet 17:399–403. https://doi.org/10.1038/ng1297-399

Gäbler J, Arnold J, Kyewski B (2007) Promiscuous gene expression and the developmental dynamics of medullary thymic epithelial cells. Eur J Immunol 37:3363–3372. https://doi.org/10.1002/eji.200737131

Gaetani M, Matafora V, Saare M et al (2012) AIRE-PHD fingers are structural hubs to maintain the integrity of chromatin-associated interactome. Nucleic Acids Res 40:11756–11768. https://doi.org/10.1093/nar/gks933

Giraud M, Taubert R, Vandiedonck C et al (2007) An IRF8-binding promoter variant and AIRE control CHRNA1 promiscuous expression in thymus. Nature 448:934–937. https://doi.org/10.1038/nature06066

Giraud M, Yoshida H, Abramson J et al (2012) Aire unleashes stalled RNA polymerase to induce ectopic gene expression in thymic epithelial cells. Proc Natl Acad Sci U S A 109:535–540. https://doi.org/10.1073/pnas.1119351109

Giraud M, Jmari N, Du L et al (2014) An RNAi screen for aire cofactors reveals a role for Hnrnpl in polymerase release and aire-activated ectopic transcription. Proc Natl Acad Sci U S A 111:1491–1496. https://doi.org/10.1073/pnas.1323535111

Gotter J, Brors B, Hergenhahn M, Kyewski B (2004) Medullary epithelial cells of the human thymus express a highly diverse selection of tissue-specific genes colocalized in chromosomal clusters. J Exp Med 199:155–166. https://doi.org/10.1084/jem.20031677

Gray DHD, Gavanescu I, Benoist C, Mathis D (2007a) Danger-free autoimmune disease in Aire-deficient mice. 104:18193–18198. https://doi.org/10.1073/pnas.0709160104

Gray D, Abramson J, Benoist C, Mathis D (2007b) Proliferative arrest and rapid turnover of thymic epithelial cells expressing aire. J Exp Med 204:2521–2528. https://doi.org/10.1084/jem.20070795

Guha M, Saare M, Maslovskaja J et al (2017) DNA breaks and chromatin structural changes enhance the transcription of autoimmune regulator target genes. J Biol Chem 292:6542–6554. https://doi.org/10.1074/jbc.M116.764704

Haljasorg U, Bichele R, Saare M et al (2015) A highly conserved NF-κB-responsive enhancer is critical for thymic expression of Aire in mice. Eur J Immunol 45:3246–3256. https://doi.org/10.1002/eji.201545928

Haljasorg U, Dooley J, Laan M et al (2017) Irf4 expression in thymic epithelium is critical for thymic regulatory T cell homeostasis. J Immunol 198:1952–1960. https://doi.org/10.4049/jimmunol.1601698

Halonen M, Kangas H, Rüppell T et al (2004) APECED-causing mutations in AIRE reveal the functional domains of the protein. Hum Mutat 23:245–257. https://doi.org/10.1002/humu.20003

Heino M, Peterson P, Kudoh J et al (1999) Autoimmune regulator is expressed in the cells regulating immune tolerance in thymus medulla. Biochem Biophys Res Commun 257:821–825. https://doi.org/10.1006/bbrc.1999.0308

Heino M, Peterson P, Sillanpaa N, et al (2000) RNA and protein expression of the murine autoimmune regulator gene (Aire) in normal, RelB-deficient and in NOD mouse. Eur J Immunol 30:1884–1893. https://doi.org/10.1002/1521-4141(200007)30:7<1884::AID-IMMU1884>3.0.CO;2-P

Heo K, Kim H, Choi SH et al (2008) FACT-mediated exchange of histone variant H2AX regulated by phosphorylation of H2AX and ADP-ribosylation of Spt16. Mol Cell 30:86–97. https://doi.org/10.1016/j.molcel.2008.02.029

Herzig Y, Nevo S, Bornstein C et al (2016) Transcriptional programs that control expression of the autoimmune regulator gene aire. Nat Immunol 18:161–172. https://doi.org/10.1038/ni.3638

Hikosaka Y, Nitta T, Ohigashi I et al (2008) The cytokine RANKL produced by positively selected thymocytes fosters medullary thymic epithelial cells that express autoimmune regulator. Immunity 29:438–450. https://doi.org/10.1016/j.immuni.2008.06.018

Hinterberger M, Aichinger M, Prazeres da Costa O et al (2010) Autonomous role of medullary thymic epithelial cells in central CD4(+) T cell tolerance. Nat Immunol 11:512–519. https://doi.org/10.1038/ni.1874

Ilmarinen T, Melen K, Kangas H et al (2006) The monopartite nuclear localization signal of autoimmune regulator mediates its nuclear import and interaction with multiple importin alpha molecules. FEBS J 273:315–324. https://doi.org/10.1111/j.1742-4658.2005.05065.x

Incani F, Serra M, Meloni A et al (2014) AIRE acetylation and deacetylation: effect on protein stability and transactivation activity. J Biomed Sci 21:85. https://doi.org/10.1186/s12929-014-0085-z

Irla M, Hugues S, Gill J et al (2008) Autoantigen-specific interactions with CD4+ thymocytes control mature medullary thymic epithelial cell cellularity. Immunity 29:451–463. https://doi.org/10.1016/j.immuni.2008.08.007

Johnnidis JB, Venanzi ES, Taxman DJ et al (2005) Chromosomal clustering of genes controlled by the aire transcription factor. Proc Natl Acad Sci 102:7233–7238. https://doi.org/10.1073/pnas.0502670102

Kanno T, Kanno Y, LeRoy G et al (2014) BRD4 assists elongation of both coding and enhancer RNAs by interacting with acetylated histones. Nat Struct Mol Biol 21:1047–1057. https://doi.org/10.1038/nsmb.2912

Khan IS, Taniguchi RT, Fasano KJ et al (2014) Canonical microRNAs in thymic epithelial cells promote central tolerance. Eur J Immunol 44:1313–1319. https://doi.org/10.1002/eji.201344079

Koh AS, Kuo AJ, Park SY et al (2008) Aire employs a histone-binding module to mediate immunological tolerance, linking chromatin regulation with organ-specific autoimmunity. Proc Natl Acad Sci U S A 105:15878–15883. https://doi.org/10.1073/pnas.0808470105

Koh AS, Kingston RE, Benoist C, Mathis D (2010) Global relevance of aire binding to hypomethylated lysine-4 of histone-3. Proc Natl Acad Sci U S A 107:13016–13021. https://doi.org/10.1073/pnas.1004436107

Kont V, Laan M, Kisand K et al (2008) Modulation of Aire regulates the expression of tissue-restricted antigens. Mol Immunol 45:25–33. https://doi.org/10.1016/j.molimm.2007.05.014

Kuroda N, Mitani T, Takeda N et al (2005) Development of autoimmunity against transcriptionally unrepressed target antigen in the thymus of aire-deficient mice. J Immunol 174:1862–1870

LaFlam TN, Seumois G, Miller CN et al (2015) Identification of a novel cis-regulatory element essential for immune tolerance. J Exp Med 212:1993–2002. https://doi.org/10.1084/jem.20151069

Liiv I, Rebane A, Org T et al (2008) DNA-PK contributes to the phosphorylation of AIRE: importance in transcriptional activity. Biochim Biophys Acta 1783:74–83. https://doi.org/10.1016/j.bbamcr.2007.09.003

Liston A, Lesage S, Wilson J et al (2003) Aire regulates negative selection of organ-specific T cells. Nat Immunol 4:350–354. https://doi.org/10.1038/ni906

Liston A, Gray DHD, Lesage S et al (2004) Gene dosage--limiting role of aire in thymic expression, clonal deletion, and organ-specific autoimmunity. J Exp Med 200:1015–1026. https://doi.org/10.1084/jem.20040581

Macedo C, Evangelista AF, Marques MM et al (2013) Autoimmune regulator (Aire) controls the expression of microRNAs in medullary thymic epithelial cells. Immunobiology 218:554–560. https://doi.org/10.1016/j.imbio.2012.06.013

Macedo C, Oliveira EH, Almeida RS et al (2015) Aire-dependent peripheral tissue antigen mRNAs in mTEC cells feature networking refractoriness to microRNA interaction. Immunobiology 220:93–102. https://doi.org/10.1016/j.imbio.2014.08.015

Malchow S, Leventhal DS, Nishi S et al (2013) Aire-dependent thymic development of tumor-associated regulatory T cells. Science 339:1219–1224. https://doi.org/10.1126/science.1233913

Mathis D, Benoist C (2009) Aire. Annu Rev Immunol 27:287–312. https://doi.org/10.1146/annurev.immunol.25.022106.141532

Meloni A, Incani F, Corda D et al (2008) Role of PHD fingers and COOH-terminal 30 amino acids in AIRE transactivation activity. Mol Immunol 45:805–809. https://doi.org/10.1016/j.molimm.2007.06.156

Meredith M, Zemmour D, Mathis D, Benoist C (2015) Aire controls gene expression in the thymic epithelium with ordered stochasticity. Nat Immunol 16:942–949. https://doi.org/10.1038/ni.3247

Metzger TC, Khan IS, Gardner JM et al (2013) Lineage tracing and cell ablation identify a post-aire-expressing thymic epithelial cell population. Cell Rep 5:166–179. https://doi.org/10.1016/j.celrep.2013.08.038

Michel C, Miller CN, Küchler R et al (2017) Revisiting the road map of medullary thymic epithelial cell differentiation. J Immunol 199:3488–3503. https://doi.org/10.4049/jimmunol.1700203

Mondal N, Parvin JD (2001) DNA topoisomerase IIalpha is required for RNA polymerase II transcription on chromatin templates. Nature 413:435–438. https://doi.org/10.1038/35096590

Murumägi A, Vähämurto P, Peterson P (2003) Characterization of regulatory elements and methylation pattern of the autoimmune regulator (AIRE) promoter. J Biol Chem 278:19784–19790. https://doi.org/10.1074/jbc.M210437200

Myhre AG, Halonen M, Eskelin P et al (2001) Autoimmune polyendocrine syndrome type 1 (APS I) in Norway. Clin Endocrinol (Oxf) 54:211–217. https://doi.org/10.1046/j.1365-2265.2001.01201.x

Nagamine K, Peterson P, Scott HS et al (1997) Positional cloning of the APECED gene. Nat Genet 17:393–398. https://doi.org/10.1038/ng1297–393

Niki S, Oshikawa K, Mouri Y et al (2006) Alteration of intra-pancreatic target-organ specificity by abrogation of aire in NOD mice. J Clin Invest 116:1292–1301. https://doi.org/10.1172/JCI26971

Nishikawa Y, Hirota F, Yano M et al (2010) Biphasic Aire expression in early embryos and in medullary thymic epithelial cells before end-stage terminal differentiation. Journal of Experimental Medicine 207:963–971. https://doi.org/10.1084/jem.20092144

Oliveira EH, Macedo C, Collares CV et al (2016) Aire downregulation is associated with changes in the posttranscriptional control of peripheral tissue antigens in medullary thymic epithelial cells. Front Immunol 7:526. https://doi.org/10.3389/fimmu.2016.00526

Org T, Chignola F, Hetényi C et al (2008) The autoimmune regulator PHD finger binds to non-methylated histone H3K4 to activate gene expression. EMBO Rep 9:370–376. https://doi.org/10.1038/sj.embor.2008.11

Org T, Rebane A, Kisand K et al (2009) AIRE activated tissue specific genes have histone modifications associated with inactive chromatin. Hum Mol Genet 18:4699–4710. https://doi.org/10.1093/hmg/ddp433

Oven I, Brdickova N, Kohoutek J et al (2007) AIRE Recruits P-TEFb for Transcriptional Elongation of Target Genes in Medullary Thymic Epithelial Cells. Mol Cell Biol 27:8815–8823. https://doi.org/10.1128/MCB.01085–07

Papadopoulou AS, Dooley J, Linterman MA et al (2011) The thymic epithelial microRNA network elevates the threshold for infection-associated thymic involution via miR-29a mediated suppression of the IFN-α receptor. Nat Immunol 13:181–187. https://doi.org/10.1038/ni.2193

Park HH, Lo Y-C, Lin S-C et al (2007) The death domain superfamily in intracellular signaling of apoptosis and inflammation. Annu Rev Immunol 25:561–586. https://doi.org/10.1146/annurev.immunol.25.022106.141656

Perheentupa J (2006) Autoimmune polyendocrinopathy-candidiasis-ectodermal dystrophy. J Clin Endocrinol Metab 91:2843–2850. https://doi.org/10.1210/jc.2005-2611

Perniola R, Musco G (2014) The biophysical and biochemical properties of the autoimmune regulator (AIRE) protein. Biochim Biophys Acta 1842:326–337. https://doi.org/10.1016/j.bbadis.2013.11.020

Peterson P, Peltonen L (2005) Autoimmune polyendocrinopathy syndrome type 1 (APS1) and AIRE gene: new views on molecular basis of autoimmunity. J Autoimmun 25(Suppl):49–55. https://doi.org/10.1016/j.jaut.2005.09.022

Peterson P, Org T, Rebane A (2008) Transcriptional regulation by AIRE: molecular mechanisms of central tolerance. Nat Rev Immunol 8:948–957. https://doi.org/10.1038/nri2450

Pinto S, Michel C, Schmidt-Glenewinkel H et al (2013) Overlapping gene coexpression patterns in human medullary thymic epithelial cells generate self-antigen diversity. Proc Natl Acad Sci U S A 110:E3497–E3505. https://doi.org/10.1073/pnas.1308311110

Pitkanen J (2000) The autoimmune regulator protein has transcriptional transactivating properties and interacts with the common coactivator CREB-binding protein. J Biol Chem 275:16802–16809. https://doi.org/10.1074/jbc.M908944199

Pitkänen J, Peterson P (2003) Autoimmune regulator: from loss of function to autoimmunity. Genes Immun 4:12–21. https://doi.org/10.1038/sj.gene.6363929

Pitkänen J, Vähämurto P, Krohn K, Peterson P (2001) Subcellular localization of the autoimmune regulator protein. characterization of nuclear targeting and transcriptional activation domain. J Biol Chem 276:19597–19602. https://doi.org/10.1074/jbc.M008322200

Pitkanen J, Rebane A, Rowell J et al (2005) Cooperative activation of transcription by autoimmune regulator AIRE and CBP. Biochem Biophys Res Commun 333:944–953. https://doi.org/10.1016/j.bbrc.2005.05.187

Ramsey C, Bukrinsky A, Peltonen L (2002a) Systematic mutagenesis of the functional domains of AIRE reveals their role in intracellular targeting. Hum Mol Genet 11:3299–3308

Ramsey C, Winqvist O, Puhakka L et al (2002b) Aire deficient mice develop multiple features of APECED phenotype and show altered immune response. Hum Mol Genet 11:397–409

Rinderle C, Christensen HM, Schweiger S et al (1999) AIRE encodes a nuclear protein co-localizing with cytoskeletal filaments: altered sub-cellular distribution of mutants lacking the PHD zinc fingers. Human Molecular Genetics 8:277–290

Roberts NA, White AJ, Jenkinson WE et al (2012) Rank signaling links the development of invariant γδ T cell progenitors and aire(+) medullary epithelium. Immunity 36:427–437. https://doi.org/10.1016/j.immuni.2012.01.016

Rossi SW, Kim MY, Leibbrandt A et al (2007) RANK signals from CD4+3 inducer cells regulate development of aire-expressing epithelial cells in the thymic medulla. J Exp Med 204:1267–1272. https://doi.org/10.1084/jem.20062497

Saare M, Rebane A, Rajashekar B et al (2012) Autoimmune regulator is acetylated by transcription coactivator CBP/p300. Exp Cell Res 318:1–12. https://doi.org/10.1016/j.yexcr.2012.04.013

Saltis M, Criscitiello MF, Ohta Y et al (2008) Evolutionarily conserved and divergent regions of the autoimmune regulator (Aire) gene: a comparative analysis. Immunogenetics 60:105–114. https://doi.org/10.1007/s00251-007-0268-9

Sansom SN, Shikama-Dorn N, Zhanybekova S et al (2014) Population and single-cell genomics reveal the aire dependency, relief from Polycomb silencing, and distribution of self-antigen expression in thymic epithelia. Genome Res 24:1918–1931. https://doi.org/10.1101/gr.171645.113

Sousa Cardoso R, Magalhães DAR, Baião AMT et al (2006) Onset of promiscuous gene expression in murine fetal thymus organ culture. Immunology 119:369–375. https://doi.org/10.1111/j.1365-2567.2006.02441.x

St-Pierre C, Trofimov A, Brochu S et al (2015) Differential Features of AIRE-Induced and AIRE-Independent Promiscuous Gene Expression in Thymic Epithelial Cells. J Immunol 195:498–506. https://doi.org/10.4049/jimmunol.1500558

Surdo PL, Bottomley MJ, Sattler M, Scheffzek K (2003) Crystal structure and nuclear magnetic resonance analyses of the SAND domain from glucocorticoid modulatory element binding protein-1 reveals deoxyribonucleic acid and zinc binding regions. Mol Endocrinol 17:1283–1295. https://doi.org/10.1210/me.2002-0409

Takaba H, Morishita Y, Tomofuji Y et al (2015) Fezf2 orchestrates a thymic program of self-antigen expression for immune tolerance. Cell 163:975–987. https://doi.org/10.1016/j.cell.2015.10.013

Tao Y, Kupfer R, Stewart BJ et al (2006) AIRE recruits multiple transcriptional components to specific genomic regions through tethering to nuclear matrix. Mol Immunol 43:335–345. https://doi.org/10.1016/j.molimm.2005.02.018

Teh CE, Daley SR, Enders A, Goodnow CC (2010) T-cell regulation by casitas B-lineage lymphoma (Cblb) is a critical failsafe against autoimmune disease due to autoimmune regulator (Aire) deficiency. Proc Natl Acad Sci USA 107:14709–14714. https://doi.org/10.1073/pnas.1009209107

Ucar O, Tykocinski L-O, Dooley J et al (2013) An evolutionarily conserved mutual interdependence between aire and microRNAs in promiscuous gene expression. Eur J Immunol 43:1769–1778. https://doi.org/10.1002/eji.201343343

Villaseñor J, Besse W, Benoist C, Mathis D (2008) Ectopic expression of peripheral-tissue antigens in the thymic epithelium: probabilistic, monoallelic, misinitiated. Proc Natl Acad Sci U S A 105:15854–15859. https://doi.org/10.1073/pnas.0808069105

Wang X, Laan M, Bichele R et al (2012) Post-aire maturation of thymic medullary epithelial cells involves selective expression of keratinocyte-specific autoantigens. Front Immunol 3:19. https://doi.org/10.3389/fimmu.2012.00019

Watanabe N, Wang Y-H, Lee HK et al (2005) Hassall's corpuscles instruct dendritic cells to induce CD4+CD25+ regulatory T cells in human thymus. Nature 436:1181–1185. https://doi.org/10.1038/nature03886

Waterfield M, Khan IS, Cortez JT et al (2014) The transcriptional regulator aire coopts the repressive ATF7ip-MBD1 complex for the induction of immunotolerance. Nat Immunol 15:258–265. https://doi.org/10.1038/ni.2820

Yang S, Bansal K, Lopes J et al (2013) Aire's plant homeodomain(PHD)-2 is critical for induction of immunological tolerance. Proc Natl Acad Sci U S A 110:1833–1838. https://doi.org/10.1073/pnas.1222023110

Yang S, Fujikado N, Kolodin D et al (2015) Immune tolerance. Regulatory T cells generated early in life play a distinct role in maintaining self-tolerance. Science 348:589–594. https://doi.org/10.1126/science.aaa7017

Yano M, Kuroda N, Han H et al (2008) Aire controls the differentiation program of thymic epithelial cells in the medulla for the establishment of self-tolerance. J Exp Med 205:2827–2838. https://doi.org/10.1084/jem.20080046

Yoshida H, Bansal K, Schaefer U et al (2015) Brd4 bridges the transcriptional regulators, aire and P-TEFb, to promote elongation of peripheral-tissue antigen transcripts in thymic stromal cells. Proc Natl Acad Sci U S A 112:E4448–E4457. https://doi.org/10.1073/pnas.1512081112

Ziv Y, Bielopolski D, Galanty Y et al (2006) Chromatin relaxation in response to DNA double-strand breaks is modulated by a novel ATM- and KAP-1 dependent pathway. Nat Cell Biol 8:870–876. https://doi.org/10.1038/ncb1446

Zuklys S, Mayer CE, Zhanybekova S et al (2012) MicroRNAs control the maintenance of thymic epithelia and their competence for T lineage commitment and thymocyte selection. J Immunol 189:3894–3904. https://doi.org/10.4049/jimmunol.1200783

Zumer K, Plemenitas A, Saksela K, Peterlin BM (2011) Patient mutation in AIRE disrupts P-TEFb binding and target gene transcription. Nucleic Acids Res 39:7908–7919. https://doi.org/10.1093/nar/gkr527

Zumer K, Low AK, Jiang H et al (2012) Unmodified histone H3K4 and DNA-dependent protein kinase recruit autoimmune regulator to target genes. Mol Cell Biol 32:1354–1362. https://doi.org/10.1128/MCB.06359-11

Chapter 8
Aire Mutations and Autoimmune Diseases

Anette S. B. Wolff and Bergithe E. Oftedal

Abstract The autoimmune regulator (*AIRE*) gene is crucial for the development of normal central immunological tolerance and prevention of autoimmunity. In this chapter, we will see how mutations in *AIRE* cause a rare autosomal recessive disease, autoimmune polyendocrine syndrome type 1 (APS-1). APS-1 patients display a variety of endocrine and ectodermal manifestations where the majority of patients develop at least two of the three main components of adrenocortical insufficiency (Addison's disease), hypoparathyroidism and chronic mucocutaneous candidiasis. Further, we will look into how the disease-causing mutations found in *AIRE* have been important in revealing the functional properties of the AIRE protein and its domains. Interestingly, a subset of specific, heterozygous AIRE mutations directly causes common organ-specific autoimmunity with propensity for pernicious anemia and vitiligo. Multiple single cases and families with heterozygous mutations in the first plant homeodomain (PHD1) zinc finger presented with dominant inheritance, later presentation, milder phenotypes, and reduced penetrance compared to the classical APS-1 caused by recessive mutations in the gene. We will also discuss how AIRE has been implicated in other, more common autoimmune diseases and look into the reported dosage response of AIRE. Hopefully, the emerging ease and lower cost of exome- and genome sequencing will aid in revealing the full involvement of AIRE in more common autoimmune diseases. As we will see, the evidence so far suggests that AIRE's role in common autoimmune disorders is still to be fully determined.

A. S. B. Wolff · B. E. Oftedal (✉)
KG Jebsen Center for Autoimmune Disorders, Department of Clinical Science, University of Bergen, Bergen, Norway

Department of Clinical Science, Haukeland University Hospital, University of Bergen, Bergen, Norway
e-mail: bergithe.oftedal@uib.no

© Springer Nature Switzerland AG 2019
G. A. Passos (ed.), *Thymus Transcriptome and Cell Biology*,
https://doi.org/10.1007/978-3-030-12040-5_8

8.1   Introduction

Autoimmune diseases were first recognised by Mackay and Macfarlane Burnet in 1962 (Burnet and Mackay 1962). It is defined as the failure of an organism to tolerate its own cells and tissue, hence resulting in an aberrant immune response by lymphocytes (T cell driven disease) and/or antibodies (B cell driven disease). The tissue that is targeted by the immune response directed against "self" undergo pathological changes and dysfunction. Autoimmune diseases can be divided into systemic autoimmune diseases (affecting the whole body) and specific organ- or body system diseases, including the endocrine, gastro-intestinal and the neurological systems. Autoimmune diseases affect hundreds of millions worldwide and cause chronic suffering, vital organ failure and premature death at the level of cancer and cardiovascular diseases. An autoimmune disease may affect one or more organs or tissues, and a person may present with more than one autoimmune disorder at the same time. Given the increasing prevalence and largely unknown nature of these disorders, the burden they represent for health care systems will only increase, emphasizing the importance of defining the mechanisms of disease and translate this knowledge into effective diagnostic tests and therapeutics. Current treatment options for autoimmune diseases are largely limited to replacement therapy and the use of unspecific anti-inflammatory agents, which do not restore full health, but rather continue to expose patients to increased morbidity, reduced quality of life, and heightened mortality. Most autoimmune diseases are most likely a result of a combination of different genetic and environmental factors. However, there are a few exemptions, like the autoimmune polyendocrine syndrome type 1 (APS-1) caused by mutations in the *Autoimmune Regulator* (*AIRE*) gene.

AIRE was identified as the underlying cause of the recessively inherited autoimmune disease APS-1 in 1997. It was then the result of many years sequencing and characterisation of the gene, and the conclusion was reached by two research groups at the same time and hence published back to back in Nature Genetics (Finnish-German APECED Consortium 1997; Nagamine et al. 1997). APS-1 is sometimes referred to an "experiment of nature" where mutations in *AIRE* typically pinpoint how immunologic tolerance is established through negative selection, and how the process of tissue restricted antigen presentation is crucial in health and disease. AIRE may also prevent autoimmunity by additional mechanisms like promoting the development of Regulatory T cells and the expression of chemokines important in mediating T cell negative selection (Lei et al. 2011; Malchow et al. 2013).

APS-1 is one of few monogenic autoimmune disorders, and when its' underlying genetic cause, *AIRE*, was detected, it held high hopes being the causative effect for other more common autoimmune diseases. But where AIRE has given us insight into the mechanisms of negative selection and the expression of tissue restricted antigens (TRAs) within the thymus, the contribution of AIRE in other autoimmune diseases than APS-1 has not been as clear. In this chapter, we will look at the clinical diagnosis and hallmark of classical APS-1. We will see that the *AIRE* mutations found in the patient cohort of APS-1 have led to characterisation of AIRE's different protein domains, and we will discuss *AIRE*'s contribution in more common autoimmune diseases as well as the dominant mutations found in *AIRE*.

8.2   Autoimmune Polyendocrine Syndrome Type I

Studies of monogenic disorders have proven to be very informative to decipher both pathogenic and normal physiological as well as biological mechanisms. Autoimmune polyendocrine syndrome type 1 (APS-1) also called APECED (autoimmune polyendocrinopathy-candidiasis ectodermal dystrophy) (online Mendelian inheritance in man (OMIM) 240300), is no exemption. APS-1 is a monogenic inherited disease caused by mutations in the *AIRE* gene. Until recently, only recessive mutations were identified as causes of APS-1 in this gene. *AIRE* occupies 13 kilobases on chromosome 21, spanning 14 exons and encodes a 545 amino acid protein [1–3]. Recessive mutations in *AIRE* lead to a serious and complex child-onset autoimmune disease with autoimmune manifestations in various, mostly endocrine, organs, as well as chronic mucocutaneous candidiasis (CMC) (Husebye et al. 2018).

The disease is rare (about 1:90,000 in Norway (Wolff et al. 2007)), but more common in certain genetic isolated populations, such as the Finnish (1:25,000), Sardinian (1:14,000) and among Iranian Jews (1:9000) (Betterle et al. 1998). The disease typically presents in childhood or early adolescence, and the patients gradually develop autoimmune manifestations throughout life. APS-1 is a devastating disease which left untreated is fatal.

8.2.1   Diagnosis of APS-1

To make a clinical diagnosis of APS-1, two of the three main components; CMC, primary hypoparathyroidism and autoimmune adrenocortical insufficiency (Addison's disease) must be present. Very often, patients also suffer from other manifestations, like primary ovarian failure, chronic diarrhoea and malabsorption, keratitis, autoimmune hepatitis, vitiligo, alopecia and enamel hypoplasia (Ahonen et al. 1990; Betterle et al. 1998; Perheentupa 2006), and early childhood onset of these manifestations could also be a reason for AIRE mutational analysis. One of the three major components is sufficient for diagnosis if (1) a sibling is already identified with APS-1 or (2) if a mutation is found in both alleles on *AIRE*. Persons who develop one of the three major components together with one of the other components before the age of 30 should be investigated for possible APS-1 (Husebye et al. 2009, 2018). It should be emphasised that because it is a rare disease, it is important to raise the awareness among general clinicians, in particular paediatricians, who often is the first medical persons to meet these patients. Enamel hypoplasia is also often seen as an early symptom of APS-1, and to better inform dentists about the disease could be a useful incentive to discover these patients.

APS-1 may have a fatal outcome if left undiagnosed and untreated. In 1962, 16 out of 23 (69.5%) patients died before the age of 30 (Gass 1962), while this number decreased to 4 out of 41 (10%) in a patient group followed from 1967 to 1996 (Betterle et al. 1998). As a devastating disease, APS-1 patients require follow up within the specialised health care system. Early diagnosis, close monitoring of new disease-components and suitable treatment are therefore essential for this group of

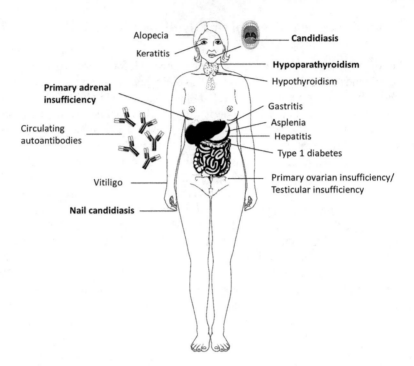

Fig. 8.1 The hallmarks of classical APS-1 are shown in the figure. The three major components composing the classical triad are marked in bold

patients, making the need for good diagnostic assays and awareness among the general physicians of outmost importance. The hallmarks of APS-1 is summarised in Fig. 8.1.

8.2.2 The Major Components of APS-1

The first sign of APS-1 is typically the diagnosis of chronic mucocutaneous candidiasis (CMC) within the first years of life. *Candida albicans* infections are seen in nails, skin, and the mucous membranes in the mouth, oesophagus and vagina. From a study of the Finnish patients, one in six patients was diagnosed with CMC by the age of 1, which increased to 98% by the age of 30 (Perheentupa 2006). The prevalence of CMC in APS-1 patients varies between populations and the severity of the infection differs between patients, but is often recurrent and difficult to treat. As a serious side effect of oral CMC, there is an increased risk of squamous cell carcinoma, and antifungal treatment and close monitoring of the oral cavity in these patients are therefore of outmost importance (Rautemaa et al. 2007).

Hypoparathyroidism is the second most prevalent feature of APS-1, and typically presents in 85% of the patients by the age of 30 (Perheentupa 2006). This is usually the first endocrine manifestation, and is more prevalent in women than in men (Gylling et al. 2003). Hypoparathyroidism is caused by autoimmune destruction of the parathyroid glands which decrease plasma calcium leading to paraesthesia and cramps. It is treated with vitamin D analogues and with calcium and magnesium supplementation, but remain the most challenging APS-1 disease-component to manage (Husebye et al. 2009). The immunological target within the parathyroid cells is the NACHT leucine-reich repeat protein 5 (NALP5), and auto-antibodies against NALP5 are excellent biomarkers of hypoparathyroidism (present in about 50% of hypoparathyroidism in connection with APS-1) (Alimohammadi et al. 2008).

Autoimmune Addison's disease (AAD) typically appears after CMC and hypoparathyroidism, as the third most common feature in APS-1, with a 78% prevalence among the Finnish patients by the age of 30 (Perheentupa 2006). In a Norwegian survey, 80% of the cohort had this manifestation, regardless of age (Myhre et al. 2001). Addison's disease was first described by Thomas Addison in 1855 (Addison 1855), and is characterised by fatigue, salt craving, weight loss and increased pigmentation of the skin and mucus membranes. The first serological sign of AAD is autoantibodies against 21-hydroxylase (21-OH) which can present up to years before any biochemical or clinical evidence of adrenal insufficiency (Winqvist et al. 1992; Soderbergh et al. 2004). Early diagnosis is important, as untreated AAD is fatal (Lovas and Husebye 2005). Replacement therapy with hydrocortisone and fludrocortisone largely normalises mortality rates (Erichsen et al. 2009).

8.2.3 Autoantibodies in APS-1

A hallmark of APS-1 is the circulating autoantibodies present in very high titres in the patients' serum. Due to their deficiency in developing tolerance to self, B cells and plasma cells from APS-1 patients produce a variety of autoantibodies, mainly directed against specific intracellular enzymes. These autoantibodies often present before any clinical symptoms and are used in the clinic to aid the diagnosis of APS-1 patients. About 10 years ago, autoantibodies against components of the immune system itself, namely against interferons and interleukins, were discovered (Meager et al. 2006). With new technological advances, several novel targets of autoantibodies have been described in APS-1 patients (Landegren et al. 2016; Meyer et al. 2016). These new results suggest that in addition to autoantibodies against specific targets, like the interferons that are found with frequencies reaching 99% in APS-1 patients there might also be a portion of "private" autoantibodies possibly reflecting the personal adaption of the immune system, or possibly the influence of HLA genotypes within the APS-1 cohort. Several of the autoantibodies that are present in patients with APS-1 are also seen in other autoimmune diseases, but usually in lower titres. These observations might indicate that there are common pathogenic and immunogenic mechanisms leading to the different diseases, independent of the mutations in *AIRE*.

8.2.4   Organ-Specific Autoantibodies in APS-1

The organ-specific autoantibodies target proteins in particular organs, and have proven to be excellent markers for autoimmune disease in the organ in which their target is expressed (Table 8.1). Since they often precede clinical symptoms, many of them are assayed at routine basis in APS-1 patients, in "APS-1 risk groups" with one syndrome component, and when APS-1 is suspected (Betterle et al. 1998, 2002; Eisenbarth and Gottlieb 2004; Boe Wolff et al. 2008; Eriksson et al. 2018b).

The adrenal steroidogenic P450 superfamily autoantigens 21-hydroxylase (21-OH), 17-α-hydroxylase (17-OH), and side-chain cleavage enzyme (SCC) catalyse chemical reactions required for the production of steroid hormones, like aldosterone and cortisol (21-OH), progesterone (17-OH), and pregnenolone (SCC) (Krohn et al. 1992; Winqvist et al. 1992, 1995; Uibo et al. 1994). Assay of autoantibodies against 21-OH is an excellent marker for Addison's disease, both in its isolated form and as part of a polyendocrine syndrome (the latter not due to mutations in *AIRE*) (Betterle et al. 1999), while 17-OH and especially SCC are the main gonadal autoantigens

Table 8.1 Common organ-specific autoantigens in APS-1, their main organ of expression relevant to APS-1, associated disease components and reported prevalence

Auto-antigen	Main expression	Disease component	Prevalence (%)[a]	References
21-OH	Adrenal cortex	Addison's disease	66	Winqvist et al. (1992), Uibo et al. (1994), Falorni et al. (1997), Peterson et al. (1997)
SCC	Adrenal cortex, gonads	Addison's disease, hypogonadism	52	Winqvist et al. (1993), Uibo et al. (1994), Peterson et al. (1997)
17-OH	Adrenal cortex, gonads	Addison's disease, hypogonadism	44	Krohn et al. (1992), Uibo et al. (1994), Peterson et al. (1997)
GAD65	Pancreas	Intestinal dysfunction	37	Velloso et al. (1994), Soderbergh et al. (2004)
CYP1A2	Liver	Autoimmune hepatitis	8	Clemente et al. (1997)
AADC	Pancreas, liver	Hepatitis, Vitiligo	51	Husebye et al. (1997)
TPH	Duodenum	Malabsorption	45	Ekwall et al. (1998)
TH	Keratinocytes	Alopecia	40	Hedstrand et al. (1999)
NALP5	Parathyroid gland, ovary	Hypoparathyroidism	41[b]	Alimohammadi et al. (2008)

21OH 21-hydroxylase, *SCC* side-chain cleavage enzyme, *17-OH* 17-α-hydroxylase, *GAD* glutamic acid decarboxylase, *CYP1A2* cytochrome P450 1A2, *AADC* aromatic ʟ-amino acid decarboxylase, *TPH* tryptophan hydroxylase, *TH* tyrosine hydroxylase, *NALP5* NACHT leucine-reich repeat protein 5
[a]Prevalence is based on detected autoantibodies in a cohort of Finnish, Swedish and Norwegian APS-1 patients, n = 90 (Soderbergh et al. 2004)
[b]Prevalence is based on detected autoantibodies in a cohort of Finnish, Swedish and Norwegian APS-1 patients, n = 87 (Alimohammadi et al. 2008)

(Chen et al. 1996). Autoantibodies directed against tryptophan hydroxylase (TPH) and aromatic L-amino acid decarboxylase (AADC) are also highly precise in their detection of APS-1 and correlate to malabsorption and hepatitis, respectively. They are rarely seen in other conditions except in a subgroup of AADC-positive Addison's patients (Husebye et al. 1997; Ekwall et al. 1999; Dal Pra et al. 2004). Cytochrome P450 IA-2 is another hepatic autoantigen, described already in 1998 (Clemente et al. 1997). Autoantibodies against tyrosine hydroxylase (TH) are correlated to alopecia in APS-1 (Hedstrand et al. 1999), and autoantibodies against TH arc also seen in patients with isolated alopecia and vitiligo (Kemp et al. 2011a, b). Autoantibodies against glutamic acid decarboxylase 65 (GAD65) are commonly found in insulin dependent diabetes mellitus (Baekkeskov et al. 1990), and are also common in APS-1, although associated with intestinal dysfunction (Soderbergh et al. 2004).

Autoantibodies against NALP5 are specific for APS-1, and positivity is a serological sign of hypoparathyroidism. NALP5 is also expressed in ovaries, and anti-NALP5 is also suggested as a marker of primary ovarian failure (Alimohammadi et al. 2008). Autoantibodies against the calcium-sensing receptor (CaSR) are also related to hypoparathyroidism (Li et al. 1996). Although controversial (Soderbergh et al. 2004), CaSR has been confirmed as a receptor-stimulating antigen in a limited number of APS-1 patients (Gavalas et al. 2007; Kemp et al. 2009).

As these antibodies are specific for either APS-1 or a specific phenotype (Table 8.1), a probable APS-1 diagnosis can be made from any disease-component together with autoantibodies against NALP5, AADC, TPH or TH.

In recent years, novel autoantibodies have been found in APS-1 patients correlated to specific, but more uncommon, manifestations, e.g. against bactericidal/permeability-increasing fold-containing B1 (BPIFB1) specific for interstitial lung disease (Shum et al. 2013), against potassium channel regulator (KCNRG) (Alimohammadi et al. 2009) in patients with respiratory symptoms and against AIE-75 and villin, correlated with gastrointestinal disease (Kluger et al. 2015). Several other antibodies to self-antigens have been identified, although without clear links to certain clinical features, e.g. endothelin-converting enzyme (ECE)-2 (ECE-2) (Smith-Anttila et al. 2017), testis specific 10 (TSGA10) (Smith et al. 2011), male specific reaction against transglutaminase 4 (TGM4) (Landegren et al. 2015), myosin-9 (MYH9) (Lindh et al. 2013; Oftedal et al. 2017), protein disulfide-isomerase-like protein of the testis (PDILT) and melanoma-associated antigen B2 (MAGEB2) (Landegren et al. 2016).

8.2.5 Cytokine Autoantibodies in APS-1

The cytokine autoantibodies recently described in APS-1 patients have changed the view on autoantibodies in autoimmune endocrine diseases, as they no longer only are directed against organ-specific antigens, but also against components of the immune system itself. The first cytokine autoantibodies described in patients with

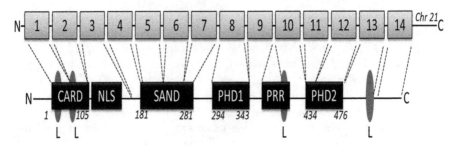

Fig. 8.2 The AIRE gene with the corresponding protein domains shown below with the amino acids for each domain outlined. CARD = Caspase activation and recruitment **domain**, NLS = nuclear localisation domain, SAND = "Sp100, AIRE-1, NucP41/75, DEAF-1" domain, PHD1 = plant homeodomain 1, PPR = proline rich region, and PHD2 = plant homeodomain 2, L = LXXLL domains

APS-1 were those against type I interferons (IFNs) (Meager et al. 2006), followed by autoantibodies against Th17-derived cytokines (Kisand et al. 2010; Puel et al. 2010). During the last years, commercialisation of these autoantibodies has been on the drawing board, with the implication of use for treatment in for example multiple sclerosis patients (immunoqure.com).

In 2006, neutralising autoantibodies against IFN-type I were detected in patients suffering from APS-1; surprisingly anti-IFN-ω was found in all the investigated patients (Meager et al. 2006). Several studies utilising different methods, control groups, and including large number of patients from different countries have confirmed the presence of anti-IFN-α or -ω in close to every patient diagnosed with APS-1 (Meager et al. 2006; Zhang et al. 2007; Meloni et al. 2008). Type I interferons is a large group of structurally similar cytokines, grouped into IFN-α (with more than 13 different members), IFN-β, IFN-ε, IFN-κ and IFN-ω in humans. They are encoded by genes clustered in one locus on the human chromosome 9. IFN-α, IFN-β and IFN-ω are expressed at very high levels in plasmacytoid DCs and monocytes following viral infection, exposure to double-stranded RNA or stimulation through Toll-like receptors, and are early key players in the innate immune response (Hertzog et al. 2003; Pestka et al. 2004). These autoantibodies also seem to precede any clinical symptoms, and have been detected as early as 7 month of age (Toth et al. 2010). Assays measuring levels of autoantibodies against IFN-ω have therefore the potential as valuable diagnostic tools and are suggested as a simpler, faster and a more cost-effective screening method preceding *AIRE* mutational analysis (Meloni et al. 2008). A probable APS-1 diagnosis can consequently rely on the presence of anti-IFN-ω autoantibodies together with any disease-components (Fig. 8.2) (Husebye et al. 2009). The origin and role of these autoantibodies are still unclear, but autoantibodies against IFN-α have been shown to down-regulate interferon-stimulated gene expression in blood cells from APS-1 patients (Kisand et al. 2008). The autoantibodies against IFN-α and IFN-ω are also found in patients suffering from Myasthenia gravis (MG) and thymoma, and their shared thymic dysfunction has been suggested as a possible connecting point.

The leukocyte-derived cytokines interleukin (IL) 17-A, IL-17F and IL-22 belong to a class of cytokines with a predominant effect on epithelial cells in various tissues. They are mainly produced by Th17 cells, which were identified as an independent linage of CD4⁺ T cells in 2005 (Harrington et al. 2005). However, also other immune cells, both adaptive and innate, may produce these interleukins.

The high occurrence of CMC in patients with APS-1 was long regarded as an immune defect and not as a part of the autoimmune reaction. The discovery of auto-antibodies against IL-17A, IL-17F and IL-22 in sera from APS-1 patients indicates an indirect autoimmune cause of CMC, where these autoantibodies inhibit elements in the defence against *C. albicans* (Kisand et al. 2010; Puel et al. 2010). Interestingly, patients suffering from thymoma and CMC in combination are also positive for these autoantibodies (Kisand et al. 2010). The Th17-response towards *C. albicans* has been studied using peripheral blood mononuclear cells (PBMC) from APS-1 patients, where the production of IL-17F and IL-22 were down-regulated, while the results for IL-17A were unequivocal (Ahlgren et al. 2011; Kisand et al. 2010).

The potential of utilising autoantibodies to detect APS-1 patients is illustrated by a recent Swedish study. Here, sera from the Swedish cohort of patients with Addison's disease (677 patients) were screened for autoantibodies targeting IL-22 and IFN-α4, and four undiagnosed APS-1 patients were detected (Eriksson et al. 2018b).

8.3   Mutations in the Different Domains of *AIRE*

AIRE is located at 21q22.3 spanning 14 exons and encodes a 545-amino-acid protein with a molecular weight of 57.5 kDa (Aaltonen et al. 1994; Nagamine et al. 1997). Over 100 mutations in the *AIRE* gene have been reported in the Human Gene Mutation Database (Liston et al. 2004), varying from single nucleotide substitutions to large deletions (Table 8.1). AIRE exhibits several different functional domains; a Caspase activation and recruitment domain, CARD (for AIRE dimerization) (amino acids 1–105), a "Sp100, AIRE 1, NucP41/75, DEAF-1" domain, SAND (for DNA interaction) (amino acids 181–281), and two plant homeodomains (PHD1 (amino acids 296–343)) for histone binding and PHD2 (amino acids 434–476), with uncertain role but probably binding to other proteins as shown in Fig. 8.2.

The different types of mutations; nonsense, insertions, deletions, duplication and frameshift mutations, are found spread out across the coding sequence, although there seem to be a pattern of clustering of missense mutation within the CARD and PHD1 domains of AIRE. Overall, there are founder effects of the disease-causing mutations found in *AIRE*; the most common mutation p.R257*, often referred to as the major Finnish mutation, is located in the SAND domain. It dominates not only in Finland, but also in many other European countries (Podkrajsek et al. 2005; Stolarski et al. 2006). A 13 base pair deletion, p.322del13, found in the PHD1 domain is the major mutation found in APS-1 patients in Norway (Bruserud et al. 2016), British Isles (Pearce et al. 1998) and North America (Heino et al. 1999). In addition, founder mutations have been described in Sardinia (p.R139X), Sicily (p. R203X), among Persian Jews (p.Y85C) and in Apulia in Italy (p.W78R) (Zlotogora and Shapiro 1992; Perniola et al. 2008; Meloni et al. 2012; Valenzise et al. 2014). Perhaps not surprisingly, most of the reported *AIRE* mutations target conserved amino acid residues (Ferguson et al. 2008). These mutations have been valuable in elucidating the different domains of AIRE and their functions.

As AIRE is functional as a homo- or multimeric complex, the **CARD domain** has been found to be crucial for this di- or multimerization (Pitkanen et al. 2000). *In vitro* studies of CARD mutations revealed that p.L28P and p.L29P, both predicted crucial for structural integrity, inhibit the nuclear localization and hence the AIRE-mediated transcription. p.K83E located on the surface of CARD had a milder effect through mediating transcription and nuclear localization, possibly by changing binding sites for protein-protein interaction (Ferguson et al. 2008). p.Y85C is located close to p.K83E (Ferguson et al. 2008) which might explain the milder phenotype observed in Persian Jews with APS-1.

The **SAND domain** is a conserved 80-amino acid sequence shown to mediate protein-protein interaction and DNA binding (Gibson et al. 1998; Waterfield et al. 2014). A number of non-sense mutations, including the major Finnish mutation (p.R257X) and Sardinian mutations are located in this domain and give rise to truncated non-functional proteins. The pathogenicity of the various SAND mutations is demonstrated *in vitro* by mutagenesis using cell assays revealing that deletions and even point mutations (e.g. p.K237A, p.K253A, p.K253E, p.R257A) result in aggregation of the polypeptide in the cytoplasm preventing proper nuclear targeting (Ramsey et al. 2002). In line with these findings, homozygosity for p.P252L has been reported in an APS-1 patient and is predicted to be damaging (Ilmarinen et al. 2005). Interestingly, p.G228W was the first reported dominant mutation in *AIRE*. It was found in an Italian APS-1 family with propensity for autoimmune thyroid disease, hence a milder phenotype than classical childhood onset APS-1 (Cetani et al. 2001). Molecular studies showed that only one p.G228W mutation was needed to cause autoimmune disease components, including diabetes mellitus and neuropathy in mice (Su et al. 2008), which can be explained by a dominant negative effect.

AIRE has two PHD-domains, of which **PHD1** is recognized as a zinc finger, while the function of the PHD2 domain is still debated. PHD1 specifically recognize demethylated lysine 4 residues (Koh et al. 2008) while **PHD2** is found to be critical for interaction with protein partners involved in chromatin binding (Yang et al. 2013). The most frequent mutation found within the PHD1 is the p.322del13 located in the 3'-end of the domain, leading to a shift of reading frame which result in a truncated protein. A number of missense mutations have been described within these two domains, often compound heterozygous for other mutations in *AIRE*. This include several of missense mutations targeting amino acids critical for zinc binding and the three dimensional structure of this domain like p.C311Y, P326Q and p.326L (Bjorses et al. 2000; Saugier-Veber et al. 2001).

Mutations are also found outside the defined protein domains of AIRE. Several gross deletions associate with APS-1(Podkrajsek et al. 2008), including a complete *AIRE* deletion (Boe Wolff et al. 2008). A large insertion of 23 base pairs is also described (Cihakova et al. 2001), as well as a number of splice site mutations. Functional testing of the splicing mutations c.463G>A and c.653-1G>A revealed an alternative splicing pattern in which parts of the intron sequence was included in the mRNA resulting in larger polypeptides and hence probably incorrect folding of the proteins leading to impaired function (Zhang et al. 2013; Mora et al. 2014).

Two single nucleotide polymorphisms (SNPs) (-655R and -230Y) are found in the promotor region of *AIRE* (Lovewell et al. 2015). Using different *AIRE* promoter reporter constructs these variants displayed difference in luciferase activity. Whether this translates into low expression of AIRE *in vivo* and susceptibility to autoimmune disease is yet to be discovered. Finally, an NFκb recognition site was described in the *AIRE* promoter, and when deleted this resulted in total inhibition of *AIRE* transcription *in vitro* (LaFlam et al. 2015). So far, there are no reports on patients with such a mutation but there might be a higher chance of detecting such mutations as whole genome sequencing are getting more common.

8.4 Non-classical APS-1

Patients harboring mono-allelic *AIRE* mutations are characterized by a later disease onset and often milder phenotype, dominated by pernicious anemia, vitiligo and thyroid disease, compared to classical APS-1 (Cetani et al. 2001; Oftedal et al. 2015). Based on a novel study, it is recognized that familial aggregation of organ-specific autoimmunity with onset at early age might in some cases be due to such heterozygous *AIRE* mutations. A family from Norway and an unrelated family from Finland presented with pernicious anemia and/or adrenal insufficiency and enamel dysplasia all had a homozygous p.C311Y mutation in AIRE, and no other disease-causing mutations could be found. Also other homozygous mutations clustering in the **PhD1 domain** were detected; p.C302Y and p.D312N were found in patients with hypoparathyroidism, and p.R303P was reported in a patient with growth hormone deficiency and premature ovarian insufficiency (Stolarski et al. 2006). Vitiligo and AAD were detected in several family members harboring a p.V301M mutation. Although these patients were diagnosed with disease-components of APS-1, none of them fulfilled the full clinical criteria. Recently, it was discovered that the majority of missense mutations in **PHD1**, including p.V301M, p.C302Y, p.R303P, p.G305S, p.D312N, and p.P326L, are all able to co-localize with wild type AIRE in nuclear speckles, and suppress gene expression driven by wild type AIRE in a dominant-negative manner (Oftedal et al. 2015). Also within the **SAND domain** a report has described a dominant mutation that do not display a complete APS-1 phenotype; p.R247C was identified in a patient with type I diabetes mellitus (T1DM) and his mother presented with rheumatoid arthritis (Abbott et al. 2018). Hence, a new set of dominant mutations in AIRE emerge as the underlying cause of autoimmune disease, but rarely as classical APS-1. Looking to public databases, the frequencies of these mutations within the general population is relatively high

(>0.0008) which could suggest a role for AIRE in more common, inherited autoimmune diseases.

Non-classical APS-1 can be challenging to discover, as these dominant mutations often present with an incomplete penetrance and the family members often display different clinical manifestations. It is further important to exclude other genetic mutations, either within *AIRE*, the promoter region or in other genes. Nevertheless, it is important to facilitate identification of these individuals and families to prevent morbidity and mortality.

8.5   *AIRE* in Other Diseases

Upon *AIRE's* identification, an obvious question was whether *AIRE* mutations or polymorphisms also could explain more common autoimmune diseases. Many *AIRE* SNPs have been studied in human disease association studies (Colobran et al. 2016), but often with conflicting results. For example the c.C961G allele (p.S278R) was studied in two different alopecia cohorts yielding opposite conclusions (Tazi-Ahnini et al. 2002; Pforr et al. 2006) and studies of vitiligo and *AIRE* variation also report conflicting findings (Jin et al. 2007; Tazi-Ahnini et al. 2008). A large scale genome-wide association study in a cohort of Japanese rheumatoid arthritis (RA) patients identified two single nucleotide polymorphisms (rs2075876 and rs760426) which strongly associated with disease (Terao et al. 2011) and this was further replicated in a Chinese population, but no associations were found in a European cohort (Stahl et al. 2010). A recent meta-analysis for rs2075876 (7145 cases and 8579 controls) and rs760426 (6696 cases and 8164 controls) showed a significantly association with an increased risk of RA in allelic, dominant, recessive, codominant heterozygous, and codominant homozygous genetic models in Asian patients (Berczi et al. 2017). Further, rs2075876 was also found to significantly increasing the risk of RA in a separate meta-analysis in Asian patient (Xu et al. 2017).

Different cohorts of patients with common autoimmune conditions such as hepatitis, inflammatory bowels disease, primary sclerosing cholangitis and primary biliary cirrhosis have been investigated for *AIRE* mutations, but no significant associations have been found. Moreover, an association between more common autoimmune endocrinopathies, such as primary adrenal insufficiency, type-1 diabetes, hypothyroidism and Graves' disease, and mutations in *AIRE* have come up negative (Nithiyananthan et al. 2000; Boe Wolff et al. 2008). However, these latter studies only analysed the respective patient cohorts for common mutations or SNPs and did not sequence the whole *AIRE gene*. A recent Swedish study analysing 1800 candidate genes by exome sequencing in 479 Addison's patients where APS-1 had been carefully ruled out, found that common variants in *AIRE* were associated with sporadic AAD (Eriksson et al. 2018a).

Single cases of patients with heterozygous mutations in *AIRE* and autoimmune disease have been reported, but the lack of biochemical studies makes it difficult to determine whether *AIRE* is the underlying cause. One example is a male teenager with common variable immunodeficiency, alopecia and onychodystrophy and a heterozygous p.S250C *AIRE* mutation (Bellacchio et al. 2014). Whether the mutation contributed to the phenotype is uncertain since he did not display the common auto-antibodies found in APS-1 (e.g. IFN-ω autoantibodies), nor did his father who carried the same mutation display any autoimmune manifestations. In patients who are heterozygous for p.S278R, an association with alopecia totalis, RA and melanoma has been reported (Tazi-Ahnini et al. 2002; Conteduca et al. 2010; Terao et al. 2011). However, p.S278R is relatively common in the general population (about 10% carry this mutation) and it remains unclear if the variant promotes autoimmune disease.

Although it does not seem like AIRE is the immediate underlying cause in more common organ-specific autoimmune disease, it contributes to private mutations inherited within families. In addition, specific combinations of suboptimal polymorphism variants of *AIRE* could potentially, together with risk variants in other genes, increase the individuals' total risk of failure of tolerance and hence development of autoimmune disease. The emerging new sequencing technologies might shed light on this challenge in the future when more extensive genetic information is more easily gathered. However, it is important that genetic findings are followed up by in vitro studies to determine the biochemical and molecular mechanisms for the causative effect.

8.6 *AIRE* in Diseases Involving the Thymus

AIRE is known to act in a dose-dependent manner (Liston et al. 2004); hence, genetic variations which directly impair AIRE activity could cause disturbances in immunological homeostasis and tolerance. Such genetic vulnerability interacting with environmental factors might then lead to autoimmune disease.

Thymomas are frequently seen in patients with MG and interestingly, expression of AIRE in thymomas is impaired in about 95% of the cases. This may contribute in the development of thymoma-related autoimmune diseases (Liu et al. 2014). Intriguingly, there are both clinical and immunological overlaps between MG and APS-1. Chronic candidiasis and other APS-1 associated manifestations are sometimes seen in MG patients with thymomas and late-onset MG disease (Liu et al. 2014). There are also similarities in autoantibody profiles, notably the presence of high-titer IFN-α and -ω autoantibodies, but also organ-specific autoantibodies typical of APS-1 such as autoantibodies against TPH and AADC (Wolff et al. 2014).

Patients with hypomorphic recombination-activating genes (*RAG*) mutations have a range of phenotypes such as Omenn syndrome, leaky severe combined immunodeficiency and combined immunodeficiency with granulomatous disease and autoimmunity (CID-G/AI). Examining thymi from Omenn patients a reduced expression of AIRE has been revealed (Cavadini et al. 2005; Somech et al. 2009), and IFN-ω autoantibodies have recently been found in both Omenn syndrome and CID-G/AI patients (Walter et al. 2015).

Thymic impairments are also found in DiGeorge syndrome (22q11.2DS) (McDonald-McGinn et al. 2015). These individuals have autoimmune characteristics, but whether this is caused by impaired AIRE expression or the altered thymus structure is not known. Individuals with Down syndrome caused by trisomy 21 also have an altered thymic microarchitecture and cell composition (Skogberg et al. 2014), and as *AIRE* is located on chromosome 21 they have three copies of the gene. The syndrome is associated with immunological dysfunction and autoimmune phenotypes are common. Interestingly, one study reported a twofold reduction in the expression of *AIRE* and tissue-restricted antigens despite having three copies of *AIRE* (Gimenez-Barcons et al. 2014). Autoantibodies hitherto regarded as unique for APS-1, like anti-21OH, anti-AADC, GAD 64 and IA2 are found among individuals with Down syndrome (Soderbergh et al. 2006). In addition, novel reports have documented IFN-autoantibodies also in patients with *FOXP3*-mutations and Tregs disturbances; the disease Immunodysregulation polyendocrinopathy enteropathy X-linked (IPEX) (Rosenberg et al. 2018). Taken together, it seems likely that various thymic defects with dysregulation of AIRE can contribute to the development of autoimmune diseases.

8.7   Expression Levels of *AIRE* as a Cause of Autoimmune Disease

From research done in mice, it is evident that there is a dosage-effect of *AIRE*, and where the complete loss of function results in APS-1, quantitative decreases in *AIRE* also predispose to autoimmune disease. This is seen with mutations that quantitatively decrease AIRE expression or function (Liston et al. 2004; Su et al. 2008), which lead to autoimmunity in mice. Secondly, expression levels of AIRE are variable through the population and correlates with the expression levels of peripheral tissue antigens (Giraud et al. 2007). Also, a quantitative decrease in the AIRE-mediated expression of peripheral antigens in the thymus might predict the development of autoimmunity (Pugliese et al. 1997; Chentoufi and Polychronakos 2002). And thirdly, polymorphisms in AIRE found in humans that decrease AIRE expression are associated with an increased development of autoimmune disease (Terao et al. 2011). This also ties back to the reduced levels of AIRE and the increased presence of autoimmune disease that we discussed in individuals with Down syndrome. In many reported cases the heterozygous family members of APS-1 patients have been investigated for traces of autoimmune disease, but so far little has been reported. However, many of these studies were biased towards looking for manifestations typically seen in APS-1 patients, and perhaps one ought to take a broader approach, as exemplified by the families with dominant *AIRE* mutations and a nonclassical phenotype.

Another intriguing aspect of the gene dosage response of AIRE is the gender inequality of autoimmune diseases. Amongst most autoimmune diseases (but not in classical APS-1) there is a gender bias where women are more affected the men.

This is seen in diseases like multiple sclerosis where the ratio between affected women to men is found to be 3:1 (Orton et al. 2006), and in Sjögrens syndrome where the ratio of women to men is as high as 9:1 (Eaton et al. 2007). While the underlying mechanisms of the gender bias is still to be determined, factors like hormones, pregnancy, the increased immune reactivity seen in women, and genetic as well as epigenetics factors have all been implicated (Ngo et al. 2014).

Intriguingly, expression levels of AIRE have been shown to vary with gender, possibly due to the expression of sex hormones. Estrogen has been shown to down regulate AIRE-mediated central tolerance in the thymus by DNA methylation. The DNA methylation by estrogen leads to silencing of the *AIRE* gene and as a result, a lower expression of tissue restricted antigens is found in the thymus, and also a decreased development of regulatory T cells. It has been suggested that androgen levels might protect males from developing autoimmunity, but the underlying mechanisms are so far unknown. Interestingly, androgen leads to enhanced levels of *AIRE*, and both in mice and humans levels of AIRE is found to be elevated in males compared to females. Hence, both administration of androgens and male gender protects against multiple sclerosis in a mice model system (Zhu et al. 2016).

Even though the dosage effect of AIRE has been suggested for quite some time now, this is still a field in development. With new and more sensitive methods of detecting both sequence variants and the effect of the variants in functional studies, this is an exciting field to follow in the future.

8.8   Concluding Remarks

Only a few autoimmune diseases follow a monogenic inheritance pattern and the study of these diseases has revealed valuable information about their molecular pathogenesis and about affected and normal physiology. As we have seen in this chapter, the characterisation of the different mutations found in patients with APS-1 has taught us not only about the AIRE proteins' functional domains, but also about negative selection mechanisms in the thymus. Of note, thymus is a challenging tissue for immunologists to work with in human patients. Not only is it protected behind the sternum, it will also undergo thymic involution with age. The development of stem cell technologies and the prospect of generating thymic cells or even tissue from patients with APS-1 is an exciting prospect, but so far there have been no reports of such efforts.

A further understanding of the genetics and the molecular mechanisms behind autoimmunity will hopefully lead to new therapeutic strategies. The discovery of non-classical APS-1 where AIRE has a dominant negative effect causing milder autoimmune phenotypes give AIRE a broader role in the pathogenesis of organ-specific autoimmune diseases. Whether AIRE is involved in more common autoimmune diseases is yet to be determined, but with the emerging ease and lower cost of whole genome sequencing, this is a prospect that should be determined in a near future.

References

Aaltonen J, Bjorses P, Sandkuijl L, Perheentupa J, Peltonen L (1994) An autosomal locus causing autoimmune disease: autoimmune polyglandular disease type I assigned to chromosome 21. Nat Genet 8(1):83–87

Abbott JK, Huoh YS, Reynolds PR, Yu LP, Rewers M, Reddy M, Anderson MS, Hur S, Gelfand EW (2018) Dominant-negative loss of function arises from a second, more frequent variant within the SAND domain of autoimmune regulator (AIRE). J Autoimmun 88:114–120

Addison T (1855) On the constitutional and local effects of disease of the suprarenal capsules. Samuel Highley, London

Ahlgren KM, Moretti S, Lundgren BA, Karlsson I, Ahlin E, Norling A, Hallgren A, Perheentupa J, Gustafsson J, Rorsman F, Crewther PE, Ronnelid J, Bensing S, Scott HS, Kampe O, Romani L, Lobell A (2011) Increased IL-17A secretion in response to Candida albicans in autoimmune polyendocrine syndrome type 1 and its animal model. Eur J Immunol 41(1):235–245

Ahonen P, Myllarniemi S, Sipila I, Perheentupa J (1990) Clinical variation of autoimmune polyendocrinopathy-candidiasis-ectodermal dystrophy (APECED) in a series of 68 patients. N Engl J Med 322(26):1829–1836

Alimohammadi M, Bjorklund P, Hallgren A, Pontynen N, Szinnai G, Shikama N, Keller MP, Ekwall O, Kinkel SA, Husebye ES, Gustafsson J, Rorsman F, Peltonen L, Betterle C, Perheentupa J, Akerstrom G, Westin G, Scott HS, Hollander GA, Kampe O (2008) Autoimmune polyendocrine syndrome type 1 and NALP5, a parathyroid autoantigen. N Engl J Med 358(10):1018–1028

Alimohammadi M, Dubois N, Skoldberg F, Hallgren A, Tardivel I, Hedstrand H, Haavik J, Husebye ES, Gustafsson J, Rorsman F, Meloni A, Janson C, Vialettes B, Kajosaari M, Egner W, Sargur R, Ponten F, Amoura Z, Grimfeld A, De Luca F, Betterle C, Perheentupa J, Kampe O, Carel JC (2009) Pulmonary autoimmunity as a feature of autoimmune polyendocrine syndrome type 1 and identification of KCNRG as a bronchial autoantigen. Proc Natl Acad Sci U S A 106(11):4396–4401

Baekkeskov S, Aanstoot HJ, Christgau S, Reetz A, Solimena M, Cascalho M, Folli F, Richter-Olesen H, De Camilli P (1990) Identification of the 64K autoantigen in insulin-dependent diabetes as the GABA-synthesizing enzyme glutamic acid decarboxylase. Nature 347(6289):151–156

Bellacchio E, Palma A, Corrente S, Di Girolamo F, Helen Kemp E, Di Matteo G, Comelli L, Carsetti R, Cascioli S, Cancrini C, Fierabracci A (2014) The possible implication of the S250C variant of the autoimmune regulator protein in a patient with autoimmunity and immunodeficiency: in silico analysis suggests a molecular pathogenic mechanism for the variant. Gene 549(2):286–294

Berczi B, Gerencser G, Farkas N, Hegyi P, Veres G, Bajor J, Czopf L, Alizadeh H, Rakonczay Z, Vigh E, Eross B, Szemes K, Gyongyi Z (2017) Association between AIRE gene polymorphism and rheumatoid arthritis: a systematic review and meta-analysis of case-control studies. Sci Rep 7(1):14096

Betterle C, Greggio NA, Volpato M (1998) Clinical review 93: Autoimmune polyglandular syndrome type 1. J Clin Endocrinol Metab 83(4):1049–1055

Betterle C, Volpato M, Pedini B, Chen S, Smith BR, Furmaniak J (1999) Adrenal-cortex autoantibodies and steroid-producing cells autoantibodies in patients with Addison's disease: comparison of immunofluorescence and immunoprecipitation assays. J Clin Endocrinol Metab 84(2):618–622

Betterle C, Dal Pra C, Mantero F, Zanchetta R (2002) Autoimmune adrenal insufficiency and autoimmune polyendocrine syndromes: autoantibodies, autoantigens, and their applicability in diagnosis and disease prediction. Endocr Rev 23(3):327–364

Bjorses P, Halonen M, Palvimo JJ, Kolmer M, Aaltonen J, Ellonen P, Perheentupa J, Ulmanen I, Peltonen L (2000) Mutations in the AIRE gene: effects on subcellular location and transactivation function of the autoimmune polyendocrinopathy-candidiasis-ectodermal dystrophy protein. Am J Hum Genet 66(2):378–392

Boe Wolff AS, Oftedal B, Johansson S, Bruland O, Lovas K, Meager A, Pedersen C, Husebye ES, Knappskog PM (2008) AIRE variations in Addison's disease and autoimmune polyendocrine syndromes (APS): partial gene deletions contribute to APS I. Genes Immun 9(2):130–136

Bruserud O, Oftedal BE, Landegren N, Erichsen M, Bratland E, Lima K, Jorgensen AP, Myhre AG, Svartberg J, Fougner KJ, Bakke A, Nedrebo BG, Mella B, Breivik L, Viken MK, Knappskog PM, Marthinussen MC, Lovas K, Kampe O, Wolff AB, Husebye ES (2016) A longitudinal follow-up of Autoimmune polyendocrine syndrome type 1. J Clin Endocrinol Metab 101(8):2975–2983

Burnet FM, Mackay IR (1962) Lymphoepithelial structures and autoimmune disease. Lancet 2(7264):1030–1033

Cavadini P, Vermi W, Facchetti F, Fontana S, Nagafuchi S, Mazzolari E, Sediva A, Marrella V, Villa A, Fischer A, Notarangelo LD, Badolato R (2005) AIRE deficiency in thymus of 2 patients with Omenn syndrome. J Clin Invest 115(3):728–732

Cetani F, Barbesino G, Borsari S, Pardi E, Cianferotti L, Pinchera A, Marcocci C (2001) A novel mutation of the autoimmune regulator gene in an Italian kindred with autoimmune polyendocrinopathy-candidiasis-ectodermal dystrophy, acting in a dominant fashion and strongly cosegregating with hypothyroid autoimmune thyroiditis. J Clin Endocrinol Metab 86(10):4747–4752

Chen S, Sawicka J, Betterle C, Powell M, Prentice L, Volpato M, Rees Smith B, Furmaniak J (1996) Autoantibodies to steroidogenic enzymes in autoimmune polyglandular syndrome, Addison's disease, and premature ovarian failure. J Clin Endocrinol Metab 81(5):1871–1876

Chentoufi AA, Polychronakos C (2002) Insulin expression levels in the thymus modulate insulin-specific autoreactive T-cell tolerance: the mechanism by which the IDDM2 locus may predispose to diabetes. Diabetes 51(5):1383–1390

Cihakova D, Trebusak K, Heino M, Fadeyev V, Tiulpakov A, Battelino T, Tar A, Halasz Z, Blumel P, Tawfik S, Krohn K, Lebl J, Peterson P (2001) Novel AIRE mutations and P450 cytochrome autoantibodies in Central and Eastern European patients with APECED. Hum Mutat 18(3):225–232

Clemente MG, Obermayer-Straub P, Meloni A, Strassburg CP, Arangino V, Tukey RH, De Virgiliis S, Manns MP (1997) Cytochrome P450 1A2 is a hepatic autoantigen in autoimmune polyglandular syndrome type 1. J Clin Endocrinol Metab 82(5):1353–1361

Colobran R, Gimenez-Barcons M, Marin-Sanchez A, Porta-Pardo E, Pujol-Borrell R (2016) AIRE genetic variants and predisposition to polygenic autoimmune disease: the case of Graves' disease and a systematic literature review. Hum Immunol 77(8):643–651

Conteduca G, Ferrera F, Pastorino L, Fenoglio D, Negrini S, Sormani MP, Indiveri F, Scarra GB, Filaci G (2010) The role of AIRE polymorphisms in melanoma. Clin Immunol 136(1):96–104

Dal Pra C, Chen S, Betterle C, Zanchetta R, McGrath V, Furmaniak J, Rees Smith B (2004) Autoantibodies to human tryptophan hydroxylase and aromatic L-amino acid decarboxylase. Eur J Endocrinol 150(3):313–321

Eaton WW, Rose NR, Kalaydjian A, Pedersen MG, Mortensen PB (2007) Epidemiology of autoimmune diseases in Denmark. J Autoimmun 29(1):1–9

Eisenbarth GS, Gottlieb PA (2004) Autoimmune polyendocrine syndromes. N Engl J Med 350(20):2068–2079

Ekwall O, Hedstrand H, Grimelius L, Haavik J, Perheentupa J, Gustafsson J, Husebye E, Kampe O, Rorsman F (1998) Identification of tryptophan hydroxylase as an intestinal autoantigen. Lancet 352(9124):279–283

Ekwall O, Sjoberg K, Mirakian R, Rorsman F, Kampe O (1999) Tryptophan hydroxylase autoantibodies and intestinal disease in autoimmune polyendocrine syndrome type 1. Lancet 354(9178):568

Erichsen MM, Lovas K, Fougner KJ, Svartberg J, Hauge ER, Bollerslev J, Berg JP, Mella B, Husebye ES (2009) Normal overall mortality rate in Addison's disease, but young patients are at risk of premature death. Eur J Endocrinol 160(2):233–237

Eriksson D, Bianchi M, Landegren N, Dalin F, Skov J, Hultin-Rosenberg L, Mathioudaki A, Nordin J, Hallgren A, Andersson G, Tandre K, Rantapaa Dahlqvist S, Soderkvist P, Ronnblom L, Hulting AL, Wahlberg J, Dahlqvist P, Ekwall O, Meadows JRS, Lindblad-Toh K, Bensing S, Rosengren Pielberg G, Kampe O (2018a) Common genetic variation in the autoimmune regulator (AIRE) locus is associated with autoimmune Addison's disease in Sweden. Sci Rep 8(1):8395

Eriksson D, Dalin F, Eriksson GN, Landegren N, Bianchi M, Hallgren A, Dahlqvist P, Wahlberg J, Ekwall O, Winqvist O, Catrina SB, Ronnelid J, Swedish Addison Registry Study Group, Hulting AL, Lindblad-Toh K, Alimohammadi M, Husebye ES, Knappskog PM, Rosengren Pielberg G, Bensing S, Kampe O (2018b) Cytokine autoantibody screening in the Swedish Addison Registry identifies patients with undiagnosed APS1. J Clin Endocrinol Metab 103(1):179–186

Falorni A, Laureti S, Nikoshkov A, Picchio ML, Hallengren B, Vandewalle CL, Gorus FK, Tortoioli C, Luthman H, Brunetti P, Santeusanio F (1997) 21-hydroxylase autoantibodies in adult patients with endocrine autoimmune diseases are highly specific for Addison's disease. Belgian Diabetes Registry. Clin Exp Immunol 107(2):341–346

Ferguson BJ, Alexander C, Rossi SW, Liiv I, Rebane A, Worth CL, Wong J, Laan M, Peterson P, Jenkinson EJ, Anderson G, Scott HS, Cooke A, Rich T (2008) AIRE's CARD revealed, a new structure for central tolerance provokes transcriptional plasticity. J Biol Chem 283(3):1723–1731

Finnish-German APECED Consortium (1997) An autoimmune disease, APECED, caused by mutations in a novel gene featuring two PHD-type zinc-finger domains. Nat Genet 17(4):399–403

Gass JD (1962) The syndrome of keratoconjunctivitis, superficial moniliasis, idiopathic hypoparathyroidism and Addison's disease. Am J Ophthalmol 54:660–674

Gavalas NG, Kemp EH, Krohn KJ, Brown EM, Watson PF, Weetman AP (2007) The calcium-sensing receptor is a target of autoantibodies in patients with autoimmune polyendocrine syndrome type 1. J Clin Endocrinol Metab 92(6):2107–2114

Gibson TJ, Ramu C, Gemund C, Aasland R (1998) The APECED polyglandular autoimmune syndrome protein, AIRE-1, contains the SAND domain and is probably a transcription factor. Trends Biochem Sci 23(7):242–244

Gimenez-Barcons M, Casteras A, Armengol Mdel P, Porta E, Correa PA, Marin A, Pujol-Borrell R, Colobran R (2014) Autoimmune predisposition in Down syndrome may result from a partial central tolerance failure due to insufficient intrathymic expression of AIRE and peripheral antigens. J Immunol 193(8):3872–3879

Giraud M, Taubert R, Vandiedonck C, Ke X, Levi-Strauss M, Pagani F, Baralle FE, Eymard B, Tranchant C, Gajdos P, Vincent A, Willcox N, Beeson D, Kyewski B, Garchon HJ (2007) An IRF8-binding promoter variant and AIRE control CHRNA1 promiscuous expression in thymus. Nature 448(7156):934–937

Gylling M, Kaariainen E, Vaisanen R, Kerosuo L, Solin ML, Halme L, Saari S, Halonen M, Kampe O, Perheentupa J, Miettinen A (2003) The hypoparathyroidism of autoimmune polyendocrinopathy-candidiasis-ectodermal dystrophy protective effect of male sex. J Clin Endocrinol Metab 88(10):4602–4608

Harrington LE, Hatton RD, Mangan PR, Turner H, Murphy TL, Murphy KM, Weaver CT (2005) Interleukin 17-producing CD4+ effector T cells develop via a lineage distinct from the T helper type 1 and 2 lineages. Nat Immunol 6(11):1123–1132

Hedstrand H, Perheentupa J, Ekwall O, Gustafsson J, Michaelsson G, Husebye E, Rorsman F, Kampe O (1999) Antibodies against hair follicles are associated with alopecia totalis in autoimmune polyendocrine syndrome type I. J Invest Dermatol 113(6):1054–1058

Heino M, Scott HS, Chen Q, Peterson P, Maebpaa U, Papasavvas MP, Mittaz L, Barras C, Rossier C, Chrousos GP, Stratakis CA, Nagamine K, Kudoh J, Shimizu N, Maclaren N, Antonarakis SE, Krohn K (1999) Mutation analyses of North American APS-1 patients. Hum Mutat 13(1):69–74

Hertzog PJ, O'Neill LA, Hamilton JA (2003) The interferon in TLR signaling: more than just antiviral. Trends Immunol 24(10):534–539

Husebye ES, Gebre-Medhin G, Tuomi T, Perheentupa J, Landin-Olsson M, Gustafsson J, Rorsman F, Kampe O (1997) Autoantibodies against aromatic L-amino acid decarboxylase in autoimmune polyendocrine syndrome type I. J Clin Endocrinol Metab 82(1):147–150

Husebye ES, Perheentupa J, Rautemaa R, Kampe O (2009) Clinical manifestations and management of patients with autoimmune polyendocrine syndrome type I. J Intern Med 265(5):514–529

Husebye ES, Anderson MS, Kampe O (2018) Autoimmune polyendocrine syndromes. N Engl J Med 378(12):1132–1141

Ilmarinen T, Eskelin P, Halonen M, Ruppell T, Kilpikari R, Torres GD, Kangas H, Ulmanen I (2005) Functional analysis of SAND mutations in AIRE supports dominant inheritance of the G228W mutation. Hum Mutat 26(4):322–331

Jin Y, Bennett DC, Amadi-Myers A, Holland P, Riccardi SL, Gowan K, Fain PR, Spritz RA (2007) Vitiligo-associated multiple autoimmune disease is not associated with genetic variation in AIRE. Pigment Cell Res 20(5):402–404

Kemp EH, Gavalas NG, Krohn KJ, Brown EM, Watson PF, Weetman AP (2009) Activating autoantibodies against the calcium-sensing receptor detected in two patients with autoimmune polyendocrine syndrome type 1. J Clin Endocrinol Metab 94(12):4749–4756

Kemp EH, Emhemad S, Akhtar S, Watson PF, Gawkrodger DJ, Weetman AP (2011a) Autoantibodies against tyrosine hydroxylase in patients with non-segmental (generalised) vitiligo. Exp Dermatol 20(1):35–40

Kemp EH, Sandhu HK, Weetman AP, McDonagh AJ (2011b) Demonstration of autoantibodies against tyrosine hydroxylase in patients with alopecia areata. Br J Dermatol 165(6):1236–1243

Kisand K, Link M, Wolff AS, Meager A, Tserel L, Org T, Murumagi A, Uibo R, Willcox N, Trebusak Podkrajsek K, Battelino T, Lobell A, Kampe O, Lima K, Meloni A, Ergun-Longmire B, Maclaren NK, Perheentupa J, Krohn KJ, Scott HS, Husebye ES, Peterson P (2008) Interferon autoantibodies associated with AIRE deficiency decrease the expression of IFN-stimulated genes. Blood 112(7):2657–2666

Kisand K, Boe Wolff AS, Podkrajsek KT, Tserel L, Link M, Kisand KV, Ersvaer E, Perheentupa J, Erichsen MM, Bratanic N, Meloni A, Cetani F, Perniola R, Ergun-Longmire B, Maclaren N, Krohn KJ, Pura M, Schalke B, Strobel P, Leite MI, Battelino T, Husebye ES, Peterson P, Willcox N, Meager A (2010) Chronic mucocutaneous candidiasis in APECED or thymoma patients correlates with autoimmunity to Th17-associated cytokines. J Exp Med 207(2):299–308

Kluger N, Jokinen M, Lintulahti A, Krohn K, Ranki A (2015) Gastrointestinal immunity against tryptophan hydroxylase-1, aromatic L-amino-acid decarboxylase, AIE-75, villin and Paneth cells in APECED. Clin Immunol 158(2):212–220

Koh AS, Kuo AJ, Park SY, Cheung P, Abramson J, Bua D, Carney D, Shoelson SE, Gozani O, Kingston RE, Benoist C, Mathis D (2008) Aire employs a histone-binding module to mediate immunological tolerance, linking chromatin regulation with organ-specific autoimmunity. Proc Natl Acad Sci U S A 105(41):15878–15883

Krohn K, Uibo R, Aavik E, Peterson P, Savilahti K (1992) Identification by molecular cloning of an autoantigen associated with Addison's disease as steroid 17 alpha-hydroxylase. Lancet 339(8796):770–773

LaFlam TN, Seumois G, Miller CN, Lwin W, Fasano KJ, Waterfield M, Proekt I, Vijayanand P, Anderson MS (2015) Identification of a novel cis-regulatory element essential for immune tolerance. J Exp Med 212(12):1993–2002

Landegren N, Sharon D, Shum AK, Khan IS, Fasano KJ, Hallgren A, Kampf C, Freyhult E, Ardesjo-Lundgren B, Alimohammadi M, Rathsman S, Ludvigsson JF, Lundh D, Motrich R, Rivero V, Fong L, Giwercman A, Gustafsson J, Perheentupa J, Husebye ES, Anderson MS, Snyder M, Kampe O (2015) Transglutaminase 4 as a prostate autoantigen in male subfertility. Sci Transl Med 7(292):292ra101

Landegren N, Sharon D, Freyhult E, Hallgren A, Eriksson D, Edqvist PH, Bensing S, Wahlberg J, Nelson LM, Gustafsson J, Husebye ES, Anderson MS, Snyder M, Kampe O (2016) Proteome-wide survey of the autoimmune target repertoire in autoimmune polyendocrine syndrome type 1. Sci Rep 6:20104

Lei Y, Ripen AM, Ishimaru N, Ohigashi I, Nagasawa T, Jeker LT, Bosl MR, Hollander GA, Hayashi Y, Malefyt Rde W, Nitta T, Takahama Y (2011) Aire-dependent production of XCL1 mediates medullary accumulation of thymic dendritic cells and contributes to regulatory T cell development. J Exp Med 208(2):383–394

Li Y, Song YH, Rais N, Connor E, Schatz D, Muir A, Maclaren N (1996) Autoantibodies to the extracellular domain of the calcium sensing receptor in patients with acquired hypoparathyroidism. J Clin Invest 97(4):910–914

Lindh E, Brannstrom J, Jones P, Wermeling F, Hassler S, Betterle C, Garty BZ, Stridsberg M, Herrmann B, Karlsson MC, Winqvist O (2013) Autoimmunity and cystatin SA1 deficiency behind chronic mucocutaneous candidiasis in autoimmune polyendocrine syndrome type 1. J Autoimmun 42:1–6

Liston A, Gray DH, Lesage S, Fletcher AL, Wilson J, Webster KE, Scott HS, Boyd RL, Peltonen L, Goodnow CC (2004) Gene dosage--limiting role of Aire in thymic expression, clonal deletion, and organ-specific autoimmunity. J Exp Med 200(8):1015–1026

Liu Y, Zhang H, Zhang P, Meng F, Chen Y, Wang Y, Yao Y, Qi B (2014) Autoimmune regulator expression in thymomas with or without autoimmune disease. Immunol Lett 161(1):50–56

Lovas K, Husebye ES (2005) Addison's disease. Lancet 365(9476):2058–2061

Lovewell TR, McDonagh AJ, Messenger AG, Azzouz M, Tazi-Ahnini R (2015) The AIRE-230Y polymorphism affects AIRE transcriptional activity: potential influence on AIRE function in the thymus. PLoS One 10(5):e0127476

Malchow S, Leventhal DS, Nishi S, Fischer BI, Shen L, Paner GP, Amit AS, Kang C, Geddes JE, Allison JP, Socci ND, Savage PA (2013) Aire-dependent thymic development of tumor-associated regulatory T cells. Science 339(6124):1219–1224

McDonald-McGinn DM, Sullivan KE, Marino B, Philip N, Swillen A, Vorstman JA, Zackai EH, Emanuel BS, Vermeesch JR, Morrow BE, Scambler PJ, Bassett AS (2015) 22q11.2 deletion syndrome. Nat Rev Dis Primers 1:15071

Meager A, Visvalingam K, Peterson P, Moll K, Murumagi A, Krohn K, Eskelin P, Perheentupa J, Husebye E, Kadota Y, Willcox N (2006) Anti-interferon autoantibodies in autoimmune polyendocrinopathy syndrome type 1. PLoS Med 3(7):e289

Meloni A, Furcas M, Cetani F, Marcocci C, Falorni A, Perniola R, Pura M, Boe Wolff AS, Husebye ES, Lilic D, Ryan KR, Gennery AR, Cant AJ, Abinun M, Spickett GP, Arkwright PD, Denning D, Costigan C, Dominguez M, McConnell V, Willcox N, Meager A (2008) Autoantibodies against type I interferons as an additional diagnostic criterion for autoimmune polyendocrine syndrome type I. J Clin Endocrinol Metab 93(11):4389–4397

Meloni A, Willcox N, Meager A, Atzeni M, Wolff AS, Husebye ES, Furcas M, Rosatelli MC, Cao A, Congia M (2012) Autoimmune polyendocrine syndrome type 1: an extensive longitudinal study in Sardinian patients. J Clin Endocrinol Metab 97(4):1114–1124

Meyer S, Woodward M, Hertel C, Vlaicu P, Haque Y, Karner J, Macagno A, Onuoha SC, Fishman D, Peterson H, Metskula K, Uibo R, Jantti K, Hokynar K, Wolff ASB, APECED Patient Collaborative, Krohn K, Ranki A, Peterson P, Kisand K, Hayday A (2016) AIRE-deficient patients harbor unique high-affinity disease-ameliorating autoantibodies. Cell 166(3):582–595

Mora M, Hanzu FA, Pradas-Juni M, Aranda GB, Halperin I, Puig-Domingo M, Aguilo S, Fernandez-Rebollo E (2014) New splice site acceptor mutation in AIRE gene in autoimmune polyendocrine syndrome type 1. PLoS One 9(7):e101616

Myhre AG, Halonen M, Eskelin P, Ekwall O, Hedstrand H, Rorsman F, Kampe O, Husebye ES (2001) Autoimmune polyendocrine syndrome type 1 (APS I) in Norway. Clin Endocrinol 54(2):211–217

Nagamine K, Peterson P, Scott HS, Kudoh J, Minoshima S, Heino M, Krohn KJ, Lalioti MD, Mullis PE, Antonarakis SE, Kawasaki K, Asakawa S, Ito F, Shimizu N (1997) Positional cloning of the APECED gene. Nat Genet 17(4):393–398

Ngo ST, Steyn FJ, McCombe PA (2014) Gender differences in autoimmune disease. Front Neuroendocrinol 35(3):347–369

Nithiyananthan R, Heward JM, Allahabadia A, Barnett AH, Franklyn JA, Gough SC (2000) A heterozygous deletion of the autoimmune regulator (AIRE1) gene, autoimmune thyroid disease, and type 1 diabetes: no evidence for association. J Clin Endocrinol Metab 85(3):1320–1322

Oftedal BE, Hellesen A, Erichsen MM, Bratland E, Vardi A, Perheentupa J, Kemp EH, Fiskerstrand T, Viken MK, Weetman AP, Fleishman SJ, Banka S, Newman WG, Sewell WA, Sozaeva LS, Zayats T, Haugarvoll K, Orlova EM, Haavik J, Johansson S, Knappskog PM, Lovas K, Wolff AS, Abramson J, Husebye ES (2015) Dominant mutations in the autoimmune regulator AIRE are associated with common organ-specific autoimmune diseases. Immunity 42(6):1185–1196

Oftedal BE, Marthinussen MC, Erichsen MM, Tveitaras MK, Kjellesvik-Kristiansen A, Hammenfors D, Jonsson MV, Kisand K, Jonsson R, Wolff ASB (2017) Impaired salivary gland activity in patients with autoimmune polyendocrine syndrome type I. Autoimmunity 50(4):211–222

Orton SM, Herrera BM, Yee IM, Valdar W, Ramagopalan SV, Sadovnick AD, Ebers GC, Canadian Collaborative Study Group (2006) Sex ratio of multiple sclerosis in Canada: a longitudinal study. Lancet Neurol 5(11):932–936

Pearce SH, Cheetham T, Imrie H, Vaidya B, Barnes ND, Bilous RW, Carr D, Meeran K, Shaw NJ, Smith CS, Toft AD, Williams G, Kendall-Taylor P (1998) A common and recurrent 13-bp deletion in the autoimmune regulator gene in British kindreds with autoimmune polyendocrinopathy type 1. Am J Hum Genet 63(6):1675–1684

Perheentupa J (2006) Autoimmune polyendocrinopathy-candidiasis-ectodermal dystrophy. J Clin Endocrinol Metab 91(8):2843–2850

Perniola R, Filograna O, Greco G, Pellegrino V (2008) High prevalence of thyroid autoimmunity in Apulian patients with autoimmune polyglandular syndrome type 1. Thyroid 18(9):1027–1029

Pestka S, Krause CD, Walter MR (2004) Interferons, interferon-like cytokines, and their receptors. Immunol Rev 202:8–32

Peterson P, Uibo R, Peranen J, Krohn K (1997) Immunoprecipitation of steroidogenic enzyme autoantigens with autoimmune polyglandular syndrome type I (APS I) sera; further evidence for independent humoral immunity to P450c17 and P450c21. Clin Exp Immunol 107(2):335–340

Pforr J, Blaumeiser B, Becker T, Freudenberg-Hua Y, Hanneken S, Eigelshoven S, Cuyt I, De Weert J, Lambert J, Kruse R, Nothen MM, Betz RC (2006) Investigation of the p.Ser278Arg polymorphism of the autoimmune regulator (AIRE) gene in alopecia areata. Tissue Antigens 68(1):58–61

Pitkanen J, Doucas V, Sternsdorf T, Nakajima T, Aratani S, Jensen K, Will H, Vahamurto P, Ollila J, Vihinen M, Scott HS, Antonarakis SE, Kudoh J, Shimizu N, Krohn K, Peterson P (2000) The autoimmune regulator protein has transcriptional transactivating properties and interacts with the common coactivator CREB-binding protein. J Biol Chem 275(22):16802–16809

Podkrajsek KT, Bratanic N, Krzisnik C, Battelino T (2005) Autoimmune regulator-1 messenger ribonucleic acid analysis in a novel intronic mutation and two additional novel AIRE gene mutations in a cohort of autoimmune polyendocrinopathy-candidiasis-ectodermal dystrophy patients. J Clin Endocrinol Metab 90(8):4930–4935

Podkrajsek KT, Milenkovic T, Odink RJ, Claasen-van der Grinten HL, Bratanic N, Hovnik T, Battelino T (2008) Detection of a complete autoimmune regulator gene deletion and two additional novel mutations in a cohort of patients with atypical phenotypic variants of autoimmune polyglandular syndrome type 1. Eur J Endocrinol 159(5):633–639

Puel A, Doffinger R, Natividad A, Chrabieh M, Barcenas-Morales G, Picard C, Cobat A, Ouachee-Chardin M, Toulon A, Bustamante J, Al-Muhsen S, Al-Owain M, Arkwright PD, Costigan C, McConnell V, Cant AJ, Abinun M, Polak M, Bougneres PF, Kumararatne D, Marodi L, Nahum A, Roifman C, Blanche S, Fischer A, Bodemer C, Abel L, Lilic D, Casanova JL (2010) Autoantibodies against IL-17A, IL-17F, and IL-22 in patients with chronic mucocutaneous candidiasis and autoimmune polyendocrine syndrome type I. J Exp Med 207(2):291–297

Pugliese A, Zeller M, Fernandez A Jr, Zalcberg LJ, Bartlett RJ, Ricordi C, Pietropaolo M, Eisenbarth GS, Bennett ST, Patel DD (1997) The insulin gene is transcribed in the human thymus and transcription levels correlated with allelic variation at the INS VNTR-IDDM2 susceptibility locus for type 1 diabetes. Nat Genet 15(3):293–297

Ramsey C, Bukrinsky A, Peltonen L (2002) Systematic mutagenesis of the functional domains of AIRE reveals their role in intracellular targeting. Hum Mol Genet 11(26):3299–3308

Rautemaa R, Hietanen J, Niissalo S, Pirinen S, Perheentupa J (2007) Oral and oesophageal squamous cell carcinoma--a complication or component of autoimmune polyendocrinopathy-candidiasis-ectodermal dystrophy (APECED, APS-I). Oral Oncol 43(6):607–613

Rosenberg JM, Maccari ME, Barzaghi F, Allenspach EJ, Pignata C, Weber G, Torgerson TR, Utz PJ, Bacchetta R (2018) Neutralizing anti-cytokine autoantibodies against interferon-alpha in immunodysregulation polyendocrinopathy enteropathy X-linked. Front Immunol 9:544

Saugier-Veber P, Drouot N, Wolf LM, Kuhn JM, Frebourg T, Lefebvre H (2001) Identification of a novel mutation in the autoimmune regulator (AIRE-1) gene in a French family with autoimmune polyendocrinopathy-candidiasis-ectodermal dystrophy. Eur J Endocrinol 144(4):347–351

Shum AK, Alimohammadi M, Tan CL, Cheng MH, Metzger TC, Law CS, Lwin W, Perheentupa J, Bour-Jordan H, Carel JC, Husebye ES, De Luca F, Janson C, Sargur R, Dubois N, Kajosaari M, Wolters PJ, Chapman HA, Kampe O, Anderson MS (2013) BPIFB1 is a lung-specific autoantigen associated with interstitial lung disease. Sci Transl Med 5(206):206ra139

Skogberg G, Lundberg V, Lindgren S, Gudmundsdottir J, Sandstrom K, Kampe O, Anneren G, Gustafsson J, Sunnegardh J, van der Post S, Telemo E, Berglund M, Ekwall O (2014) Altered expression of autoimmune regulator in infant down syndrome thymus, a possible contributor to an autoimmune phenotype. J Immunol 193(5):2187–2195

Smith CJ, Oscarson M, Ronnblom L, Alimohammadi M, Perheentupa J, Husebye ES, Gustafsson J, Nordmark G, Meloni A, Crock PA, Kampe O, Bensing S (2011) TSGA10 - a target for autoantibodies in autoimmune polyendocrine syndrome type 1 and systemic lupus erythematosus. Scand J Immunol 73(2):147–153

Smith-Anttila CJA, Bensing S, Alimohammadi M, Dalin F, Oscarson M, Zhang MD, Perheentupa J, Husebye ES, Gustafsson J, Bjorklund P, Fransson A, Nordmark G, Ronnblom L, Meloni A, Scott RJ, Hokfelt T, Crock PA, Kampe O (2017) Identification of endothelin-converting enzyme-2 as an autoantigen in autoimmune polyendocrine syndrome type 1. Autoimmunity 50(4):223–231

Soderbergh A, Myhre AG, Ekwall O, Gebre-Medhin G, Hedstrand H, Landgren E, Miettinen A, Eskelin P, Halonen M, Tuomi T, Gustafsson J, Husebye ES, Perheentupa J, Gylling M, Manns MP, Rorsman F, Kampe O, Nilsson T (2004) Prevalence and clinical associations of 10 defined autoantibodies in autoimmune polyendocrine syndrome type I. J Clin Endocrinol Metab 89(2):557–562

Soderbergh A, Gustafsson J, Ekwall O, Hallgren A, Nilsson T, Kampe O, Rorsman F, Anneren G (2006) Autoantibodies linked to autoimmune polyendocrine syndrome type I are prevalent in Down syndrome. Acta Paediatr 95(12):1657–1660

Somech R, Simon AJ, Lev A, Dalal I, Spirer Z, Goldstein I, Nagar M, Amariglio N, Rechavi G, Roifman CM (2009) Reduced central tolerance in Omenn syndrome leads to immature self-reactive oligoclonal T cells. J Allergy Clin Immunol 124(4):793–800

Stahl EA, Raychaudhuri S, Remmers EF, Xie G, Eyre S, Thomson BP, Li Y, Kurreeman FA, Zhernakova A, Hinks A, Guiducci C, Chen R, Alfredsson L, Amos CI, Ardlie KG, Barton A, Bowes J, Brouwer E, Burtt NP, Catanese JJ, Coblyn J, Coenen MJ, Costenbader KH, Criswell LA, Crusius JB, Cui J, de Bakker PI, De Jager PL, Ding B, Emery P, Flynn E, Harrison P, Hocking LJ, Huizinga TW, Kastner DL, Ke X, Lee AT, Liu X, Martin P, Morgan AW, Padyukov L, Posthumus MD, Radstake TR, Reid DM, Seielstad M, Seldin MF, Shadick NA, Steer S, Tak PP, Thomson W, van der Helm-van Mil AH, van der Horst-Bruinsma IE, van der Schoot CE, van Riel PL, Weinblatt ME, Wilson AG, Wolbink GJ, Wordsworth BP, Wijmenga C, Karlson EW, Toes RE, de Vries N, Begovich AB, Worthington J, Siminovitch KA, Gregersen PK, Klareskog L, Plenge RM (2010) Genome-wide association study meta-analysis identifies seven new rheumatoid arthritis risk loci. Nat Genet 42(6):508–514

Stolarski B, Pronicka E, Korniszewski L, Pollak A, Kostrzewa G, Rowinska E, Wlodarski P, Skorka A, Gremida M, Krajewski P, Ploski R (2006) Molecular background of polyendocrinopathy-

ctorial?

candidiasis-ectodermal dystrophy syndrome in a Polish population: novel AIRE mutations and an estimate of disease prevalence. Clin Genet 70(4):348–354

Su MA, Giang K, Zumer K, Jiang H, Oven I, Rinn JL, Devoss JJ, Johannes KP, Lu W, Gardner J, Chang A, Bubulya P, Chang HY, Peterlin BM, Anderson MS (2008) Mechanisms of an autoimmunity syndrome in mice caused by a dominant mutation in Aire. J Clin Invest 118(5):1712–1726

Tazi-Ahnini R, Cork MJ, Gawkrodger DJ, Birch MP, Wengraf D, McDonagh AJ, Messenger AG (2002) Role of the autoimmune regulator (AIRE) gene in alopecia areata: strong association of a potentially functional AIRE polymorphism with alopecia universalis. Tissue Antigens 60(6):489–495

Tazi-Ahnini R, McDonagh AJ, Wengraf DA, Lovewell TR, Vasilopoulos Y, Messenger AG, Cork MJ, Gawkrodger DJ (2008) The autoimmune regulator gene (AIRE) is strongly associated with vitiligo. Br J Dermatol 159(3):591–596

Terao C, Yamada R, Ohmura K, Takahashi M, Kawaguchi T, Kochi Y, Human Disease Genomics Working Group, RA Clinical and Genetic Study Consortium, Okada Y, Nakamura Y, Yamamoto K, Melchers I, Lathrop M, Mimori T, Matsuda F (2011) The human AIRE gene at chromosome 21q22 is a genetic determinant for the predisposition to rheumatoid arthritis in Japanese population. Hum Mol Genet 20(13):2680–2685

Toth B, Wolff AS, Halasz Z, Tar A, Szuts P, Ilyes I, Erdos M, Szegedi G, Husebye ES, Zeher M, Marodi L (2010) Novel sequence variation of AIRE and detection of interferon-omega antibodies in early infancy. Clin Endocrinol 72(5):641–647

Uibo R, Aavik E, Peterson P, Perheentupa J, Aranko S, Pelkonen R, Krohn KJ (1994) Autoantibodies to cytochrome P450 enzymes P450scc, P450c17, and P450c21 in autoimmune polyglandular disease types I and II and in isolated Addison's disease. J Clin Endocrinol Metab 78(2):323–328

Valenzise M, Fierabracci A, Cappa M, Porcelli P, Barcellona R, De Luca F, Barollo S, Garelli S, Betterle C (2014) Autoimmune polyendocrinopathy-candidiasis-ectodermal dystrophy: report of seven additional sicilian patients and overview of the overall series from sicily. Horm Res Paediatr 82(2):127–132

Velloso LA, Winqvist O, Gustafsson J, Kampe O, Karlsson FA (1994) Autoantibodies against a novel 51 kDa islet antigen and glutamate decarboxylase isoforms in autoimmune polyendocrine syndrome type I. Diabetologia 37(1):61–69

Walter JE, Rosen LB, Csomos K, Rosenberg JM, Mathew D, Keszei M, Ujhazi B, Chen K, Lee YN, Tirosh I, Dobbs K, Al-Herz W, Cowan MJ, Puck J, Bleesing JJ, Grimley MS, Malech H, De Ravin SS, Gennery AR, Abraham RS, Joshi AY, Boyce TG, Butte MJ, Nadeau KC, Balboni I, Sullivan KE, Akhter J, Adeli M, El-Feky RA, El-Ghoneimy DH, Dbaibo G, Wakim R, Azzari C, Palma P, Cancrini C, Capuder K, Condino-Neto A, Costa-Carvalho BT, Oliveira JB, Roifman C, Buchbinder D, Kumanovics A, Franco JL, Niehues T, Schuetz C, Kuijpers T, Yee C, Chou J, Masaad MJ, Geha R, Uzel G, Gelman R, Holland SM, Recher M, Utz PJ, Browne SK, Notarangelo LD (2015) Broad-spectrum antibodies against self-antigens and cytokines in RAG deficiency. J Clin Invest 126(11):4389

Waterfield M, Khan IS, Cortez JT, Fan U, Metzger T, Greer A, Fasano K, Martinez-Llordella M, Pollack JL, Erle DJ, Su M, Anderson MS (2014) The transcriptional regulator Aire coopts the repressive ATF7ip-MBD1 complex for the induction of immunotolerance. Nat Immunol 15(3):258–265

Winqvist O, Karlsson FA, Kampe O (1992) 21-Hydroxylase, a major autoantigen in idiopathic Addison's disease. Lancet 339(8809):1559–1562

Winqvist O, Gustafsson J, Rorsman F, Karlsson FA, Kampe O (1993) Two different cytochrome P450 enzymes are the adrenal antigens in autoimmune polyendocrine syndrome type I and Addison's disease. J Clin Invest 92(5):2377–2385

Winqvist O, Gebre-Medhin G, Gustafsson J, Ritzen EM, Lundkvist O, Karlsson FA, Kampe O (1995) Identification of the main gonadal autoantigens in patients with adrenal insufficiency and associated ovarian failure. J Clin Endocrinol Metab 80(5):1717–1723

Wolff AS, Erichsen MM, Meager A, Magitta NF, Myhre AG, Bollerslev J, Fougner KJ, Lima K, Knappskog PM, Husebye ES (2007) Autoimmune polyendocrine syndrome type 1 in Norway:

phenotypic variation, autoantibodies, and novel mutations in the autoimmune regulator gene. J Clin Endocrinol Metab 92(2):595–603

Wolff AS, Karner J, Owe JF, Oftedal BE, Gilhus NE, Erichsen MM, Kampe O, Meager A, Peterson P, Kisand K, Willcox N, Husebye ES (2014) Clinical and serologic parallels to APS-I in patients with thymomas and autoantigen transcripts in their tumors. J Immunol 193(8):3880–3890

Xu YS, Jiang XJ, Chen JM (2017) A single nucleotide polymorphism of AIRE gene located in the 21q22.3 increases the risk of rheumatoid arthritis. Oncotarget 8(42):71556–71562

Yang S, Bansal K, Lopes J, Benoist C, Mathis D (2013) Aire's plant homeodomain (PHD)-2 is critical for induction of immunological tolerance. Proc Natl Acad Sci U S A 110(5):1833–1838

Zhang L, Barker JM, Babu S, Su M, Stenerson M, Cheng M, Shum A, Zamir E, Badolato R, Law A, Eisenbarth GS, Anderson MS (2007) A robust immunoassay for anti-interferon autoantibodies that is highly specific for patients with autoimmune polyglandular syndrome type 1. Clin Immunol 125(2):131–137

Zhang J, Liu H, Liu Z, Liao Y, Guo L, Wang H, He L, Zhang X, Xing Q (2013) A functional alternative splicing mutation in AIRE gene causes autoimmune polyendocrine syndrome type 1. PLoS One 8(1):e53981

Zhu ML, Bakhru P, Conley B, Nelson JS, Free M, Martin A, Starmer J, Wilson EM, Su MA (2016) Sex bias in CNS autoimmune disease mediated by androgen control of autoimmune regulator. Nat Commun 7:11350

Zlotogora J, Shapiro MS (1992) Polyglandular autoimmune syndrome type I among Iranian Jews. J Med Genet 29(11):824–826

Chapter 9
The Thymus as a Mirror of the Body's Gene Expression

Geraldo A. Passos, Adriana B. Genari, Amanda F. Assis, Ana C. Monteleone-Cassiano, Eduardo A. Donadi, Ernna H. Oliveira, Max J. Duarte, Mayara V. Machado, Pedro P. Tanaka, and Romário Mascarenhas

Abstract The thymus is a complex organ formed by different cell types that establish close interaction. The role played by the thymic stroma is very intriguing, since it is not only a connective tissue or a support structure. The stromal thymic epithelial cells (TECs), establish physical and functional interaction with developing thymocytes culminating in a unique function of this organ, the induction of central tolerance. The role played by the medullary thymic epithelial cells (mTECs) is noteworthy and is being the focus of many studies. The transcriptome of mTEC cells is also very complex. These cells express nearly the entire functional genome without altering their morphological and functional features. Among thousand mRNAs expressed, a particular set encodes together all peripheral tissue antigens (PTAs), which represent the different tissues and organs of the body. The consequence of ectopic proteins translated from these mRNAs in the thymus is immunological and is associated with self-non-self discrimination and induction of central tolerance. Due to the wide variety of PTAs, this process was termed *promiscuous gene expression* (PGE), whose control is shared between autoimmune regulator (human *AIRE*/murine *Aire*), a transcriptional modulator and forebrain-expressed zinc finger 2 (*FEZF2*/*Fezf2*), a transcription factor. Therefore, this molecular-genetic process is closely linked to the elimination of autoreactive thymocytes in the

G. A. Passos (✉)
Laboratory of Genetics and Molecular Biology, Department of Basic and Oral Biology, School of Dentistry of Ribeirão Preto, University of São Paulo, Ribeirão Preto, SP, Brazil

Molecular Immunogenetics Group, Department of Genetics, Ribeirão Preto Medical School, University of São Paulo, Ribeirão Preto, SP, Brazil
e-mail: passos@usp.br

A. B. Genari · A. F. Assis · A. C. Monteleone-Cassiano · E. H. Oliveira · M. J. Duarte
M. V. Machado · P. P. Tanaka · R. Mascarenhas
Molecular Immunogenetics Group, Department of Genetics, Ribeirão Preto Medical School, University of São Paulo, Ribeirão Preto, SP, Brazil

E. A. Donadi
Department of Clinical Medicine, Ribeirão Preto Medical School, University of São Paulo, Ribeirão Preto, SP, Brazil

© Springer Nature Switzerland AG 2019   215
G. A. Passos (ed.), *Thymus Transcriptome and Cell Biology*,
https://doi.org/10.1007/978-3-030-12040-5_9

thymus through negative selection. In this chapter, we review PGE in mTECs and its immunologic implication, the role of the *Aire* and *Fezf2* genes, the role of *Aire* on the expression of miRNAs in mTECs, its consequence on PGE and the manipulation of the *Aire* expression either by siRNA or by genome editing using the Crispr-Cas9 system.

9.1   The Role of Promiscuous Gene Expression in the Induction of Central Tolerance

The central tolerance to self-antigens occurs in the thymus through a sophisticated process (Gallegos and Bevan 2004; Mouchess and Anderson 2014; Passos et al. 2015, 2018). For this to occur, the developing thymocytes sequentially interact in a three-dimensional mesh within the thymus passing through two histologically distinct structures, the cortex and then the medulla. The double-negative (DN) CD4⁻ CD8⁻ and double-positive (DP) CD4⁺ CD8⁺ thymocytes initially interact with the cortical structure adhering to the cortical thymic epithelial cells (cTECs). This allows MHC-mediated self-peptide presentation to DP thymocytes with intermediate affinity/avidity TCR α/β^+. The result of this interaction is termed *positive selection* (PS), which triggers DP thymocytes to differentiate into mature single-positive (SP) CD4⁺ CD8⁻ or CD4⁻ CD8⁺ thymocytes (Klein et al. 2009; Groettrup et al. 2010; Passos et al. 2015; Takaba and Takayanagi 2017; Abramson and Anderson 2017; James et al. 2018). Those DP thymocyte clones that do not go through PS are neglected and die (death by neglect).

After that, the remaining SP and DP thymocytes through the corticomedullary junction migrate to the medullar compartment where they interact with medullary thymic epithelial cells (mTECs) (Halkias et al. 2013; Abramson and Anderson 2017). The mTEC-thymocyte adhesion is a crucial process that enables the exposure of the diverse population of thymocytes (in terms of their TCR α/β specificities) to a vast diversity of peripheral tissue antigens (PTAs) that are expressed and presented by mTECs (Koble and Kyewski 2009).

The mutual interactions between the thymic epithelial cells with thymocytes is referred to *thymic crosstalk* (Savino et al. 2002, 2004; Klein et al. 2009; Groettrup and Kirk 2010; Zuklys et al. 2012; Hu et al. 2015; Lopes et al. 2015, 2018) a subject of main importance during the induction of central tolerance, which is extensively discussed in the Chaps. 5 and 6 of this book.

Of note, mTEC cells represent an unusual set of cells since in a population they ectopically express virtually all self-PTAs through *promiscuous gene expression* (PGE) (Mizuochi et al. 1992; Klein and Kyewski 2000; Derbinski et al. 2001; Kyewski et al. 2002; Gotter et al. 2004; Magalhães et al. 2006; Tykocinski et al. 2008; Derbinski and Kyewski 2010; Ucar and Rattay 2015; Passos et al. 2015, 2018).

The mRNA transcriptome sequencing (RNA-Seq) demonstrated that of the nearly 24,000 functional genes in the mouse, the full population of mTECs in the thymus express approximately 19,000, which represent approximately 80% of the murine genome (St-Pierre et al. 2013, 2015; Sansom et al. 2014). The mTEC cells appear to follow a stochastic pattern of gene expression (Derbinski et al. 2008; Meredith et al. 2015), that means that a given mTEC cell expresses a set of PTAs that may differ from another mTEC cell at same maturation stage. As already interpreted by St-Pierre et al. (2015) collectively the mTEC cells express almost the entire functional genome.

Regarding the phenotypic characterization of TEC cells, cTECs express cyto keratin 8, EpCAM (CD36), Ly51, CD205 and MHC-I and MHC-II molecules. These cells form a reticular network in the thymic cortex and acts supporting PS (Nakagawa et al. 2012). On the other hand, mTECs are marked by the expression of cytokeratin 5, MHC-II and CD80 and these cells are divided into two groups i.e., mTEC^{lo} that expresses low levels of MHC-II and CD80 and mTEC^{hi} that expresses high levels of these molecules. The phenotypic profile of mTECs is CD45$^-$, EpCam$^+$, Ly51$^-$, UEA1$^+$ (extensively reviewed in the Chaps. 2, 3 and 6 of this book). An additional subset of TECs are the "thymic nurse cells" (TNC), which is a large epithelial cell that envelops lymphoid cells (Kyewski and Kaplan 1982). The TNC is in fact a complex that includes clusters cTEC-thymocyte interactions. The cTECs express β5t, a component of the thymoproteasome, which is essential for PS of CD8$^+$ T cells. The TNC complex represents a persistent interaction between cTECs and DP thymocytes that undergo secondary TCRα rearrangement (Nakagawa et al. 2012).

As the random V(D)J recombination generates TCR α/β specificities that recognize self-PTAs, those thymocyte clones expressing high affinity receptors directed to autoantigens are eliminated by apoptosis, a process termed *negative selection* (NS) or clonal deletion, thus preventing that autoreactive thymocyte clones reach the periphery. This includes the self-reactive regulatory T cells (T$_{reg}$), which also are negatively selected within the thymus, even if they could play a role in the periphery (Kyewski and Derbinski 2004; Kyewski and Klein 2006; Takahama 2006; Holländer 2007; Irla et al. 2008; Villaseñor et al. 2008; Takaba and Takayanagi 2017; Passos et al. 2015, 2018).

The thymic dendritic cells (DCs) also participate in the process of elimination of autoreactive thymocytes but only after the transfer of PTAs from mTECs (Mizuochi et al. 1992; Klein and Kyewski 2000; Derbinski et al. 2001; Aschenbrenner et al. 2007; Tykocinski et al. 2008; Koble and Kyewski 2009; Hubert et al. 2011; Aichinger et al. 2013; Oh and Shin 2015; Lopes et al. 2015; Hasegawa and Matsumoto 2018).

Accordingly, the main implication of PGE, which is heterogeneous and ectopic, is linked to immune homeostasis and controlling the auto-reactivity and consequently autoimmune diseases (Passos et al. 2015, 2018). The cTECs and mTECs although located in distinct compartments of the thymus, act in synergism. These cells are essential, but alone do not run into selection events (Gäbler et al. 2007). The cTECs regulate the PS of thymocytes that recognize the MHC-peptide complexes themselves and mTECs that express autoimmune regulator (AIRE) protein help ensure tolerance to self-antigens through NS (Anderson et al. 2007; Passos et al. 2015, 2018).

9.2   The Thymus Transcriptome

As previously conceptualized, the transcriptome is referred as the total set of RNA species, including coding and non-coding RNAs (ncRNAs), that are transcribed in a given cell type, tissue or organ at any given time under normal physiological or pathological conditions (Assis et al. 2014). Accordingly, the thymus should exhibit its own transcriptome pattern, which has been shown to be peculiar due to the enormous diversity of mRNAs encoding ectopic proteins. The conceptual challenge was to give a plausible interpretation to this phenomenon.

However, the ectopic expression of genes or proteins in the thymus was initially determined by using small-scale biochemistry, histology or molecular biology methods as Sephadex column chromatography for the separation of neurohypophyseal hormones (Geenen et al. 1986), *in situ* hybridization histochemistry for the localization of the mRNAs that encode insulin-like growth factors (IGFs) (Han et al. 1987), indirect immunoperoxidase with monoclonal antibodies for the localization of acetylcholine receptors (Kirchner et al. 1987) and the use of antisense RNA probes to demonstrate the expression of the somatostatin gene in the rat thymus (Fuller and Verity 1989). The significance of these pioneering studies was in the field of neuroimmunoendocrinology, which evidenced the role of the thymus in the connection of the neuronal, immune and endocrine systems.

Perniola (2018) review that the immunologic interpretation of the ectopic expression of proteins in the thymus begins later at the end of 1990s when independent research groups observed that insulin gene was expressed in thymi of aborted fetuses and children dead at various ages. It was from these observations that there was the suggestion that thymic insulin could be implicated in the elimination of related autoreactive thymocytes (Vafiadis et al. 1996, 1997; Pugliese et al. 1997; Heath et al. 1997; Hanahan 1998; Werdelin et al. 1998; Sospedra et al. 1998).

The understanding of this phenomenon as *promiscuous gene expression* (PGE) appears when Derbinski et al. (2001) assessed the expression of a larger set of genes in thymic stromal cells including cTECs and mTECs besides dendritic cells (DCs) and macrophages. These authors observed that the expression of a greater number of autoantigen-encoding genes was restricted to mTEC cells presenting UEA1+ CD80+ phenotype concomitantly with the expression of autoimmune regulator (*Aire*) gene (Derbinski et al. 2001; Gray et al. 2008; Sawanobori et al. 2014). Since the need to study the transcriptional expression of mTEC cells based on the concept of the transcriptome, researchers use large-scale methods as microarrays and/or RNA-sequencing (RNA-Seq) whose data were analyzed through bioinformatics pipelines and made freely available to the scientific community through public databases such as ArrayExpress (https://www.ebi.ac.uk/arrayexpress/) and/or Gene Expression Omnibus, GEO, (https://www.ncbi.nlm.nih.gov/geo/) (Sousa Cardoso et al. 2006; Derbinski et al. 2008; Giraud et al. 2012; Brennecke et al. 2015; Sansom et al. 2014; St-Pierre et al. 2015; Oliveira et al. 2016; Speck-Hernandez et al. 2018).

As discussed by Perniola (2018) the observation that PGE was restricted to mTEC cells stimulated researchers examine for a possible role of *Aire* in its control and in thymic self-representation and tolerance. In fact, *Aire* triggers PGE in mTEC cells (Derbinski et al. 2001; Anderson et al. 2002) whose property was confirmed in whole thymi using either murine fetal thymus organ culture (FTOC) (Sousa Cardoso et al. 2006) or murine adult thymus organ culture (ATOC) (Chen et al. 2008).

9.3   *Aire* Is a Transcriptional Controller

The history of the autoimmune regulator gene (*AIRE* for the human gene or *Aire* for the murine counterpart) begins with the genetic dissection of the autoimmune polyendocrinopathy-candidiasis-ectodermal dystrophy syndrome (APECED)/auto-immune polyglandular syndrome 1 (APS-1), OMIM entry # 240300 (Passos et al. 2018). This syndrome is characterized by immunological manifestations involving lymphocytic infiltration in the affected organs and the presence of specific autoanti-bodies (Björses et al. 1998).

The human DNA segment where there could be a gene implicated with APECED/APS-1 had its positioning in an unbiased way. Researchers noted that there was a link between this syndrome and mutations on chromosome 21q22.3, a region that had not yet been associated with autoimmune diseases (Aaltonen et al. 1994). Detailed sequencing from that region identified a new gene harboring two PHD finger motifs, which suggested that their role could be linked in some way to transcriptional regulation. The identification of mutations in this new gene validated its association with APECED/APS-1 and finally the gene was termed autoimmune regulator (*Aire*) due to the role in the prevention of aggressive autoimmunity (Aaltonen et al. 1994; Björses et al. 1998; Nagamine et al. 1997). As the Chaps. 7 and 8 of this book are dedicated to *AIRE*/*Aire*, here we pinpoint some of its main features.

In the mouse *Mus musculus* the *Aire* gene is located at chromosome 10 position 39.72 cM and the majority of the studies on the its mechanism was made with murine cells. *Aire* knockout mice (*Aire*$^{-/-}$ KO) feature destructuring in thymic architecture as well as changes in the expression of PTAs in mTEC cells besides the presence of specific autoantibodies, the decreasing in regulatory T cells and activation of self-reactive T cells. These characteristics promote general autoimmune reactions comparable to that observed in APECED/APS-1 patients (Anderson et al. 2002; Yano et al. 2008). Comparative analysis of the transcriptome profile of mTEC cells from wild-type mice versus *Aire* KO allowed a strong association between the expression of this gene and the control of PGE (Derbinski et al. 2001; Anderson et al. 2002).

Murine mTEC cells are peculiar since they express about 80% of their functional genome, i.e. from the 24,000 functional genes in the mouse, these cells express about 19,000 (Sansom et al. 2014; St-Pierre et al. 2015) from which many encode tissue-specific proteins, an exclusive characteristic of this cell type. The ectopic gene expression found in the thymus does not correspond to a fault in the cell's transcriptional control but rather to an inherent ability of these cells to promote PGE

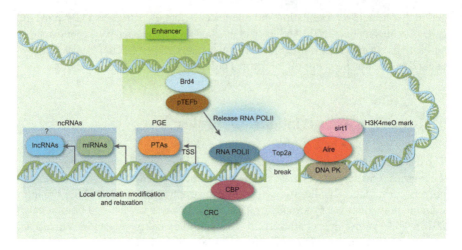

Fig. 9.1 Mechanism of action of AIRE protein and its partners. AIRE is a transcriptional controller of promiscuous gene expression in mTEC cells rather than a classical transcription factor (TF) due to its mode of action. Instead, interact directly with DNA on chromatin like TFs, AIRE associates with several other proteins to form the AIRE protein complex. In turn, this complex, binds to the transcription start site (TSS) of genes that encode peripheral tissue antigens (PTAs) or other proteins as those involved with cell adhesion in mTECs. Deacetylase sirtuin 1 (SIRT1) activates AIRE through deacetylation of its lysine residues. The PHD1 domain is involved with recognition of repressive epigenetic signatures that includes H3K4me0 of AIRE-dependent genes. TOP2a at the TSSs of AIRE-dependent genes initiates breaks on DNA that activates DNA-PK and other partners resulting in local chromatin relaxation. The pTEFb/BRD4 complex then liberates stalled RNA Pol II on chromatin to proceed transcription elongation of AIRE-dependent genes, which include the peripheral tissue antigens (PTAs) and non-PTA genes. As RNA Pol II transcribes miRNAs, AIRE protein complex is also involved in the transcriptional control of this RNA species (and possibly of lncRNAs) in mTECs. Finally, AIRE is repressed by acetylation of its lysine residues by CREB-binding protein (CBP). This figure was taken from Passos et al. (2018) with permission of John Wiley & Sons Ltd/Immunology

(Derbinski et al. 2001; Kyewski et al. 2002; Ucar and Rattay 2015; Passos et al. 2015, 2018; Magalhães et al. 2006).

Aire controls a set of about 3900 down-stream genes in mTECs, which are termed *Aire*-dependent genes. But rather associating directly to DNA, as found in the classical transcription factors, Aire protein play its role within a large transcriptional complex formed by partners proteins (Anderson and Su 2016; St-Pierre et al. 2015; Meredith et al. 2015; Abramson and Goldfarb 2016; Passos et al. 2018) (Fig. 9.1). Within the transcription machinery of mTECs, Aire protein directly binds to RNA Pol II molecules that although have started transcription are stalled at the gene promoters. This facilitates the pushing of RNA Pol II to proceed with gene transcription (Giraud et al. 2012).

As RNA Pol II is not specific and acts during the transcription of most, if not all, functional genes of a mammalian cell, this might explains how *Aire* controls PGE in mTECs (Passos et al. 2015, 2018). This has a tremendous implication in the thymic self-representation in which self-peptides originated from thymoproteasome processing of PTAs are presented, via MHC-II, to developing thymocytes during the negative selection process (Ucar and Rattay 2015; Passos et al. 2015, 2018).

However, not every mTEC cell expresses all these ectopic genes at any given time. The phenomenon of PGE occurs at the population level of mTECs, i.e. the diversity of PTAs occurs in a set of these cells. This type of gene expression was interpreted as "ordered" and "stochastic" since only 1–3% of mTECs expresses a certain PTA at any given time although *Aire* can controls the expression of a large set of PTAs at a same time. The expression of a given PTA in an mTEC cell is then probabilistic and the term "ordered" refers to an increased chance of particular PTA expression (Passos et al. 2018).

As already mentioned, Aire protein is in fact associated with several others when the transcriptional complex is formed, which fall into four functional groups: (1) proteins involved in nuclear transport, (2) proteins that interact directly with chromatin, (3) proteins that can participate in both transcription and mRNA processing and (4) proteins that participate in the transcriptional complex of PTAs (reviewed in Passos et al. 2018). In turn, some of these proteins regulate Aire protein itself (Herzig et al. 2017). Some Aire protein partners, e.g. those associated with DNA-damage response, localize on chromatin super-enhancers (Bansal et al. 2017).

Briefly, the Aire protein activates PGE in the transcriptional start sites (TSS) of most functional genes in mTEC cells. When Aire protein associates with RNA Pol II and, through histone modifications and deacetylation mediated by Sirt-1, the complex releases Top2a which generates disruptions in the regions of transcription initiation, induces epigenetic changes and chromatin relaxation. All this causes recruitment of the pTefb/Brd4 protein complex releasing the RNA Pol II to elongate the transcription (Fig. 9.1).

9.4 *Aire* as a Pleiotropic Gene

Since Aire protein, through association with RNA Pol II and other partner proteins (Fig. 9.1), may controls many downstream genes that are involved in different biological functions, this gene can be classified as pleiotropic, i.e., implicated in different functions or phenotypes. In addition to PTAs, *Aire* can control the tolerance of B cells, a process observed outside the thymic compartment (Meyer et al. 2016), regulates de expression of genes influencing the recruitment of cells through induction of chemokine expression (Lei et al. 2011) or even regulate the expression of miRNAs in the thymus. This last process has important implications, since the miRNAs exert post-transcriptional control of the gene expression and consequently they can control the PGE (Macedo et al. 2013; Passos et al. 2015; Oliveira et al. 2016) as well as the thymic architecture (Ucar and Rattay 2015).

Alternative processing of mRNAs is a way that ends up increasing the diversity of PTAs and consequently the range of self-representation in the thymus. Recent results showed that *Aire+* mTECs have an alternative splicing rate higher than that other cell populations (St-Pierre et al. 2015; Keane et al. 2015; Danan-Gotthold et al. 2016) whose role of *Aire* in the control of mRNA isoforms was confirmed through Crispr-Cas9-mediated gene disruption (Speck-Hernandez et al. 2018).

As already mentioned, thymic microenvironment is composed by several cell types including Aire$^+$ (protein) and Aire$^-$ mTECs, DCs and thymocytes that are continuously interacting during the process of induction of central tolerance of T cells to self-antigens (Lopes et al. 2015). Aire$^+$ mTECs do not only express PTAs but also proteins involved with cell-cell interaction highlighting adhesion molecules (St-Pierre et al. 2015).

Experiments using the small interfering RNA (siRNA) to silencing *Aire* (Pezzi et al. 2016) or the Crispr-Cas9 system to disrupt this gene in mTECs (Speck-Hernandez et al. 2018) demonstrated that either its reduction (siRNA) or its annulation (Crispr-Cas9) leads reduction in not only the PTA mRNAs but also those that encode cell-adhesion proteins. This reduced the adhesion property of mTECs to thymocytes.

Regarding the *Aire* silencing experiments, researchers tested the hypothesis that partial reduction (down-regulation) of *Aire* could impair the property of mTEC cells to adhere to thymocytes that is crucial for negative selection. As a model system they used the murine mTEC 3.10 cell line (CD45$^-$, EpCam$^+$, Lyn51$^-$, UEA1$^+$) (Pezzi et al. 2016). As this cell line constitutively express *Aire* mRNA and protein, it was considered adequate for siRNA-mediated silencing. Results demonstrating the reduction of *Aire* mRNA were confirmed by both methods RT-qPCR and RNA-fluorescent *in situ* hybridization (RNA-FISH). The RNA-FISH technique for *Aire* worked so well that it was possible to visualize, for the first time, the stochasticity of *Aire* expression in mTEC 3.10 cells. Moreover, they explored the large-scale gene expression profiling (mRNA transcriptome profiling) comparing control vs. *Aire* silenced mTEC cells and have found a set of mRNAs that encode cell-adhesion molecules, i.e., the extracellular matrix constituent Lama 1, the CAM family adhesion molecules Vcam1 and Icam4 as well as those that encode PTAs was concomitantly down-regulated under *Aire* reduction. This supported the authors to demonstrate that variations in the expression of *Aire* causes mTEC cells to reduce their ability to adhesion with thymocytes.

However, the question whether deleterious mutations of *Aire* would be implicated in the adhesive property of mTECs was still elusive. Following this question, Speck-Hernandez et al. (2018) employed for the first time, the Crispr-Cas9 system to induce mutations in the *Aire* exon 3 of mTEC cells in culture. Exon 3 encodes the NLS domain of the Aire protein that is involved in its transport to the nucleus (Fig. 9.1). The authors demonstrated that *Aire* KO mTECs did not present Aire protein in the nucleus and, consequently, the loss of function of this gene reduced the ability of mTECs to adhere to thymocytes. This contributes to the concept that *Aire* is actually implicated in the adhesion of mTECs, but it is still open whether the mutations found in APECED/APS-1 patients would have such an effect. Since mTEC-thymocyte adhesion is crucial for the negative selection of autoreactive thymocytes and induction of immunological tolerance, these results demonstrated how *Aire* might participates in this process, i.e. controlling the expression of PTAs and cell adhesion proteins.

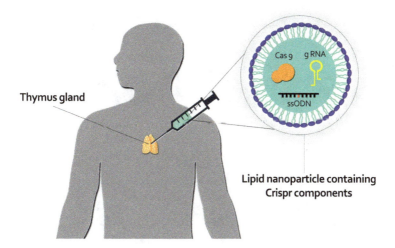

Thymus gland

Cas 9   g RNA

ssODN

Lipid nanoparticle containing
Crispr components

Fig. 9.2 Schematic representation for the use of Crispr-Cas9 system (or even making use of another more efficient Cas enzyme) for correction of Aire gene mutations directly into the thymus of newly diagnosed APECED/APS1 syndrome patients with pathogenic mutations in the Aire gene. Lipid nanoparticles containing Crispr components [Cas9 enzyme, gRNA directed to the target Aire mutant exon and a single-stranded oligodeoxynucleotide (ssODN) carrying the respective wild-type sequence] could be injected *in vivo* and directly into the thymus. The presence of Aire wild-type ssODN will serve to stimulate homology directed repair (HDR) to correct Aire mutation(s). A part of the mTEC cells of these patients could have the corrected Aire sequence, would express more efficiently and present autoantigens and participate in the negative selection of auto aggressive thymocytes. This type of treatment could be a useful adjuvant and help reduce the symptoms of autoimmune syndrome

Regarding the use of the Crispr-Cas9 system in TEC cells, as performed in this pioneering work, we point out that this opened up new perspectives for using this gene-editing system in an attempt to correct those mutations of *Aire* found in the APECED/APS-1 syndrome patients directly in the thymus (Fig. 9.2).

9.5   *Fezf2*, a Second Controller of Promiscuous Gene Expression

As previously stated, mTEC cells collectively express almost all (~80%) of the murine genome, i.e. more than 19,000 protein-coding genes, which includes all PTAs and non-PTA proteins (Sansom et al. 2014; St-Pierre et al. 2015). To date, no other normal cell type has shown such transcriptional activity. In the mouse, *Aire* controls around 3900 genes in mTECs, which are *Aire*-dependent and represent approximately 20% of the PGE. Clearly, *Aire* is an important controller of PGE and mutations in this gene represent the cause of severe autoimmunity.

Over the past 20 years, research in this area has focused on the *Aire* gene about which there is still much to do and understand. However, since this gene controls

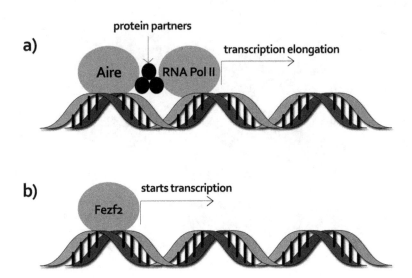

Fig. 9.3 Comparison between Aire and Fezf2 protein mode of action in mTEC cells. While Aire protein binds to several other protein partners to liberate stalled RNA Pol II on chromatin (**a**), Fezf2 protein binds directly to DNA and plays their role like a classical transcription factor (**b**)

only a part of the PGE, others must also function. This long-standing suspicion was reinforced with the observation that under *Aire* abrogation, PGE is not fully cancelled in mTECs (Derbinski et al. 2005; reviewed in Passos et al. 2018). Hence one question remained; what are the elements controlling the expression of *Aire*-independent genes?

Progress in this regard has appeared with the work of Takaba et al. (2015) in which the authors tested the hypothesis that if there is any other transcriptional PGE controller it must be differentially expressed between cTECs and mTECs. While the finding of *Aire* was based on phenotype-genotype association studies, the identification of this second PGE controller i.e. forebrain-expressed zinc finger 2 (*Fezf2*) was based on the transcriptomics of the TEC cells. The reason for the late discovery of a further PGE controller, in addition to *Aire*, was in part that the researchers focused on *Aire* to unravel its mechanism of action. The other reason is inherent to the intrinsic properties of *Fezf2*. In fact, this gene had already been identified and its function was associated with corticospinal motor neuron differentiation (Lodato et al. 2014). Since *Fezf2*-deficient mice do not survive after weaning, the function of this gene in the immune system cannot be revealed. It was only when they used the concept and tools of transcriptomics, comparing the differential expression between mTECs and cTECs that pointed to *Fezf2*.

In fact, *Aire* and *Fezf2* function independently, but synergistically. While the Aire protein associates with several partners and functions in the transcription elongation step, the *Fezf2* protein acts as a classical transcription factor, i.e. it binds directly to DNA (Takaba et al. 2015; Klein 2015) (Fig. 9.3). As previously discussed (Passos et al. 2018), the discovery of *Fezf2* as a second PGE controller in mTECs undoubtedly contributed to increase the knowledge and raised further questions as follow:

(1) is there any function of this gene in the three-dimensional organization of the thymus? (Klein 2015), (2) does it control genes associated with mTECs-thymocyte adhesion and/or thymocyte migration? (3) does it control the expression of non-coding RNAs (miRNAs or lncRNAs)? These are open questions that will surely guide interesting work in this area.

Perniola (2018) makes another interesting observation when referring to the *Aire*-dependent and *Aire*-independent genes. According to this author, this "dichotomy" needs to be revised and perhaps this is an inappropriate simplification. Observing gene networking data previously published by Macedo et al. (2009) and Donate et al. (2011), this author noted that genes that participate in the two categories are connected through networks that recognize a hierarchy. He then considered that this could explain how *Aire* regulates genes indirectly (or hierarchically) and that the participation of further transcription factors cannot be excluded.

9.6 The Link Between *Aire* and miRNAs

As previously discussed by Perniola (2018) and Passos et al. (2018), researchers were soon interested in see if there were some sort of link between *Aire* and miRNAs in the thymus. The first observations came between 2012 and 2014 with works in which the deletion of genes involved with the miRNA pathways in TEC cells disturbed the thymic microenvironment to the point of not sustaining the thymocyte maturation, though with not so obvious *Aire* deregulation (Zuklys et al. 2012; Khan et al. 2014). Soon there came the reports that the miRNA expression pattern changes during the differentiation of mTEC cells in which they identified a subset of miRNAs that would potentially affect the translation of *Aire* mRNA (Papadopoulou et al. 2011; Ucar et al. 2013; Ucar and Rattay 2015).

In the work by Macedo et al. (2013), researchers suspected that if *Aire* interacts with RNA Pol II and this enzyme besides mRNAs also transcribes miRNAs then *Aire* could control this RNA species. They tested this hypothesis through the loss of function strategy in whose work they used siRNA to silencing *Aire* in mTEC cells. The expression profile of the total set of miRNAs, comparing control cells with *Aire*-silenced cells they observed a set of miRNAs that was concomitantly downregulated with the decrease of *Aire*. Among the downregulated miRNAs, the miR-376 family was focused because it harbors the members miR-376a, miR-376b and miR-376c that are located within the Gm2922 open-reading frame (ORF) at chromosome 12F1. This ORF features T-boxes (TTATTA) and G-boxes (GATTGG), which represent putative RNA Pol II promoter motifs. Interestingly, Gm2922 ORF encodes an mRNA that also was downregulated upon *Aire* silencing. Authors infer that *Aire* can control miRNAs located within genomic regions encompassing ORF and/or mRNA genes.

The investigation continued to find out what the consequences of this process would be on PGE. *Aire* could, at the same time, favor the transcriptional expression of PTAs, but also indirectly control them at post-transcriptional level? To test this hypothesis, Oliveira et al. (2016) used *in vivo* intra thymic electrotransfection of siRNA to induce transient changes in the *Aire* transcript levels and then to profile the *Aire*-dependent mRNAs and miRNAs. The reconstruction of miRNA-mRNA networks enabled authors to observe that reduction in *Aire* imbalance the interaction between the mRNAs that encode PTAs with miRNAs and consequently altering PGE profile in mTECs. Remarkably, none of the considered *Aire*-dependent PTAs interacted with miRNAs (Macedo et al. 2015), remaining the question whether *Aire*, during the transcription of PTAs, could alter the structure of mRNAs making them somehow refractory to the action of miRNAs (Macedo et al. 2015; Perniola 2018).

Chen et al. (2017) reported the role of miR-449a in regulating the development and function of TEC cells. The authors demonstrated that the *in vitro* overexpression of miR-449a stimulated the differentiation of progenitor epithelial cells into mature mTEC cells. In addition, the ectopic overexpression of miR-449a in FTOC via lentiviral transduction induced expression of *Aire* and *Aire*-dependent PTAs.

This shows that miRNAs might directly or indirectly control the expression of *Aire*, but there are still no conclusive results in this regard.

Intriguingly, the normal thymus involves as age passes, and this age-dependent involution begins as early as in adolescence. However, pathological processes such as infections, autoimmunity and neoplasia also influence thymus involution. (Taub and Longo 2005; Lynch et al. 2009; Rezzani et al. 2014; Linhares-Lacerda et al. 2015; Palmer et al. 2018). Studies have shown the participation of miRNAs in age-dependent thymus involution (Dooley and Liston 2012; Xu et al. 2017) and in infectious processes (Papadopoulou et al. 2011; Linhares-Lacerda et al. 2015).

Recently, Assis et al. (2018) demonstrated the participation of miRNAs in the differentiation between the thoracic (TT) and cervical thymus (CT) in the mouse. The CT appears in both humans and mice (Dooley et al. 2006; Terszowski et al. 2006; Li et al. 2013). It is an independent organ implicated in the generation of the thymocyte repertoire and positive and negative selection. To study the molecular differences that still exist between TT and CT, the authors analyzed the respective expression profiles of mRNAs and miRNAs of these two organs. Through the reconstruction of post-transcriptional interaction networks, they observed significant differences between these two organs suggesting that functional uniqueness found in CT might be under post-transcriptional control involving miRNAs.

9.7   Would There Be a Role for Long Non-coding RNAs in the Thymus?

Large-scale transcriptomic analysis have recently revealed a large number of long non-coding RNAs (lncRNAs) in mammalian cells. These non-coding RNAs are endogenous transcripts longer than 200 nucleotides that are predict do carry no functional ORFs. They are poorly conserved and represent a very heterogeneous group of RNAs, which allows covering a wide spectrum of cellular and molecular functions, comprising transcriptional regulation, organization of nuclear domains and regulation of proteins or RNA molecules (Kapranov et al. 2007; Wilusz et al. 2009; Chen and Carmichael 2010; Wang and Chang 2011; Derrien et al. 2012a).

Increasing studies have implicated lncRNAs as key players of cellular differentiation, cell lineage choice and organogenesis and tissue homeostasis. Their deregulation is associated with various pathologic states in humans and animal models. Together lncRNAs present low level of expression, however, they are expressed with high level depending on cell and/or developmental stage specificity (Khalil et al. 2009; Derrien et al. 2012b; Fang and Fullwood 2016; Delás and Hannon 2017; Kopp and Mendell 2018).

Bogu et al. (2015) set up a catalog of lncRNAs across eight different tissues in mouse including thymus. Using RNA-Seq and Chip-Seq data from ENCODE Project (https://www.encodeproject.org/), authors addressed specific lncRNAs from the studied tissues and characterized those that might play a role as regulatory molecules that associate with promoter or enhancer chromatin states. This opens up perspectives for novel experiments seeking to demonstrate the possible role of lncRNAs in the thymus and/or TEC cells.

9.8 Thymocyte Abnormalities and the Emergence of Autoimmunity

As explained above, thymic stroma is not solely a scaffold structure. It plays a crucial role in the control of immune tolerance. However, what was happening to the "other side of the coin" that is, to thymocytes? When immune tolerance fails, developing autoreactive thymocytes that should have been eliminated through NS remain viable and may reach the periphery. The thymocyte TCR receptor specificity determine then the target for the autoimmune attack. For example, if a particular thymocyte clone exhibiting autoreactive TCR α/β against pancreatic antigens it may attack the pancreas and destroy insulin-producing beta cells through inflammatory insulitis. The emergence of type 1 diabetes mellitus (DM-1) is a consequence of this and pancreas-infiltrating T lymphocytes (PILs) represent the main effector cells in this process (reviewed in Mendes-da-Cruz et al. 2018).

The emergence of DM-1 in non-obese diabetic mice (NOD) is currently the most important experimental model of aggressive autoimmunity (Pearson et al. 2016) and alterations in the thymus of these mice were observed during the pathogenesis of the disease. The main thymic alterations are those related to the migration and accumulation of Foxp3$^+$ mature thymocytes (Mendes-da-Cruz et al. 2018). Studies in our group have shown that during the course of DM-1 in NOD mice, genes involved with immune reactivity/inflammation is modulated in mature CD3$^+$ T lymphocytes and during all stages of lymphocyte development, i.e. from the thymocyte stage, through the peripheral T CD3$^+$ lymphocytes to PILs phase (Fornari et al. 2011, 2015).

To better understand how this gene modulation is controlled we explore the transcriptome, involving mRNAs and miRNAs, of these cells in all their phases. It was observed that these species of RNA interact with each other through post-transcriptional control during the transition from thymocytes to peripheral T CD3+ lymphocytes to PILs. The miRNA-mRNA interaction networks demonstrated the participation of miRNAs that regulate target mRNAs that encode proteins associated with apoptosis, cell adhesion, cell regulation and immune system development, among others. The miR-202-3p drew attention as it regulates both the Ccr7 chemokine receptor and Cd3 zeta chain mRNAs. This type of post-transcriptional regulation was observed in NOD mouse PILs and helped us to better understand how the lack of immune homeostasis and the emergence of autoimmunity are associated with decreasing levels of the respective mRNAs encoding Ccr7 or Cd3 zeta chain that are regulated by miR-202-3p (Fornari et al. 2015).

9.9   Concluding Remarks and Perspectives

It is conceptual that self-non-self discrimination is crucial for maintenance of immune homeostasis. This is the basis of central immune tolerance and the thymus is the organ involved in this process. Within the thymus occurs two of the most intriguing molecular-genetic processes; the generation of the T cell receptor (TCR) repertoire through the random recombination of the V (D) J segments and the promiscuous gene expression (PGE). Although these processes are apparently independent and occur in distinct types of cells [V (D) J recombination occurs in developing thymocytes and PGE in mTEC cells], they are complementary in the establishment of central tolerance. The meaning of PGE is immunological and enables the presentation of the myriad of autoantigens to the developing thymocytes, culminating in the negative selection of potentially dangerous self-reactive clones. The mTEC cells, which exhibit PGE, are very peculiar since they express almost 80% of their functional genome. *Aire* and *Fezf2* are two genes that control PGE in mTECs and mutations of *Aire* causes APECED/APS-1 syndrome, a severe autoimmune disease. *Aire* is actually a pleiotropic gene and is therefore involved with several functions of mTEC cells, among which we highlight the control of autoantigens, miRNAs and mTEC-thymocyte adhesion. Devoid of these processes, the NS of auto-reactive thymocytes is disturbed, which allows the emergence of autoimmune diseases. For these reasons, research into the molecular genetics of thymic cells is progressing. Increasing number of researchers consider that this is fundamental to better understanding autoimmune diseases and possibly in the near future to establish safe protocols for a more effective treatment, looking at the use of the emerging Crispr-Cas genome editing technology to correct *Aire* mutations directly in the thymus.

Acknowledgments We thank São Paulo Research Foundation (FAPESP), São Paulo, Brazil, through grant 13/17481-1 and 17/10780-4 to GAP, National Council for Scientific and Technological Development (CNPq), Brasília, Brazil, through grant 305787/2017-9 to GAP and grant 304931/2014-1 to EAD. This study was financed in part by the Coordenação de Aperfeiçoamento de Pessoal de Nível Superior (CAPES)—Brazil—Financial code 001. We also thank Ms Janine Bottosso Passos for the Figs. 9.2 and 9.3.

References

Aaltonen J, Björses P, Sandkuijl L, Perheentupa J, Peltonen L (1994) An autosomal locus causing autoimmune disease: autoimmune polyglandular disease type I assigned to chromosome 21. Nat Genet 8:83–87

Abramson J, Anderson G (2017) Thymic epithelial cells. Annu Rev Immunol 35:85–118

Abramson J, Goldfarb Y (2016) AIRE: from promiscuous molecular partnerships to promiscuous gene expression. Eur J Immunol 46:22–33

Aichinger M, Wu C, Nedjic J, Klein L (2013) Macroautophagy substrates are loaded onto MHC class II of medullary thymic epithelial cells for central tolerance. J Exp Med 210:287–300

Anderson MS, Su MA (2016) AIRE expands: new roles in immune tolerance and beyond. Nat Rev Immunol 16:247–258

Anderson MS, Venanzi ES, Klein L, Chen Z, Berzins SP, Turley SJ, von Boehmer H, Bronson R, Dierich A, Benoist C, Mathis D (2002) Projection of an immunological self shadow within the thymus by the aire protein. Science 298:1395–1401

Anderson G, Lane PJ, Jenkinson EJ (2007) Generating intrathymic microenvironments to establish T-cell tolerance. Nat Rev Immunol 7:954–963

Aschenbrenner K, D'Cruz LM, Vollmann EH, Hinterberger M, Emmerich J, Swee LK, Rolink A, Klein L (2007) Selection of Foxp3+ regulatory T cells specific for self antigen expressed and presented by Aire+ medullary thymic epithelial cells. Nat Immunol 8:351–358

Assis AF, Oliveira EH, Donate PB, Giuliatti S, Nguyen C, Passos GA (2014) What is the transcriptome and how it is evaluated? In: Passos GA (ed) Transcriptomics in health and disease. Springer International Publishing, Switzerland, pp 3–48

Assis AF, Li J, Donate PB, Dernowsek JA, Manley NR, Passos GA (2018) Predicted miRNA-mRNA-mediated posttranscriptional control associated with differences in cervical and thoracic thymus function. Mol Immunol 99:39–52

Bansal K, Yoshida H, Benoist C, Mathis D (2017) The transcriptional regulator Aire binds and activates super-enhancers. Nat Immunol 18:63–273

Björses P, Aaltonen J, Horelli-Kuitunen N, Yaspo ML, Peltonen L (1998) Gene defect behind APECED: a new clue to autoimmunity. Hum Mol Genet 7:1547–1553

Bogu GA, Vizan P, Stanton LW, Beato M, Di Croce L, Marti-Renom MA (2015) Chromatin and RNA maps reveal regulatory long noncoding RNAs in mouse. Mol Cell Biol 36:809–819

Brennecke P, Reyes A, Pinto S, Rattay K, Nguyen M, Küchler R, Huber W, Kyewski B, Steinmetz LM (2015) Single-cell transcriptome analysis reveals coordinated ectopic gene-expression patterns in medullary thymic epithelial cells. Nat Immunol 16:933–941

Chen LL, Carmichael GG (2010) Long noncoding RNAs in mammalian cells: what, where, and why? Wiley Interdiscip Rev RNA 1:2–21

Chen J, Yang W, Yu C, Li Y (2008) Autoimmune regulator initiates the expression of promiscuous genes in thymic epithelial cells. Immunol Investig 37:203–214

Chen P, Zhang H, Sun X, Hu Y, Jiang W, Liu Z, Liu S, Zhang X (2017) microRNA-449a modulates medullary thymic epithelial cell differentiation. Sci Rep 7:15915

Danan-Gotthold M, Guyon C, Giraud M, Levanon EY, Abramson J (2016) Extensive RNA editing and splicing increase immune self-representation diversity in medullary thymic epithelial cells. Genome Biol 17:219

Delás MJ, Hannon GJ (2017) lncRNAs in development and disease: from functions to mechanisms. Open Biol 7(7):170121

Derbinski J, Kyewski B (2010) How thymic antigen presenting cells sample the body's self-antigens. Curr Opin Immunol 22:592–600

Derbinski J, Schulte A, Kyewski B, Klein L (2001) Promiscuous gene expression in medullary thymic epithelial cells mirrors the peripheral self. Nat Immunol 2:1032–1039

Derbinski J, Gäbler J, Brors B, Tierling S, Jonnakuty S, Hergenhahn M, Peltonen L, Walter J, Kyewski B (2005) Promiscuous gene expression in thymic epithelial cells is regulated at multiple levels. J Exp Med 202:33–45

Derbinski J, Pinto S, Rösch S, Hexel K, Kyewski B (2008) Promiscuous gene expression patterns in single medullary thymic epithelial cells argue for a stochastic mechanism. Proc Natl Acad Sci U S A 105:657–662

Derrien T, Johnson R, Bussotti G, Tanzer A, Djebali S, Tilgner H, Guernec G, Martin D, Merkel A (2012a) The GENCODE v7 catalog of human long noncoding RNAs: analysis of their gene structure, evolution, and expression. Genome Res 22:1775–1789

Derrien T, Guigó R, Johnson R (2012b) The long non-coding RNAs: a new (p)layer in the "dark matter". Front Genet 2:107

Donate PB, Fornari TA, Junta CM, Magalhães DA, Macedo C, Cunha TM, Nguyen C, Cunha FQ, Passos GA (2011) Collagen induced arthritis (CIA) in mice features regulatory transcriptional network connecting major histocompatibility complex (MHC H2) with autoantigen genes in the thymus. Immunobiology 2016:591–603

Dooley J, Liston A (2012) Molecular control over thymic involution: from cytokines and microRNA to aging and adipose tissue. Eur J Immunol 42:1073–1079

Dooley J, Erickson M, Gillard GO, Farr AG (2006) Cervical thymus in the mouse. J Immunol 176:6484–6490

Fang Y, Fullwood MJ (2016) Roles, functions, and mechanisms of long non-coding RNAs in cancer. Genomics Proteomics Bioinformatics 14:42–54

Fornari TA, Donate PB, Macedo C, Sakamoto-Hojo ET, Donadi EA, Passos GA (2011) Development of type 1 diabetes mellitus in nonobese diabetic mice follows changes in thymocyte and peripheral T lymphocyte transcriptional activity. Clin Dev Immunol 2011:158735

Fornari TA, Donate PB, Assis AF, Macedo C, Sakamoto-Hojo ET, Donadi EA, Passos GA (2015) Comprehensive survey of miRNA-mRNA interactions reveals that Ccr7 and Cd247 (CD3 zeta) are posttranscriptionally controlled in pancreas infiltrating T lymphocytes of non-obese diabetic (NOD) mice. PLoS One 10:e0142688

Fuller PJ, Verity K (1989) Somatostatin gene expression in the thymus gland. J Immunol 143:1015–1017

Gäbler J, Arnold J, Kyewski B (2007) Promiscuous gene expression and the developmental dynamics of medullary thymic epithelial cells. Eur J Immunol 37:3363–3372

Gallegos AM, Bevan MJ (2004) Central tolerance to tissue-specific antigens mediated by direct and indirect antigen presentation. J Exp Med 200:1039–1049

Geenen V, Legros JJ, Franchimont P, Baudrihaye M, Defresne MP, Boniver J (1986) The neuroendocrine thymus: coexistence of oxytocin and neurophysin in the human thymus. Science 232:508–511

Giraud M, Yoshida H, Abramson J, Rahl PB, Young RA, Mathis D, Benoist C (2012) Aire unleashes stalled RNA polymerase to induce ectopic gene expression in thymic epithelial cells. Proc Natl Acad Sci U S A 109:535–540

Gotter J, Brors B, Hergenhahn M, Kyewski B (2004) Medullary epithelial cells of the human thymus express a highly diverse selection of tissue-specific genes colocalized in chromosomal clusters. J Exp Med 199:155–166

Gray DH, Fletcher AL, Hammett M, Seach N, Ueno T, Young LF, Barbuto J, Boyd RL, Chidgey AP (2008) Unbiased analysis, enrichment and purification of thymic stromal cells. J Immunol Methods 329:56–66

Groettrup M, Kirk CJ, Basler M (2010) Proteasomes in immune cells: more than peptide producers? Nat Rev Immunol 10:73–77

Halkias J, Melichar HJ, Taylor KT, Ross JO, Yen B, Cooper SB, Winoto A, Robey EA (2013) Opposing chemokine gradients control human thymocyte migration in situ. J Clin Invest 123:2131–2142

Han VK, D'Ercole AJ, Lund PK (1987) Cellular localization of somatomedin (insulin-like growth factor) messenger RNA in the human fetus. Science 236:19319–19317

Hanahan D (1998) Peripheral-antigen-expressing cells in thymic medulla: factors in self-tolerance and autoimmunity. Curr Opin Immunol 10:656–662

Hasegawa H, Matsumoto T (2018) Mechanisms of tolerance induction by dendritic cells in vivo. Front Immunol 9:350

Heath V, Mason D, Ramirez F, Seddon B (1997) Homeostatic mechanisms in the control of auto-immunity. Semin Immunol 9:375–380

Herzig Y, Nevo S, Bornstein C, Brezis MR, Ben-Hur S, Shkedy A, Eisenberg-Bord M, Levi B, Delacher M et al (2017) Transcriptional programs that control expression of the autoimmune regulator gene Aire. Nat Immunol 18:161–172

Holländer GA (2007) Claudins provide a breath of fresh Aire. Nat Immunol 8:234–246

Hu Z, Lancaster JN, Ehrlich L (2015) The contribution of chemokines and migration to the induction of central tolerance in the thymus. Front Immunol 6:398

Hubert FX, Kinkel SA, Davey GM, Phipson B, Mueller SN, Liston A, Proietto AI, Cannon PZ, Forehan S, Smyth GK, Wu L, Goodnow CC, Carbone FR, Scott HS, Heath WR (2011) Aire regulates the transfer of antigen from mTECs to dendritic cells for induction of thymic tolerance. Blood 118:2462–2472

Irla M, Hugues S, Gill J, Nitta T, Hikosaka Y, Williams IR, Hubert FX, Scott HS, Takahama Y, Holländer GA, Reith W (2008) Autoantigen-specific interactions with CD4+ thymocytes control mature medullary thymic epithelial cell cellularity. Immunity 29:451–463

James KD, Jenkinson WE, Anderson G (2018) T-cell egress from the thymus: should I stay or should I go? J Leukoc Biol 104(2):275–284

Kapranov P, Cheng J, Dike S, Nix DA, Duttagupta R, Willingham AT, Stadler PF, Hertel J, Hackermüller J et al (2007) RNA maps reveal new RNA classes and a possible function for pervasive transcription. Science 316(5830):1484–1488

Keane P, Ceredig R, Seoighe C (2015) Promiscuous mRNA splicing under the control of AIRE in medullary thymic epithelial cells. Bioinformatics 31:986–990

Khalil AM, Guttman M, Huarte M, Garber M, Raj A, Rivea Morales D, Thomas K, Presser A, Bernstein BE et al (2009) Many human large intergenic noncoding RNAs associate with chromatin-modifying complexes and affect gene expression. Proc Natl Acad Sci U S A 106:11667–11672

Khan IS, Tanigushi RT, Fasano KJ, Anderson MS, Jeker LT (2014) Canonical microRNAs in thymic epithelial cells promote central tolerance. Eur J Immunol 44:1313–1309

Kirchner T, Hoppe F, Müller-Hermelink HK, Schalke B, Tzartos S (1987) Acetylcholine receptor epitopes on epithelial cells of thymoma in myasthenia gravis. Lancet 1:218

Klein L (2015) Aire gets company for immune tolerance. Cell 163:794–795

Klein L, Kyewski B (2000) Self-antigen presentation by thymic stromal cells: a subtle division of labor. Curr Opin Immunol 12:179–186

Klein L, Hinterberger M, Wirnsberger G, Kyewski B (2009) Antigen presentation in the thymus for positive selection and central tolerance induction. Nat Rev Immunol 9:833–844

Koble C, Kyewski B (2009) The thymic medulla: a unique microenvironment for intercellular self-antigen transfer. J Exp Med 206:1505–1513

Kopp F, Mendell JT (2018) Functional classification and experimental dissection of long noncoding RNAs. Cell 172:393–407

Kyewski B, Derbinski J (2004) Self-representation in the thymus: an extended view. Nat Rev Immunol 4:688–698

Kyewski B, Kaplan HS (1982) Lymphoepithelial interactions in the mouse thymus: phenoptypic and kinetic studies on thymic nurse cells. J Immunol 128:2287–2294

Kyewski B, Klein L (2006) A central role for central tolerance. Annu Rev Immunol 24:571–606

Kyewski B, Derbinski J, Gotter J, Klein L (2002) Promiscuous gene expression and central T-cell tolerance: more than meets the eye. Trends Immunol 23:364–371

Lei Y, Ripen AM, Ishimaru N, Ohigashi I, Nagasawa T, Jeker LT, Bösl MR, Holländer GA, Hayashi Y (2011) Aire-dependent production of XCL1 mediates medullary accumulation of thymic dendritic cells and contributes to regulatory T cell development. J Exp Med 208:383–394

Li J, Liu Z, Xiao S, Manley NR (2013) Transdifferentiation of parathyroid cells into cervical thymi promotes atypical T-cell development. Nat Commun 4:2959–2966

Linhares-Lacerda L, Palu CC, Ribeiro-Alves M, Paredes BD, Morrot A, Garcia-Silva MR, Cayota A, Savino W (2015) Thymic epithelial cells from *Trypanosoma cruzi* acutely infected mice: role in thymic atrophy. Front Immunol 6:428

Lodato S, Molyneaux BJ, Zuccaro E, Goff LA, Chen HH, Yuan W, Meleski A, Takahashi E, Mahony S, Rinn JL, Gifford DK, Arlotta P (2014) Gene co-regulation by Fezf2 selects neurotransmitter identity and connectivity of corticospinal neurons. Nat Neurosci 17:1046–1054

Lopes N, Sergé A, Ferrier P, Irla M (2015) Thymic crosstalk coordinates medulla organization and T-cell tolerance induction. Front Immunol 6:365

Lopes N, Charaix J, Cédile O, Sergé A, Irla M (2018) Lymphotoxin α fine-tunes T cell clonal deletion by regulating thymic entry of antigen-presenting cells. Nat Commun 9:1262

Lynch HE, Goldberg GL, Chidgey A, Van den Brink MR, Boyd R, Sempowski GD (2009) Thymic involution and immune reconstitution. Trends Immunol 30:366–373

Macedo C, Evangelista AF, Magalhães DA, Fornari TA, Linhares LL, Junta CM, Silva GL, Sakamoto-Hojo ET, Donadi EA, Savino W, Passos GA (2009) Evidence for a network transcriptional control of promiscuous gene expression in medullary thymic epithelial cells. Mol Immunol 46:3240–3244

Macedo C, Evangelista AF, Marques MM, Octacílio-Silva S, Donadi EA, Sakamoto-Hojo ET, Passos GA (2013) Autoimmune regulator (Aire) controls the expression of microRNAs in medullary thymic epithelial cells. Immunobiology 218:554–560

Macedo C, Oliveira EH, Almeida RS, Donate PB, Fornari TA, Pezzi N (2015) Aire-dependent peripheral tissue antigen mRNAs in mTEC cells feature network refractoriness to microRNA interaction. Immunobiology 220:93–102

Magalhães DA, Silveira EL, Junta CM, Sandrin-Garcia P, Fachin AL, Donadi EA, Sakamoto-Hojo ET, Passos GA (2006) Promiscuous gene expression in the thymus: the root of central tolerance. Clin Dev Immunol 13:81–99

Mendes-da-Cruz DA, Lemos JP, Passos GA, Savino W (2018) Abnormal T-cell development in the thymus of non-obese diabetic mice: possible relationship with the pathogenesis of Type 1 autoimmune diabetes. Front Endocrinol 9:381

Meredith M, Zemmour D, Mathis D, Benoist C (2015) Aire controls gene expression in the thymic epithelium with ordered stochasticity. Nat Immunol 16:942–949

Meyer S, Woodward M, Hertel C, Vlaicu P, Haque Y, Kärner J, Macagno A, Onuoha SC, Fishman D et al (2016) AIRE-deficient patients harbor unique high-affinity disease-ameliorating autoantibodies. Cell 166:582–595

Mizuochi T, Kasai M, Kokuho T, Kakiuchi T, Hirokawa K (1992) Medullary but not cortical thymic epithelial cells present soluble antigens to helper T cells. J Exp Med 175:1601–1605

Mouchess ML, Anderson M (2014) Central tolerance induction. Curr Top Microbiol Immunol 373:69–86

Nagamine K, Peterson P, Scott HS, Kudoh J, Minoshima S, Heino M, Krohn KJ, Lalioti MD, Mullis PE, Antonarakis SE, Kawasaki K, Asakawa S, Ito F, Shimizu N (1997) Positional cloning of the APECED gene. Nat Genet 17:393–398

Nakagawa Y, Ohigashi I, Nitta T, Sakata M, Tanaka K, Murata S, Kanagawa O, Takahama Y (2012) Thymic nurse cells provide microenvironment for secondary T cell receptor α rearrangement in cortical thymocytes. Proc Natl Acad Sci U S A 109(50):20572–20577

Oh J, Shin JS (2015) The role of dendritic cells in central tolerance. Immune Netw 15:111–120

Oliveira EH, Macedo C, Collares CV, Freitas AC, Donate PB, Sakamoto-Hojo ET, Donadi EA, Passos GA (2016) Aire downregulation is associated with changes in the posttranscriptional control of peripheral tissue antigens in medullary thymic epithelial cells. Front Immunol 7:526

Palmer S, Albergante L, Blackburn CC, Newman TJ (2018) Thymic involution and rising disease incidence with age. Proc Natl Acad Sci U S A 115:1883–1888

Papadopoulou AS, Dooley J, Linterman MA, Pierson W, Ucar O, Kyewsky B, Hollander GA, Matthys P, Gray DH et al (2011) The thymic epithelial microRNA network elevates the threshold for infection-associated thymic involution via miR-29a mediated suppression of the IFN-α receptor. Nat Immunol 13:181–187

Passos GA, Mendes-da-Cruz DA, Oliveira EH (2015) The thymic orchestration involving Aire, miiRNAs and cell-cell interactions during the induction of central tolerance. Front Immunol 6:352

Passos GA, Speck-Hernandez CA, Assis AF, Mendes-da-Cruz DA (2018) Update on Aire and thymic negative selection. Immunology 153:10–20

Pearson JA, Wong FS, Wen L (2016) The importance of the Non Obese Diabetic (NOD) mouse model in autoimmune diabetes. J Autoimmun 66:76–88

Perniola R (2018) Twenty years of AIRE. Front Immunol 9:98

Pezzi N, Assis AF, Cotrim-Sousa L, Lopes GS, Mosella MS, Lima DS, Bombonato-Prado KF, Passos GA (2016) Aire knockdown in medullary thymic epithelial cells affects Aire protein, deregulates cell adhesion genes and decreases thymocyte interaction. Mol Immunol 77:157–173

Pugliese A, Zeller M, Fernandez A Jr, Zalcberg LJ, Bartlett RJ, Ricordi C, Pietropaolo M, Eisenbarth GS, Bennett ST, Patel DD (1997) The insulin gene is transcribed in the human thymus and transcription levels correlated with allelic variation at the INS VNTR-IDDM2 susceptibility locus for type 1 diabetes. Nat Genet 15:293–297

Rezzani R, Nardo L, Favero G, Peroni M, Rodella LF (2014) Thymus and aging: morphological, radiological, and functional overview. Age 36:313–351

Sansom SN, Shikama-Dorn N, Zhanybekova S, Nusspaumer G, Macaulay IC, Deadman ME, Heger A, Ponting CP, Hollander GA (2014) Population and single-cell genomics reveal the Aire dependency relief from Polycomb silencing, and distribution of self-antigen expression in thymic epithelia. Genome Res 24:1918–1931

Savino W, Mendes-da-Cruz DA, Silva JS, Dardenne M, Cotta-de-Almeida V (2002) Intrathymic T-cell migration: a combinatorial interplay of extracellular matrix and chemokines? Trends Immunol 23:305–313

Savino W, Mendes-Da-Cruz DA, Smaniotto S, Silva-Monteiro E, Villa-Verde DM (2004) Molecular mechanisms governing thymocyte migration: combined role of chemokines and extracellular matrix. J Leukoc Biol 75:951–961

Sawanobori Y, Ueta H, Dijkstra CD, Park CG, Satou M, Kitazawa Y, Matsuno K (2014) Three distinct subsets of thymic epithelial cells in rats and mice defined by novel antibodies. PLoS One 9(10):e109995

Sospedra M, Ferrer-Francesch X, Domínguez O, Juan M, Foz-Sala M, Pujol-Borrell R (1998) Transcription of a broad range of self-antigens in human thymus suggests a role for central mechanisms in tolerance toward peripheral antigens. J Immunol 161:5918–5929

Sousa Cardoso R, Magalhães DA, Baião AM, Junta CM, Macedo C, Marques MM, Sakamoto-Hojo ET, Donadi EA, Passos GA (2006) Onset of promiscuous gene expression in murine fetal thymus organ culture. Immunology 119:369–375

Speck-Hernandez CA, Assis AF, Felicio RF, Cotrim-Sousa L, Pezzi N, Lopes GS, Bombonato-Prado KF, Giuliatti S, Passos GA (2018) Aire disruption influences the medullary thymic epithelial cell transcriptome and interaction with thymocytes. Front Immunol 9:964

St-Pierre C, Brochu S, Vanegas JR, Dumont-Lagacé M, Lemieux S, Perreault C (2013) Transcriptome sequencing of neonatal thymic epithelial cells. Sci Rep 3:1860

St-Pierre C, Trofimov A, Brochu S, Lemieux S, Perreault C (2015) Differential features of AIRE-induced and AIRE-independent promiscuous gene expression in thymic epithelial cells. J Immunol 195:498–506

Takaba H, Takayanagi H (2017) The mechanisms of T cell selection in the thymus. Trends Immunol 38:805–816

Takaba H, Morishita Y, Tomofuji Y, Danks L, Nitta T, Komatsu N, Kodama T, Takayanagi H (2015) Fezf2 orchestrates a thymic program of self-antigen expression for immune tolerance. Cell 163:975–987

Takahama Y (2006) Journey through the thymus: stromal guides for T-cell development and selection. Nat Rev Immunol 6:127–135

Taub DD, Longo DL (2005) Insights into thymic aging and regeneration. Immunol Rev 205:72–93

Terszowski G, Müller SM, Bleul CC, Blum C, Schirmbeck R, Reimann J, Pasquier LD, Amagai T, Boehm T et al (2006) Evidence for a functional second thymus in mice. Science 312:284–287

Tykocinski LO, Sinemus A, Kyewski B (2008) The thymus medulla slowly yields its secrets. Ann N Y Acad Sci 1143:105–122

Ucar O, Rattay K (2015) Promiscuous gene expression in the thymus: a matter of epigenetics, miRNA, and more? Front Immunol 6:93

Ucar O, Tykocinski LO, Dooley J, Liston A, Kyewski B (2013) An evolutionarily conserved mutual interdependence between Aire and microRNAs in promiscuous gene expression. Eur J Immunol 43:1769–1778

Vafiadis P, Bennett ST, Colle E, Grabs R, Goodyer CG, Polychronakos C (1996) Imprinted and genotype-specific expression of genes at the IDDM2 locus in pancreas and leucocytes. J Autoimmun 9:397–403

Vafiadis P, Bennett ST, Todd JA, Nadeau J, Grabs R, Goodyer CG, Wickramasinghe S, Colle E, Polychronakos C (1997) Insulin expression in human thymus is modulated by INS VNTR alleles at the IDDM2 locus. Nat Genet 15:289–292

Villaseñor J, Besse W, Benoist C, Mathis D (2008) Ectopic expression of peripheral-tissue antigens in the thymic epithelium: probabilistic, monoallelic, misinitiated. Proc Natl Acad Sci U S A 105:15854–15859

Wang KC, Chang HY (2011) Molecular mechanisms of long noncoding RNAs. Mol Cell 43:904–914

Werdelin O, Cordes U, Jensen T (1998) Aberrant expression of tissue-specific proteins in the thymus: a hypothesis for the development of central tolerance. Scand J Immunol 47:95–100

Wilusz JE, Sunwoo H, Spector DL (2009) Long noncoding RNAs: functional surprises from the RNA world. Genes Dev 1(23):1494–1504

Xu M, Zhang X, Hong R, Su D-M, Wang L (2017) MicroRNAs regulate thymic epithelium in age-related thymic involution via down- or upregulation of transcription factors. J Immunol Res 2017:2528957

Yano M, Kuroda N, Han H, Meguro-Horike M, Nishikawa Y, Kiyonari H, Maemura K, Yanagawa Y, Obata K, Takahashi S, Ikawa T, Satoh R, Kawamoto H, Mouri Y, Matsumoto M (2008) Aire controls the differentiation program of thymic epithelial cells in the medulla for the establishment of self-tolerance. J Exp Med 205:2827–2838

Zuklys S, Mayer CE, Zhanybekova S, Stefanski HE, Nusspaumer G, Gill J, Barthlott T, Chappaz S, Nitta T et al (2012) MicroRNAs control the maintenance of thymic epithelia and their competence for T lineage commitment and thymocyte selection. J Immunol 189:3894–3904

Chapter 10
Functional Genomics of the Infant Human Thymus: AIRE and Minipuberty

Carlos Alberto Moreira-Filho, Silvia Yumi Bando,
Fernanda Bernardi Bertonha, and Magda Carneiro-Sampaio

Abstract Sexual dimorphism in the immune system is well documented in humans and it encompasses sex differences in responses to self and foreign antigens. Indeed, women usually mount stronger immune responses to infections and vaccination but have higher susceptibility to autoimmune diseases than men (Klein and Flanagan 2016). Regarding autoimmune diseases, it is striking that 80% of autoimmune patients are women (Rubtsova et al. 2015). Autoimmunity results from a tolerance breakdown and essentially involves the thymus, the site of T cell selection (Cheng and Anderson 2018). T cell selection depends on the ectopic thymic transcription of thousands of genes coding for tissue-specific antigens, which is induced by the autoimmune regulator gene *AIRE* (Passos et al. 2018; Perniola 2018). In spite of our incomplete knowledge on the biological processes responsible for autoimmunity, it is reasonable to assume that sex hormones impact the genomic mechanisms governing *AIRE* functions. An important experimental evidence supporting this assumption came from the work of Dumont-Lagacé et al. (2015), who showed in a murine model that sex hormones have pervasive effects on thymic epithelial cells (TEC)—antigen presenting cells that regulate T cell repertoire and tolerance—and that androgens have a greater impact on TEC transcriptome than estrogens. In this study, the authors observed that sex steroids repressed the expression of tissue-restricted antigens but did not alter the expression of *Aire*. Just after this work, Dragin et al. (2016) demonstrated that estrogen mediates the downregulation of *AIRE* in human pubescent and adult thymic tissues, thus indicating that the reduced expression of AIRE protein in women may be related to autoimmunity susceptibility. However, this study did not cover infants along the first 6 months of age, i.e. during minipuberty (Kuiri-Hänninen et al. 2014), a period when sex hormones conceivably act on thymic tissue.

C. A. Moreira-Filho (✉) · S. Y. Bando · F. B. Bertonha · M. Carneiro-Sampaio
Department of Pediatrics, Faculdade de Medicina da Universidade de São Paulo,
São Paulo, SP, Brazil
e-mail: carlos.moreira@hc.fm.usp.br; silvia.bando@hc.fm.usp.br;
fernanda.bernardi@fm.usp.br; magdacs@usp.br

© Springer Nature Switzerland AG 2019
G. A. Passos (ed.), *Thymus Transcriptome and Cell Biology*,
https://doi.org/10.1007/978-3-030-12040-5_10

10.1   Introduction

Sexual dimorphism in the immune system is well documented in humans and it encompasses sex differences in responses to self and foreign antigens. Indeed, women usually mount stronger immune responses to infections and vaccination but have higher susceptibility to autoimmune diseases than men (Klein and Flanagan 2016). Regarding autoimmune diseases, it is striking that 80% of autoimmune patients are women (Rubtsova et al. 2015). Autoimmunity results from a tolerance breakdown and essentially involves the thymus, the site of T cell selection (Cheng and Anderson 2018). T cell selection depends on the ectopic thymic transcription of thousands of genes coding for tissue-specific antigens, which is induced by the auto-immune regulator gene *AIRE* (Passos et al. 2018; Perniola 2018). In spite of our incomplete knowledge on the biological processes responsible for autoimmunity, it is reasonable to assume that sex hormones impact the genomic mechanisms govern-ing *AIRE* functions. An important experimental evidence supporting this assumption came from the work of Dumont-Lagacé et al. (2015), who showed in a murine model that sex hormones have pervasive effects on thymic epithelial cells (TEC)—antigen presenting cells that regulate T cell repertoire and tolerance—and that androgens have a greater impact on TEC transcriptome than estrogens. In this study, the authors observed that sex steroids repressed the expression of tissue-restricted antigens but did not alter the expression of *Aire*. Just after this work, Dragin et al. (2016) demon-strated that estrogen mediates the downregulation of *AIRE* in human pubescent and adult thymic tissues, thus indicating that the reduced expression of AIRE protein in women may be related to autoimmunity susceptibility. However, this study did not cover infants along the first 6 months of age, i.e. during minipuberty (Kuiri-Hänninen et al. 2014), a period when sex hormones conceivably act on thymic tissue.

In order to further investigate the presumptive sexual dimorphism induced by minipuberty on infant thymus, we performed comparative genomic and immunohis-tochemical studies on thymic surgical explants (corticomedullar sections) obtained at cardiac surgery from karyotypically normal male (M) and female (F) infants dur-ing minipuberty, here termed MM and MF groups. The same studies were con-ducted on thymic explants obtained from karyotypically normal M and F non-puberty (N) infants aged 7–31 months, the NM and NF groups. Analyses included gene coexpression networks for global expressed genes, abundantly expressed miRNA analyses, *AIRE*-centered gene-gene interaction networks encompassing the genes coding for AIRE interactors and miRNA, and quantitative RT-qPCR and immuno-histochemical measurements of AIRE expression. The methodologies here men-tioned are detailed described in Moreira-Filho et al. (2018).

We employed a network-based approach for global gene coexpression—weighted gene coexpression network analysis, or WGCNA (Langfelder and Horvath 2008)—that allows the identification of modular transcriptional repertoires (communities) and the interactions between all the system's constituents through community detection (Chaussabel and Baldwin 2014; Moreira-Filho et al. 2016). Abundantly expressed miRNA comparative analysis was performed between minipuberty and non-puberty groups. *AIRE*-centered networks allowed the measuring of gender-related differences in AIRE-interactors gene-gene expression correlation.

10.2   Functional Genomics Studies by WGCNA

Networks were constructed using the WGCNA R software package (Langfelder and Horvath 2008). Pearson's correlation coefficient was used for obtaining gene coexpression similarity measures and for the subsequent construction of an adjacency matrix using soft power and topological overlap matrix (TOM). TOM was used to filter weak connections during network construction. Module identification was based on TOM and in average linkage hierarchical clustering. Keeping to the scale-free topology criterion, it is necessary to choose a soft power threshold (β). Soft-thresholding process transformed the correlation matrix to mimic the scale-free topology. Finally, dynamic cut-tree algorithm was used for dendrogram's branch selection. The module eigengene is defined as the first principal component of a given module (ME), which can be considered a representative of the gene expression profiles in a module.

Module-trait association analysis was obtained by the correlation between the gene expression values and the phenotypic trait, such as clinical and demographic data (here represented by age and gender). This correlation value is defined as Gene Significance (GS). The mean GS for a particular module is considered as a measure of module significance (MS). The GS values were obtained using Pearson's correlation and to assign a p-value to the module significance, we used Student's t test. The modules, which presented $p < 0.05$ were considered significantly associated with the trait.

Three global gene expression networks were constructed by WGCNA using 6662 genes for three comparative analyses of module-trait association: (1) minipuberty (age < 7 months, 16 samples) and non-puberty (7 months $\leq$ age $\leq$ 31 months, 24 samples) groups; (2) minipuberty male (MM, 9 samples) and female (MF, 6 samples) groups; and (3) male (NM, 11 samples) and female (NF, 6 samples) non-puberty groups. Figure 10.1 depicts all three module-trait association analyses.

In the minipuberty and non-puberty groups (Fig. 10.1a), we obtained 23 modules and two of these modules—tan and light green—were positively and significantly associated with the minipuberty group, whereas two other modules—purple and green—were positively and significantly associated with the non-puberty group. Tan and light green modules encompass, respectively, 53 out of 191 and 58 out of 124 differentially expressed genes ($p < 0.05$), where 37 genes of the tan module and 40 genes of the light green module are hyper-expressed in the minipuberty group. In the non-puberty group, purple and green modules contain, respectively, 118 out of 237 and 212 out of 378 differentially expressed genes ($p < 0.05$), where 66 gene of the purple module and 185 genes of the green module are hyper-expressed. These results showed that minipuberty and non-puberty age groups have different gene

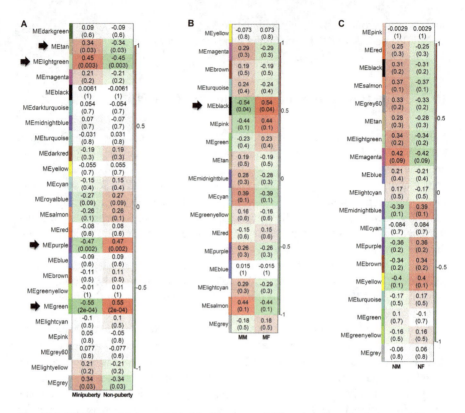

Fig. 10.1 WGCNA modules and Module Eigengene (ME) correlations. In the rows, MEs are named by their module colors. The columns indicate the traits of interest: minipuberty and non-puberty (**a**) and gender for the minipuberty (**b**) and non-puberty (**c**) groups. Numbers inside each colored box are the correlation coefficients between the ME and the specific trait (MS), and p-value is indicated between brackets. The more intense the color of the box, the more negatively (green) or positively (red) correlated is the module with the trait. None correlation presented p < 0.05. Keeping to the scale-free topology criterion, soft power β values 8, 12 and 10 were considered for **a**, **b** and **c** networks respectively

expression profiles, thus reflecting the well-known hormonal differences between these two groups (Kuiri-Hänninen et al. 2014).

We then investigated gender differences in the minipuberty and in the non-puberty groups. A total of 16 modules were obtained for the MM and MF groups (Fig. 10.1b), of which only one—the black module—was significantly associated with gender (positively with the MF group). This module contains 79 out of 422 differentially expressed genes, from which 37 genes were hyper-expressed in MF group. Finally, no differences were found between the NM and NF groups (Fig. 10.1c). As a whole, these results indicated that the gene expression profiles here obtained reflect different gene expression regulation patterns in boys and girls during minipuberty. These differences are probably due to the joint influence of genomic and hormonal mechanisms.

Next, we investigated how the distinct gene expression regulation patterns found for MM and MF groups could affect AIRE expression in minipuberty.

10.3 *AIRE* Expression Assessment in Minipuberty

AIRE expression values in MM and MF groups showed no significant difference in microarray data (p = 0.50) and in subsequent RT-qPCR analysis (p = 0.35). Microarray absolute values for *AIRE* mRNA expression were normalized to those of two thymic mTEC markers, keratin 5 (*KRT5*) and keratin 14 (*KRT14*), and no significant differences between male and female groups (p = 0.14) were found in both comparisons (Fig. 10.2a–c). The total number of thymic AIRE-positive cells and of medullary thymic epithelial cells (mTECs) expressing AIRE—positive for

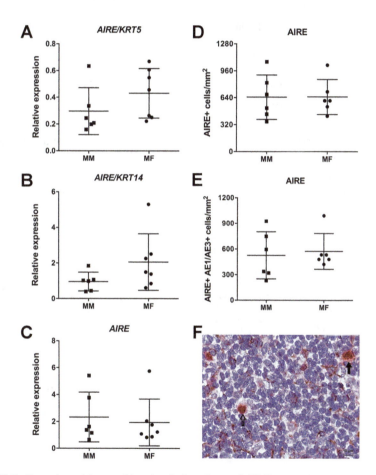

Fig. 10.2 Genomic and immunohistochemical analyses of AIRE expression. DNA microarray relative expression of *AIRE* mRNA normalized to *KRT5* (**a**) and *KRT14* (**b**); RT-qPCR relative expression of *AIRE* mRNA (**c**); Scatter plot of total AIRE positive thymic cells/mm^2 (**d**); Scatter plot of AIRE/cytokeratin positive, mTEC cells/mm^2 (**e**); Double-staining immunohistochemistry for AIRE (red color) and cytokeratin (AE1/AE3 brown color) showing AIRE/cytokeratin positive, mTEC cell (filled arrow) and AIRE-positive/cytokeratin negative, non-mTEC cells (empty arrows) cells in thymic medullary area, 400× (**f**). Unpaired Mann Whitney test was used for the comparisons shown in **a** and **b**. Unpaired Student's t-test was used for the comparisons shown in **c**, **d** and **e**. Error bars represent s.d. Statistical significance was considered with p values less than 0.05

AIRE and positive for the cytokeratin markers AE1/AE3—were comparatively assessed by IHC in thymic samples from six male and six female donors aged <6 months (Fig. 10.2d–f). Statistical analysis showed no significant difference between male and female samples for total AIRE expression (p = 0.49) and for AIRE expression in mTECs (p = 0.37). A detailed description of our studies on *AIRE* expression in minipuberty appeared in Moreira-Filho et al. (2018).

On the other hand, if no differences in AIRE expression were detected between boys and girls in minipuberty, could the androgen or estrogen rich environments cause other effects on *AIRE* functioning in thymus? This question prompted us to study: (1) the miRNA expression differences between minipuberty and non-puberty groups; (2) the *AIRE* interactors' gene-gene and gene-miRNA expression relationships in the minipuberty and non-puberty groups.

10.4 MiRNA Expression Analysis in Minipuberty and Non-puberty Groups

Abundantly expressed miRNAs were identified for minipuberty and non-puberty groups. The abundantly expressed miRNAs for these two groups—16 in the minipuberty and 17 in the non-puberty—were selected after analyzing miRNA expression value distribution through a scatter dot plot and adopting a cut-off for abundant expression values for minipuberty and non-puberty groups. In the minipuberty *vs* non-puberty comparison two miRNAs were found to be abundantly expressed only in the minipuberty group—*miR-494-3p* and *miR-6869-5p*—and three others in the non-puberty group—*let-7f-5p*, *miR-205-5p*, and *miR-125-5p* (Fig. 10.3).

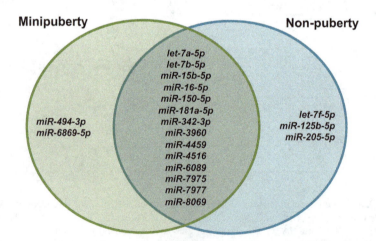

Fig. 10.3 Abundantly expressed miRNAs identified for minipuberty and non-puberty groups. In the minipuberty *vs.* non-puberty comparison two miRNAs are exclusive to the minipuberty group and three to the non-puberty group. Fourteen miRNAs are common to both groups

10.5   *AIRE* Interactors' Gene-Gene and Gene-miRNA Expression Relationships

In order to investigate these relationships, we firstly selected a set of genes coding for the AIRE-targeted proteins previously identified in TECs by Abramson et al. (2010). In the human thymus *AIRE* is almost exclusively expressed in thymic epithelial cells (TECs): only a small fraction of thymic B cells, around 5%, express *AIRE* and B cells constitute just 1% of thymic lymphocytes (Gies et al. 2017). Therefore, regarding *AIRE* expression there is no artifact in our data caused by thymocyte background. On the other hand, only genes known to be expressed in mice and/or human thymic epithelial cells (TECs)—and whose coded proteins were shown to physically associate with AIRE in TECs—were included in our *AIRE*-interactors network analysis. Additionally, we included in our analysis the abundantly expressed miRNAs found in the minipuberty and non-puberty groups. The gene expression matrix of each group was used for gene-gene expression network construction based on Pearson's correlation coefficient. Visualization was achieved through Cytoscape v. 3.0.0 (Shannon et al. 2003). We adopted an absolute correlation value of 0.80 for all AIRE-interactors, but for miRNA-gene we considered only inverse correlations. All 16 abundantly expressed miRNA have many interactions with genes in minipuberty group, differently of what was found for the non-puberty group, where about only half of the abundantly expressed miRNA have interactions—and few interactions—with genes in the networks.

These AIRE-interactors networks included *AIRE* and other 34 genes (34 genes in the minipuberty group or 33 genes in the non-puberty group), which code for proteins that are associated, directly or indirectly, with AIRE (Fig. 10.4) and exert impact on its functions. *AIRE* interactors were classified according to their molecular function and represented by different node colors in the networks. There are 14 high-interactors in MM, only 1 in MF, 8 in NM, and 6 in NF. Consequently, distinctive profiles of *AIRE* interactors' gene-gene relationships were found for each group. Altogether, these data suggest that sex hormones and genomic background exert their influence on *AIRE* interactors' gene-gene expression relationship during and after minipuberty.

It is relevant to mention that in minipuberty male network (MM) all miRNAs have links with genes interacting in first-degree with AIRE (first neighbours, thick lines). Differently, in the minipuberty female network (MF) all miRNAs have links mostly with *RANBP9*. In the non-puberty networks (NM and NF) AIRE has first-degree interaction with miR-150-5p in NM and with miR-7977 in NF. Moreover, *miR-494-3p* and *miR-6869-5p* interact with the same genes of the MM group, with exception of *LMNB1* and *CAND1*, which interact only with *miR-494-3p*. The same pattern is observed in the MF group, where both miRNAs interact with *RANBP9* and *IPO7*, but only *miR-6869-5p* interacts with *RANBP2* (Fig. 10.4). It is worth to note that *RANBP2* and *RANBP9* are involved in the shuttling of AIRE into the nucleus (Abramson and Goldfarb 2016).

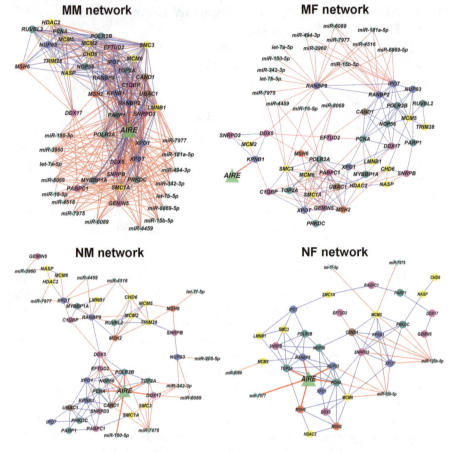

Fig. 10.4 AIRE interactors' gene-gene and gene-miRNA expression relationships. Gene-gene and gene-miRNA expression relationship networks for Minipuberty Male—MM, Minipuberty Female—MF, Non-puberty Male—NM, and Non-puberty Female—NF groups. Nodes are colored according to their molecular function (GO): green for transcription, yellow for chromatin binding/ structure, blue for nuclear transport, brown for ubiquitination, pink for pre-mRNA processing, red for DNA repair, and purple for *AIRE*. miRNAs are showed as turquoise vee nodes. Gene-gene or gene-miRNA expression correlation values >|0.80| are depicted in blue (positive values) and in red (negative values) links. Links between *AIRE* and its first neighbors (genes or miRNA) are depicted with a thick line. The node size is not proportional to the number of links

10.6   Minipuberty and Sexual Dimorphism in the Infant Human Thymus

The effects of sex steroids on thymic tissue constitute a matter of great interest since these hormones could act on the mechanisms of immune tolerance (Rubtsova et al. 2015; Dragin et al. 2016). Here we focused the effects of the transient post-natal sex steroids surge of infancy, or minipuberty, on thymic global gene expression, AIRE expression and *AIRE* interactors' gene-gene and gene-miRNA expression relationships.

Significant differences were firstly observed regarding global gene expression between boys and girls in minipuberty, as detected by WGCNA, thus indicating that the sex steroids surge in minipuberty modifies thymic global gene expression. Differential gene expression between male and female groups was found in minipuberty only and vanished in non-puberty groups. The significant sex-related thymic differential gene expression in minipuberty is probably triggered by the transient hormonal surge. However, its effects on gene functioning may extend beyond minipuberty, as shown by the differences found in AIRE-interactors networks.

AIRE expression was comparatively assessed in minipuberty male (MM) and female (MF) groups by DNA microarray, RT-qPCR, and immunohistochemistry and no significant differences were found. These results corroborate the previous findings of Dumont-Lagacé et al. (2015) in mice: in spite of gene expression differences in male and female thymic epithelial cells, the expression of *Aire* was found to be quantitatively the same in male and female mice thymuses. On the other hand, our data do not support the claim by Zhu et al. (2016) of a consistently higher expression of *AIRE* in male infants along minipuberty. Gender differences in AIRE expression were well established for prepubescent, pubescent, and adult individuals by Dragin et al. (2016).

A relevant finding of our work is derived from the analysis of gene-gene expression relationships between *AIRE* and its interactors depicted in networks constructed for minipuberty (MM and MF) and non-puberty groups (NM and NF). It was possible to obtain distinctive profiles of *AIRE* interactors' gene-gene relationships for each minipuberty and non-puberty groups. These results clearly evidence that sex hormones and XY and XX genomic backgrounds exert their influence on *AIRE* interactors' gene-gene expression relationship during and after minipuberty. Interestingly, as we previously shown (Moreira-Filho et al. 2018), neither the sex steroids surge during minipuberty, nor the XY or XX background, seem to promote any significant gender-related histomorphometric changes in the infant thymus, corroborating previous data (Steinmann et al. 1985; Steinmann 1986).

The fact that the gender-specific *AIRE*-interactors gene-gene relationships profiles were found for all minipuberty and non-puberty groups must be further considered here. Although sex hormones are thought to be major mediators of sexual dimorphism in the immune system, sex differences in immune response arise from a complex interplay of genomic, hormonal and environmental mechanisms, whose molecular bases remain to be fully determined (Markle and Fish 2014; Arnold 2017; Edwards et al. 2018). Indeed, sex differences in the susceptibility to infectious diseases (Guerra-Silveira and Abad-Franch 2013; Muenchhoff and Goulder 2014) and in response to vaccines (Klein and Flanagan 2016) are quite evident in infancy (below 1 year of age) and in early childhood (1–4 years of age), i.e. well before puberty, thus evidencing that genetic and epigenetic factors may have a role in shaping immune system sexual dimorphism.

The results presented here indicate that genomic mechanisms and postnatal hormonal influences probably act synergistically in shaping thymic sexual dimorphism along the first 6 months of life, but this process does not involve changes in *AIRE* expression, although may involve differences—perhaps long-lasting differences—in the interactions of *AIRE* with its partners.

Acknowledgments *Financial Support* This work was funded by Fundação de Amparo à Pesquisa do Estado de São Paulo (FAPESP) research grants 2015/22308-2 (CAM-F) and 2014/50489-9 (MC-S); and Conselho Nacional de Desenvolvimento Científico e Tecnológico (CNPq) grant 307626/2014-8 (CAM-F).

References

Abramson J, Goldfarb Y (2016) AIRE: from promiscuous molecular partnerships to promiscuous gene expression. Eur J Immunol 46(1):22–33

Abramson J, Giraud M, Benoist C, Mathis D (2010) Aire's partners in the molecular control of immunological tolerance. Cell 140(1):123–135

Arnold AP (2017) Y chromosome's roles in sex differences in disease. Proc Natl Acad Sci U S A 114(15):3787–3789

Chaussabel D, Baldwin N (2014) Democratizing systems immunology with modular transcriptional repertoire analyses. Nat Rev Immunol 14(4):271–280

Cheng M, Anderson MS (2018) Thymic tolerance as a key brake on autoimmunity. Nat Immunol 19(7):659–664

Dragin N, Bismuth J, Cizeron-Clairac G, Biferi MG, Berthault C, Serraf A et al (2016) Estrogen-mediated downregulation of AIRE influences sexual dimorphism in autoimmune diseases. J Clin Invest 126(4):1525–1537

Dumont-Lagacé M, St-Pierre C, Perreault C (2015) Sex hormones have pervasive effects on thymic epithelial cells. Sci Rep 5:12895

Edwards M, Dai R, Ahmed SA (2018) Our environment shapes us: the importance of environment and sex differences in regulation of autoantibody production. Front Immunol 9:478

Gies V, Guffroy A, Danion F, Billaud P, Keime C, Fauny JD et al (2017) B cells differentiate in human thymus and express AIRE. J Allergy Clin Immunol 139(3):1049–52.e12

Guerra-Silveira F, Abad-Franch F (2013) Sex bias in infectious disease epidemiology: patterns and processes. PLoS One 8(4):e62390

Klein SL, Flanagan KL (2016) Sex differences in immune responses. Nat Rev Immunol 16(10):626–638

Kuiri-Hänninen T, Sankilampi U, Dunkel L (2014) Activation of the hypothalamic-pituitary-gonadal axis in infancy: minipuberty. Horm Res Paediatr 82(2):73–80

Langfelder P, Horvath S (2008) WGCNA: an R package for weighted correlation network analysis. BMC Bioinformatics 9:559

Markle JG, Fish EN (2014) SeXX matters in immunity. Trends Immunol 35(3):97–104

Moreira-Filho CA, Bando SY, Bertonha FB, Silva FN, Costa LF, Ferreira LR et al (2016) Modular transcriptional repertoire and MicroRNA target analyses characterize genomic dysregulation in the thymus of Down syndrome infants. Oncotarget 7(7):7497–7533

Moreira-Filho CA, Bando SY, Bertonha FB, Ferreira LR, Vinhas CF, Oliveira LHB et al (2018) Minipuberty and sexual dimorphism in the infant human thymus. Sci Rep 8(1):13169

Muenchhoff M, Goulder PJ (2014) Sex differences in pediatric infectious diseases. J Infect Dis 209(Suppl 3):S120–S126

Passos GA, Speck-Hernandez CA, Assis AF, Mendes-da-Cruz DA (2018) Update on Aire and thymic negative selection. Immunology 153(1):10–20

Perniola R (2018) Twenty years of AIRE. Front Immunol 9:98

Rubtsova K, Marrack P, Rubtsov AV (2015) Sexual dimorphism in autoimmunity. J Clin Invest 125(6):2187–2193

Shannon P, Markiel A, Ozier O, Baliga NS, Wang JT, Ramage D et al (2003) Cytoscape: a software environment for integrated models of biomolecular interaction networks. Genome Res 13(11):2498–2504

Steinmann GG (1986) Changes in the human thymus during aging. Curr Top Pathol 75:43–88

Steinmann GG, Klaus B, Müller-Hermelink HK (1985) The involution of the ageing human thymic epithelium is independent of puberty. A morphometric study. Scand J Immunol 22(5):563–575

Zhu ML, Bakhru P, Conley B, Nelson JS, Free M, Martin A et al (2016) Sex bias in CNS autoimmune disease mediated by androgen control of autoimmune regulator. Nat Commun 7:11350

Chapter 11
Thymus Transcriptome of TGF-β Superfamily

Arnon Dias Jurberg and Vinicius Cotta-de-Almeida

Abstract Members of the TGF-β superfamily play pleiotropic activities in the thymus, which ultimately involve epithelial cell homeostasis and proper maturation of many T cell subsets. This chapter performs a second-level analysis of thymic transcriptomes to highlight key factors and functional evidence that may be used to guide putative medical interventions for rescuing the production of a given specific T cell type.

11.1 Introduction

The production of multiple T cell subsets as diverse as conventional CD4$^+$ and CD8$^+$ αβTCR T cells, γδTCR T cells, Foxp3$^+$ regulatory T (Treg) cells, and invariant natural killer T (NKT) cells occurs once common bone-marrow derived early thymic precursors (ETPs) enter the thymus and progressively mature as they migrate from the corticomedullary region towards the cortex and then to the thymic medulla. Thymic regions can be recognized based on their distinct cell densities caused by an initial cell expansion upon β-selection in the cortex and a later strong decline in cell numbers by negative selection, as cells move to the medulla, where central T cell tolerance emerges (Anderson et al. 2005; Pearse 2006). This cellular saga for developing T lymphocytes involves several processes, such as gene recombination, cell migration, proliferation and apoptosis, which are regulated by numerous complex molecular cascades. Some of these signaling pathways are part of the transforming

A. D. Jurberg (✉)
Laboratory on Thymus Research, Oswaldo Cruz Institute, Oswaldo Cruz Foundation, Pavilhão Leônidas Deane, Rio de Janeiro, RJ, Brazil
e-mail: ajurberg@ioc.fiocruz.br

V. Cotta-de-Almeida (✉)
Laboratory on Thymus Research, Oswaldo Cruz Institute, Oswaldo Cruz Foundation, Pavilhão Leônidas Deane, Rio de Janeiro, RJ, Brazil

Estácio de Sá University (UNESA), Rio de Janeiro, RJ, Brazil
e-mail: vca@ioc.fiocruz.br

© Springer Nature Switzerland AG 2019
G. A. Passos (ed.), *Thymus Transcriptome and Cell Biology*,
https://doi.org/10.1007/978-3-030-12040-5_11

growth factor-β (TGF-β) superfamily, which play important roles on thymus development and function (Jurberg et al. 2015).

In this chapter, we revisited the components of the TGF-β superfamily to perform a second-level analysis on transcriptomic data sets, retrieved from the Immunological Genome Project (ImmGen) Consortium, for both thymic epithelial cells and thymocytes (Heng et al. 2008). Our analysis sought to provide additional insights on the physiology of thymic epithelial cells (TECs) and thymopoiesis, in an attempt to direct further studies using conditional gene targeting systems and complement the understanding of bidirectional lymphoid-epithelial communication. Precise modulation of signaling cascades may ultimately contribute to improve thymus function, revert its age-related involution, or mitigate autoimmunity.

11.2   Signaling by Members of the TGF-β Superfamily in the Thymus

The superfamily of TGF-β ligands gathers about 33 proteins with a similar overall basic structure, which includes a N-terminal signal peptide involved in cell secretion, a regulatory latency-associated peptide (LAP) that may act in protein sequestration, and a mature peptide at the C-terminus that must be released by proprotein convertases to exert biological functions. These factors are subdivided based on sequence similarity and downstream cascade activation into families, such as TGF-β, bone morphogenetic proteins (BMPs), growth and differentiation factors (GDFs), activins/inhibins, and glial cell line-derived neurotrophic factor (GDNF), in addition to other ligands such as Nodal, Lefty proteins, anti-Müllerian hormone (AMH), and SCUBE3 (Wu et al. 2011; Budi et al. 2017). In general, they bind single-pass transmembrane type II receptors, like ActRII (*Acvr2a*) and ActRIIB (*Acvr2b*), which then recruit type I activin receptor-like kinases (ALKs) to form a heterotetramer which phosphorylates Smad proteins. However, this sequence of receptor recruitment may vary depending on the ligand (de Caestecker 2004). Upon phosphorylation, these intracellular effectors move into the nucleus to regulate the expression of numerous target genes (Wrana 2013; David and Massagué 2018). The impact of TGF-β superfamily components on thymus physiology and aging is not fully understood, partially because of the formation of active heterodimers, the utilization of multiple receptor combinations and shared downstream components, as well as similar outcome responses and the existence of many selective inhibitory proteins (Walker et al. 2016).

The differential expression of TGF-β superfamily ligands and corresponding receptors in the thymus clustered cells into lymphoid and stromal compartments. In general, TECs showed higher levels of both TGF-β factors and receptors than lymphoid cells. The inhibitory Smad7 was also found highly expressed not only in TECs, but also in developing thymocytes (Fig. 11.1). This Smad protein targets type I receptors for proteasome-dependent degradation and causes severe thymic atrophy when overexpressed in TECs under the control of a keratin 5 promoter (He et al. 2002). Furthermore, TGF-β-associated gene expression profiles specifically subdi-

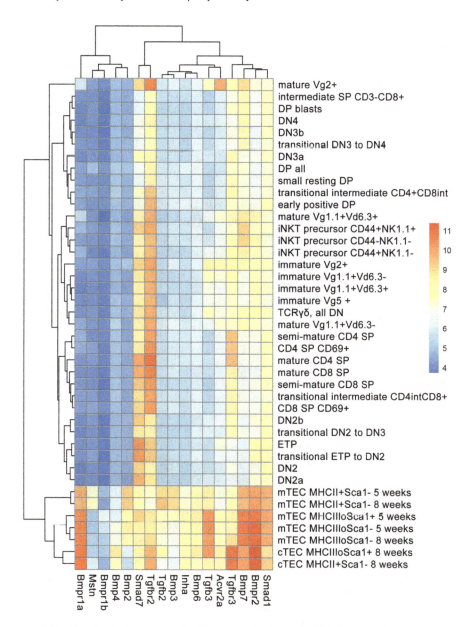

Fig. 11.1 Signaling by TGF-β superfamily members in thymus. TGF-β ligands, their receptors and downstream Smad effectors (columns) were differentially expressed in distinct subsets of thymic epithelial cells and thymocytes (rows). Hierarchical clustering heatmap of microarray datasets from ImmGen using a coefficient of variation of 10%. Colors represent levels of expression as indicated

vided the stromal compartment into mTECs and cTECs, with *Bmpr1a*, *Bmpr2*, and *Tgfbr3* receptor genes and the *Bmp7* ligand gene exhibiting the highest levels of expression (Fig. 11.1). Indeed, Bmp7 signals through Alk3 (*Bmpr1a*) and BmprII (*Bmpr2*) receptors to phosphorylate Smad1/5/8 and inhibits TGF-β signaling in many distinct systems (Weiskirchen and Meurer 2013). The expression of *Bmpr2* was found to be higher in α6^{hi}Sca-1^{hi}CD45$^-$EpCAM$^+$Foxn1$^+$ thymic epithelial progenitor cells (TEPCs) and in immature CD45$^-$EpCAM$^+$UEA1$^-$Ly51$^+$MHCII^{lo} cTECs (Barsanti et al. 2017). This receptor also binds Bmp2 and Bmp4 ligands with low affinity, which are expressed in TECs at intermediate levels along with *Bmp3* and *Bmp6*, as well as the TGF-β isoform gene *Tgfb3* and the activin-related gene *Inha* (Fig. 11.1). Among these Bmp ligands, Bmp4 has a major role in TEC homeostasis by stimulating the expression of *Foxn1*, TEPC self-renewal and immature TEC survival (Barsanti et al. 2017). In addition, the presence of the type III co-receptor betaglycan (*Tgfbr3*) in TECs, which is also expressed by most developing thymocytes, can potentiate ligand activity upon binding (López-Casillas et al. 1993, 1994; Eickelberg et al. 2002). Accordingly, both *Acvr2a* and *Tgfbr2* receptor genes are found at mid-levels of expression in TECs (Fig. 11.1). Whereas the ActRII (*Acvr2a*) participates in many pathways, including TGF-βs and activins, the TβRII (*Tgfbr2*) is exclusive for relaying TGF-β signals (reviewed by Massagué and Gomis 2006; Mueller and Nickel 2012). In the epithelial compartment, as revealed by its conditional gene inactivation, the TβRII seems to restrict only the proliferation of mTECs (Hauri-Hohl et al. 2014). In turn, *Bmpr1b* and *Mstn* (myostatin/Gdf8) exhibited mid to low expressions (Fig. 11.1), but their roles in thymus biology are vastly unknown. When taken together, none of the aforementioned genes was uniquely expressed by a single thymic epithelial cell type. This discourages the possibility that they can be used as cell specific markers in the thymus, although both *Bmpr1a* and *Bmpr2* may be used to discriminate between the epithelial and lymphoid compartments. Moreover, these expression profiles suggest the existence of multiple TGF-β superfamily-associated autocrine loops of signaling in TECs that may also play paracrine roles during thymopoiesis. However, it is not possible to exclude the contribution of distant sources of TGF-β superfamily signals in the thymus, since they can circulate through the bloodstream (Loffredo et al. 2013), or the participation of other stromal cells, such as macrophages and endothelial cells. For instance, thymic endothelial cells secrete Bmp4 upon acute damage to stimulate Foxn1-dependent cTEC multiplication and revert organ atrophy (Wertheimer et al. 2018). Although we observed no sharp age-related influence on gene expression due to the small age difference among TEC samples, age-associated modulation of TGF-β superfamily members and their receptors cannot be disregarded in future studies.

Similar to the epithelial compartment, almost all lymphoid cell types examined exhibited high expression of *Tgfbr2* and *Smad7* (Fig. 11.1). The role of Smad7 in lymphocytes was demonstrated in peripheral CD4$^+$CD25$^-$ T cells through the TGF-β-mediated induction of Foxp3, which directly inhibited the expression of *Smad7* and led to the generation of Treg cells (Fantini et al. 2004). It is therefore possible that Smad7 also takes part in the production of Treg cells in the thymus. In turn, TβRII seems to participate in thymic negative selection, as the inactivation of *Tgfbr2* in developing thymocytes promoted apoptosis of TCRβ^{high}CD4$^+$ and TCRβ^{high}CD8$^+$

mature SP cells upon activation by anti-CD3 or transgenic TCRβ^highOT-II T cells upon antigen-dependent stimulation (Ouyang et al. 2010). However, unlike TECs, the general expression of *Bmpr1a* was low, whereas the levels of *Bmp7*, *Bmpr2*, *Tgfbr3* and *Smad1* were more intermediate. By conditionally inactivating *Bmpr1a* in hematopoietic cells, Hager-Theodorides et al. (2014) have shown that Bmp2/4 signaling play a dual role during thymopoiesis, first by controlling the pace of cell differentiation to DN3 and later restraining cell expansion at the DP stage. Indeed, treatment of fetal thymic organ culture (FTOC) at E15.0 with recombinant BMP4 resulted in cell cycle arrest of DN1 cells, a phenotype that was also achieved upon BMP2, but not with BMP7 incubation (Graf et al. 2002; Tsai et al. 2003; Cejalvo et al. 2007). Possibly due to redundant Bmp7 secretion by TECs, no phenotype was observed on thymopoiesis upon conditional gene inactivation in the hematopoietic lineage (Passa et al. 2011). On the other hand, the genetic inactivation of *Tgfbr3* in FTOC led to impaired production of DP cells due to increased thymocyte apoptosis (Aleman-Muench et al. 2012).

Overall, TGF-β-associated gene expression profiles in the lymphoid compartment clustered roughly by cell developmental stage, except by some particularities. Early thymic precursors (ETP) clustered with both DN2a and DN2b cells, but instead of clustering with DN3a, DN3b, DN3-to-DN4, and DN4 thymocytes, this group was closely related to distinct stages of both CD4^+ and CD8^+ SP cells. In turn, the DN3/DN4 group was close to DP cells and CD3^−CD8^+ intermediate SP cells (Fig. 11.1). At the protein level, however, DN cells exhibited elevated Alk4 (*Acvr1b*) and increased Smad2 phosphorylation (pSmad2), whereas Alk5/TβRI (*Tgfbr1*) and TβRII (*Tgfbr2*) receptors showed only moderate quantities (Rosendahl et al. 2003). Curiously, distinct subsets of innate-like natural killer T (iNKT) cells and γδ T cells cluster together based on the expression profile of TGF-β superfamily members (Fig. 11.1). Their proximity to ETP/SP cells may be explained by shifts in the conventional αβ TCR differentiation program. In particular, recent evidence points out that iNKT cells have at least two developmental origins in the thymus. Whereas classical iNKT cells arise from DP cells undergoing positive selection through interaction with CD1d, a non-classical MHC-related protein, Th1-biased iNKT cells differentiate earlier from late DN cells (Bendelac 1997; Dashtsoodol et al. 2017; White et al. 2018).

The importance of TGF-β superfamily signaling on the differentiation of iNKT cells was unveiled by genetic investigation of TβRI and TβRII, although a few mature iNKT cells have escaped these perturbations (Doisne et al. 2009). It is possible, however, that this evasion may be explained by the alternative differentiation from DN cells instead of the classical CD4^+ DP thymocytes, a hypothesis that awaits further experimental investigation. By using mixed bone marrows from CD4-Cre;*Tgfbr2*^fl/fl mice and *CD1d* mutant mice transferred into irradiated *Rag2*-deficient mice, Doisne et al. (2009) have shown that unresponsiveness to TGF-β signals disrupted the generation of DP-derived iNKT cells due to impaired precursor survival. Furthermore, the conditional activation of a constitutively active form of TβRI (TβRI^CA) in CD4^+ thymocytes blocked the progression of iNKT precursors at the CD44^loNK1.1^− stage through a T-bet/CD122-dependent mechanism, which led to an increased circulation of immature iNKT cells. In addition, *survival* of early CD44^loNK1.1^− *precursors depended on the TIF1γ branch of* TGF-β *signaling*, whereas the Smad4 branch was responsible to maintain T-bet/CD122 levels and sustain NK1.1 expression in late CD44^{hi}NK1.1$^+$ precursors. However, the transition from CD44^{hi}NK1.1$^-$ to CD44^{hi}NK1.1$^+$ relied on a brief TGF-β inhibition as revealed by the additive phenotype of compound CD4-Cre;*Tif1γ*$^{fl/fl}$;*Smad4*$^{fl/fl}$ mice (Doisne et al. 2009). In accordance with this finding, the downregulation of TGF-β signaling at this stage may result from an increase in the levels of miR-17 and miR-19 families that target *Tgfbr2* mRNA (Fedeli et al. 2016). Although signaling by Bmp ligands influence iNKT effector activities in the periphery, there is no information to our knowledge on their effects in thymic cell maturation.

Likewise, evidence of TGF-β superfamily activity on γδ T cells in the thymus is also more restricted at present. These cells arise from DN progenitors receiving stronger TCR signaling, which further produce distinct tissue-specific cell subsets based on VγVδ combination and cytokine production (reviewed by Muñoz-Ruiz et al. 2017). Although not fully understood, each subset requires TCR-dependent and TCR-independent stimulation for proper differentiation into IFNγ-producing γδ T cells or IL-17-producing γδ T cells. Among TCR-independent signals, Do et al. (2010) have demonstrated that inactivation of either *Tgfb1* or *Smad3* abrogates the expression of *Il17* without interfering with IFN-γ levels. Other TGF-β superfamily-associated stimuli are largely unknown and further studies are still required to elucidate their participation on γδ T cell development.

11.3   Concluding Remarks

Our discussion presented herein reveals that a collection of thymus metadata from comprehensive gene expression databases, along with detailed phenotypic information, might be correlated with functional data on TGF-β superfamily members. In fact, although gene expression information should be cautiously considered, since mRNA levels might correlate poorly with protein abundance (Vogel and Marcotte 2012), we have presented many instances supporting functional roles of TGF-β superfamily components for thymopoiesis and TEC homeostasis. The broad and particular expression of several TGF-β superfamily ligands and receptors in the thymus reinforces their importance for proper development and production of both lymphoid and stromal cell subsets. In addition, it is important to point out that further variability in gene expression may arise from infection, injury, aging, or other conditions that should not be disregarded at the time to evaluate putative candidate interventions.

Acknowledgments This work benefited from data assembled by the ImmGen consortium. A.D.J. is supported by a fellowship from the Rio de Janeiro State Research Council/FAPERJ (E-26/202.683/2016). The work was financially supported by the Oswaldo Cruz Institute/Fiocruz, the Estacio de Sá University (Productivity in Research Program), the Brazilian National Institute of Science and Technology on Neuroimmunomodulation/INCT-NIM, MERCOSUL Fund for Structural Convergence/FOCEM, the Brazilian Research Council/CNPq, and the Rio de Janeiro State Research Council (FAPERJ).

References

Aleman-Muench GR, Mendoza V, Stenvers K et al (2012) Betaglycan (TβRIII) is expressed in the thymus and regulates T cell development by protecting thymocytes from Apoptosis. PLoS One 7:1–10. https://doi.org/10.1371/journal.pone.0044217

Anderson MS, Venanzi ES, Chen Z et al (2005) The cellular mechanism of Aire control of T cell tolerance. Immunity 23:227–239. https://doi.org/10.1016/j.immuni.2005.07.005

Barsanti M, Lim JMC, Hun ML et al (2017) A novel *Foxn1eGFP*/+ mouse model identifies Bmp4-induced maintenance of *Foxn1* expression and thymic epithelial progenitor populations. Eur J Immunol 47:291–304. https://doi.org/10.1002/eji.201646553

Bendelac A (1997) Selection and adaptation of cells expressing major histocompatibility complex class I-specific receptors of the natural killer complex. J Exp Med 186:349–351

Budi EH, Duan D, Derynck R (2017) Transforming growth factor-β receptors and Smads: regulatory complexity and functional versatility. Trends Cell Biol 27:658–672. https://doi.org/10.1016/j.tcb.2017.04.005

Cejalvo T, Sacedón R, Hernández-López C et al (2007) Bone morphogenetic protein-2/4 signalling pathway components are expressed in the human thymus and inhibit early T-cell development. Immunology 121:94–104. https://doi.org/10.1111/j.1365-2567.2007.02541.x

Dashtsoodol N, Shigeura T, Aihara M et al (2017) Alternative pathway for the development of Vα14+ NKT cells directly from CD4−CD8− thymocytes that bypasses the CD4+CD8+ stage. Nat Immunol 18:274–282. https://doi.org/10.1038/ni.3668

David CJ, Massagué J (2018) Contextual determinants of TGFβ action in development, immunity and cancer. Nat Rev Mol Cell Biol 19:419–435. https://doi.org/10.1038/s41580-018-0007-0

de Caestecker M (2004) The transforming growth factor-beta superfamily of receptors. Cytokine Growth Factor Rev 15:1–11. https://doi.org/10.1002/stem.504

Do J, Fink PJ, Li L et al (2010) Cutting Edge: Spontaneous development of IL-17-producing γδ T cells in the thymus occurs via a TGF-β1-dependent mechanism. J Immunol 184:1675–1679. https://doi.org/10.4049/jimmunol.0903539

Doisne J-M, Bartholin L, Yan K-P et al (2009) iNKT cell development is orchestrated by different branches of TGF-β signaling. J Exp Med 206:1365–1378. https://doi.org/10.1084/jem.20090127

Eickelberg O, Centrella M, Reiss M et al (2002) Betaglycan inhibits TGF-β signaling by preventing type I-type II receptor complex formation: glycosaminoglycan modifications alter betaglycan function. J Biol Chem 277:823–829. https://doi.org/10.1074/jbc.M105110200

Fantini MC, Becker C, Monteleone G et al (2004) Cutting Edge: TGF-β induces a regulatory phenotype in CD4+CD25− T cells through Foxp3 induction and down-regulation of Smad7. J Immunol 172:5149–5153. https://doi.org/10.4049/jimmunol.172.9.5149

Fedeli M, Riba M, Garcia Manteiga JM et al (2016) miR-17~92 family clusters control iNKT cell ontogenesis via modulation of TGF-β signaling. Proc Natl Acad Sci U S A 113:E8286–E8295. https://doi.org/10.1073/pnas.1612024114

Graf D, Nethisinghe S, Palmer DB et al (2002) The developmentally regulated expression of Twisted gastrulation reveals a role for bone morphogenetic proteins in the control of T cell development. J Exp Med 196:163–171. https://doi.org/10.1084/jem.20020276

Hager-Theodorides A, Ross S, Sahni H et al (2014) Direct BMP2/4 signaling through BMP receptor IA regulates fetal thymocyte progenitor homeostasis and differentiation to CD4+CD8+ double-positive cell. Cell Cycle 13:324–333. https://doi.org/10.4161/cc.27118

Hauri-Hohl M, Zuklys S, Holländer GA, Ziegler SF (2014) A regulatory role for TGF-β signaling in the establishment and function of the thymic medulla. Nat Immunol 15:554–561. https://doi.org/10.1038/ni.2869

He W, Li AG, Wang D et al (2002) Overexpression of Smad7 results in severe pathological alterations in multiple epithelial tissues. EMBO J 21:2580–2590. https://doi.org/10.1093/emboj/21.11.2580

Heng TSP, Painter MW, Immunological Genome Project Consortium (2008) The Immunological Genome Project: networks of gene expression in immune cells. Nat Immunol 9:1091–1094. https://doi.org/10.1038/ni1008-1091

Jurberg AD, Vasconcelos-Fontes L, Cotta-de-Almeida V (2015) A tale from TGF-β superfamily for thymus ontogeny and function. Front Immunol 6:1–15. https://doi.org/10.3389/fimmu.2015.00442

Loffredo FS, Steinhauser ML, Jay SM et al (2013) Growth differentiation factor 11 is a circulating factor that reverses age-related cardiac hypertrophy. Cell 153:828–839. https://doi.org/10.1016/j.cell.2013.04.015

López-Casillas F, Wrana JL, Massagué J (1993) Betaglycan presents ligand to the TGF-β signaling receptor. Cell 73:1435–1444. https://doi.org/10.1016/0092-8674(93)90368-Z

López-Casillas F, Payne HM, Andres JL, Massagué J (1994) Betaglycan can act as a dual modulator of TGF-β access to signaling receptors: mapping of ligand binding and GAG attachment sites. J Cell Biol 124:557–568. https://doi.org/10.1083/jcb.124.4.557

Massagué J, Gomis RR (2006) The logic of TGFβ signaling. FEBS Lett 580:2811–2820

Mueller TD, Nickel J (2012) Promiscuity and specificity in BMP receptor activation. FEBS Lett 586:1846–1859. https://doi.org/10.1016/j.febslet.2012.02.043

Muñoz-Ruiz M, Sumaria N, Pennington DJ, Silva-Santos B (2017) Thymic determinants of γδ T cell differentiation. Trends Immunol 38:336–344. https://doi.org/10.1016/j.it.2017.01.007

Ouyang W, Beckett O, Ma Q, Li MO (2010) Transforming growth factor-β signaling curbs thymic negative selection promoting regulatory T cell development. Immunity 32:642–653. https://doi.org/10.1016/j.immuni.2010.04.012

Passa O, Tsalavos S, Belyaev NN et al (2011) Compartmentalization of bone morphogenetic proteins and their antagonists in lymphoid progenitors and supporting microenvironments and functional implications. Immunology 134:349–359. https://doi.org/10.1111/j.1365-2567.2011.03495.x

Pearse G (2006) Normal structure, function and histology of the thymus. Toxicol Pathol 34:504–514. https://doi.org/10.1080/01926230600865549

Rosendahl A, Speletas M, Leandersson K et al (2003) Transforming growth factor-β- and Activin-Smad signaling pathways are activated at distinct maturation stages of the thymopoeisis. Int Immunol 15:1401–1414. https://doi.org/10.1093/intimm/dxg139

Tsai PT, Lee RA, Wu H (2003) BMP4 acts upstream of FGF in modulating thymic stroma and regulating thymopoiesis. Blood 102:3947–3953. https://doi.org/10.1182/blood-2003-05-1657

Vogel C, Marcotte EM (2012) Insights into the regulation of protein abundance from proteomic and transcriptomic analyses. Nat Rev Genet 13:227–232. https://doi.org/10.1038/nrg3185

Walker RG, Poggioli T, Katsimpardi L et al (2016) Biochemistry and biology of GDF11 and myostatin (Response to Walker et al). Circ Res 118:1125–1142. https://doi.org/10.1161/CIRCRESAHA.116.308391

Weiskirchen R, Meurer SK (2013) BMP-7 counteracting TGF-β1 activities in organ fibrosis. Front Biosci (Landmark Ed) 18:1407–1434

Wertheimer T, Velardi E, Tsai J et al (2018) Production of BMP4 by endothelial cells is crucial for endogenous thymic regeneration. Sci Immunol 3:eaal2736. https://doi.org/10.1126/sciimmunol.aal2736

White AJ, Lucas B, Jenkinson WE, Anderson G (2018) Invariant NKT cells and control of the thymus medulla. J Immunol 200:3333–3339. https://doi.org/10.4049/jimmunol.1800120

Wrana JL (2013) Signaling by the TGFβ superfamily. Cold Spring Harb Perspect Biol 5:a011197. https://doi.org/10.1101/cshperspect.a011197

Wu YY, Peck K, Chang YL et al (2011) SCUBE3 is an endogenous TGF-β receptor ligand and regulates the epithelial-mesenchymal transition in lung cancer. Oncogene 30:3682–3693. https://doi.org/10.1038/onc.2011.85

Chapter 12
Development of Thymic Regulatory T Lymphocytes

Larissa Vasconcelos-Fontes, Rafaella Ferreira-Reis, João Ramalho Ortigão-Farias, Arnon Dias Jurberg, and Vinicius Cotta-de-Almeida

Abstract A healthy immune system should maintain a balance between the ability to respond to infectious agents and tumor cells at the same time that sustains self-tolerance. For this purpose, the immune system must be capable of restraining foreign antigens, stopping the immune response after the resolution of a problem and blocking autoreactivity of immune cells that have escaped negative selection. For the maintenance of homeostasis and peripheral tolerance, a group of T lymphocytes named regulatory T (Treg) cells is produced from the CD4$^+$ T cells in the thymus or in the periphery, where they have the ability to control the immune response. This chapter reviews the development of Treg cells within the thymus, detailing their markers and involved factors, including Foxp3 and other recently discovered transcription factors, as well as some noncoding RNAs. We ultimately discuss disorders associated with Treg cell deficiency and the modulation of thymic Treg cells to treat diseases.

12.1 A Brief History of the Discovery of Treg Cells

The first description of specialized cells in regulating the immune response occurred in 1969, with the identification of a suppressive subset of T cells in autoimmune oophoritis (Nishizuka and Sakakura 1969). The authors observed that mice thymectomyzed at the third day of life developed an organ-specific autoimmune disease in the ovaries. Later reports demonstrated that depletion of certain thymic T cells in

L. Vasconcelos-Fontes (✉) · R. Ferreira-Reis · J. R. Ortigão-Farias · A. D. Jurberg
Laboratory on Thymus Research, Oswaldo Cruz Institute, Oswaldo Cruz Foundation, Rio de Janeiro, RJ, Brazil
e-mail: lvfbiomed@gmail.com

V. Cotta-de-Almeida (✉)
Laboratory on Thymus Research, Oswaldo Cruz Institute, Oswaldo Cruz Foundation, Rio de Janeiro, RJ, Brazil

Estácio de Sá University (UNESA), Rio de Janeiro, RJ, Brazil
e-mail: vca@ioc.fiocruz.br

© Springer Nature Switzerland AG 2019   255
G. A. Passos (ed.), *Thymus Transcriptome and Cell Biology*,
https://doi.org/10.1007/978-3-030-12040-5_12

normal animals induced autoimmune diseases, whereas the reconstitution of these cells prevented and suppressed them (Gershon and Kondo 1970). However, despite two decades of study, the information on this suppressor cell population was yet incomplete. A central observation was that a small population of CD4+ T lymphocytes had a suppressive action and somehow inhibited CD4 and CD8 T-cell responses in grafts (Hall et al. 1990). A subpopulation of CD4+ T lymphocytes expressing the α chain of the IL-2 receptor (CD25) was further characterized and, via cell transfer assays, was demonstrated to be responsible for suppressing autoimmune diseases (Sakaguchi et al. 1995).

Later, the Foxp3 transcription factor (Forkhead box P3), a central protein for the differentiation process of Treg cells, was discovered in a murine model (scurfy mice). It is a member of the forkhead/winged-helix family of transcription factors and highly conserved between mice and humans (Bennett et al. 2001; Wildin et al. 2001). This transcription factor is a marker preferentially expressed in thymic and peripheral CD4+CD25+ Treg cells, which distinguishes them from effector CD4+ T cells. Mutations in the *Foxp3* gene, causing loss of gene function, lead to a fatal disease in humans known as IPEX (polyendocrinopathy, enteropathy, X-linked syndrome), which is similar to what happens in scurfy mice. Both present immune dysregulation caused by the absence of Treg cells, demonstrating that Foxp3 is critical for their development (Bennett et al. 2001). Therefore, Foxp3 is defined as the main regulator for the differentiation of Treg cells, a central factor for their intrathymic development and immune regulatory functions. This discovery defined them as a bona fide lineage and opened the field to understand their mechanisms of action and clinical use.

12.2 Subpopulations of Treg Cells

Treg cell populations are defined according to their origin of development. Some studies have shown that early Treg cell development happens in the thymus, whereas others have shown that they can also develop in the periphery. For standardization purposes, the nomenclature is defined as: (a) thymic Treg (tTreg) cells emerge on the thymus (previously called nTreg cells based on their natural origin); (b) pTreg cells differentiate in the periphery (formerly called adaptive Treg cells or simply aTreg cells), and include *in vitro* induced Treg cells (iTreg cells) (Abbas et al. 2013). Nevertheless, there are discussions about the existence of a unique marker that could differentiate both populations. For some time it was believed that the transcription factor Helios, which belongs to the Ikaros family, was a potential marker to distinguish tTreg and pTreg, since only tTreg cells would present this factor. However, Helios could be induced in CD4 T cells with transgenic TCR under stimulation and presence of IL-2 and TGF-β (Verhagen and Wraith 2010). Helios also could be induced during the activation and proliferation of T cells, but recedes under resting conditions (Akimova et al. 2011). Therefore, these data suggest that the expression of Helios in iTreg cells can be observed upon particular *in vitro* activation methods.

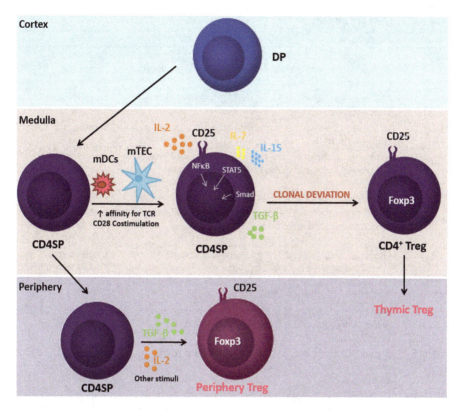

Fig. 12.1 Signaling in Treg cell differentiation processes. When CD4[+] SP cells bind with high affinity to MHC class II presented by medullary dendritic cells (mDCs) and mTECs, in the presence of costimulatory B7-1/B7-2 signals, they begin to express the IL-2 receptor α-chain. Together with co-stimulation with IL-2, IL-7, IL-15 and TGF-β, these cells upregulate distinct transcription factors, such as NFκB, STAT5 and Smads, driving clonal deviation toward tTreg cells, which escape apoptosis and express Foxp3. CD4[+] T cells that reach the periphery and may get in contact with TGF-β and IL-2, in conjunction with other tissue-specific or inflammation-associated stimuli, can also differentiate into pTreg cells

12.2.1 tTreg Cells

tTreg cells are generated in the thymus and escape negative selection during lymphocyte maturation (Fig. 12.1). These cells express Foxp3 as the main developmental regulator, which is critical for their suppressive function. The tTreg cells are selected following TCR recognition of self-peptides/MHC with high affinity along with the costimulatory signal of B7 molecules and other cytokines, such as IL-2, TGF-β, IL-7 and IL-15 (Josefowicz et al. 2012; Salomon et al. 2000). Mutant mice for TGF-β or IL-2 continued to have tTreg cells, which suggests that these molecules are important for the effector function of the cell, but are not essential for its generation (Furtado et al. 2002). However, loss of the TGF-β receptor II in thymocytes caused a reduction in the production of tTreg cells in newborn mice, likely due to a decreased negative

Table 12.1 tTreg cell markers

Protein	Function	References
CD25	IL-2 receptor α-chain; It activates the signaling pathway that will culminate with the activation of STAT5. It is important for Treg cell development and maintenance.	Sakaguchi et al. (1995)
CD122	IL-2 receptor β-chain. It forms the high affinity receptor complex for IL-2.	Thornton et al. (2004)
CD132	IL-2 receptor γ-chain. It forms the high affinity receptor complex for IL-2.	Thornton et al. (2004)
FOXP3	Transcription factor critical to the development of Treg cells. It represses the transcription of Th1 and Th2 genes.	Fontenot et al. (2003), Hori et al. (2003), Lin et al. (2007)
HELIOS	Transcription factor from the Ikaros family. It binds to the Foxp3 promoter, sustains the expression of Foxp3 and enhances its suppressive activity.	Sugimoto et al. (2006), Thornton et al. (2010), Getnet et al. (2010)
GITR	Glucocorticoid-induced TNF receptor. It is important for the generation of tTreg cells by increasing the responsiveness to IL-2 and STAT5. It is expressed after TCR activation upon TAK1 kinase activation and CD28 co-stimulation.	Kanamaru et al. (2004), Mahmud et al. (2014), Shimizu et al. (2002)
C-REL	Member of the nuclear factor κB (NF-κB) transcription factor family; important for the development of Foxp3⁺Helios⁺Treg cells.	Richards et al. (2015), Luu et al. (2017)
STAT5	Signal transducer and activator of transcription 5; member of the STAT family of transcription factors. It induces *Foxp3* expression.	Burchill et al. (2007), Soper et al. (2007)
CTLA-4	Cytotoxic T-lymphocyte-associated protein 4; it binds to CD80 and CD86 on antigen presenting cells and prevents binding of effector T cells.	Takahashi et al. (2000), Zheng et al. (2006)
THPOK	Zinc finger-containing transcription factor for precursor cells prior the expression of *Foxp3*.	Carpenter et al. (2017)
OX40 (CD134)	Member of the TNF Receptor Superfamily important in the generation of Treg cells by increasing the responsiveness to IL-2 and STAT5. It is expressed upon TCR activation upon TAK1 kinase activation and CD28 co-stimulation.	So et al. (2008), Mahmud et al. (2014)

selection, further confirming that TGF-β signaling is crucial for the inhibition of Bim-dependent apoptosis in autoreactive thymocytes (Ouyang et al. 2010).

The tTreg cells express high levels of CD25 on its surface and the Foxp3 transcription factor, thus being defined as CD4⁺CD25⁺Foxp3⁺ cells. The β-chain (CD122) and the γ-chain (CD132) of the IL-2R shape the receptor together with the α-chain (CD25), thus constituting a high affinity complex for IL-2 (Thornton and Shevach 1998; Thornton et al. 2004). In addition, several other molecules have often been added to the lists of tTreg cell markers (Table 12.1) and play functional roles during cell development.

It is important to note that human tTreg cells can be divided into functionally and phenotypically distinct subpopulations based on the levels of Foxp3 expression, cell surface markers as CD45RA, the DNA methylation profile at the FOXP3 regulatory regions and the physiological state of the cell, among other factors. In this context, resting Treg cells can be defined as CD45RA$^+$FOXP3^{lo}, whereas activated Treg cells are CD45RA$^-$FoxP3^{hi} and CD45RA$^-$FoxP3^{lo} cells are cytokine-secreting non-Treg lymphocytes (Miyara et al. 2009).

12.2.2 pTreg Cells

Induction of Smad3-dependent Foxp3 expression in the periphery leads to a pTreg phenotype (Fig. 12.1), but the involved mechanisms are not completely understood due to the lack of reliable methods to discriminate distinct Treg cell types in lymphoid tissues and in the blood (Thornton et al. 2010). Generation of pTreg cells is triggered by TCR activation and by TGF-β signaling, in addition to a retinoic acid-associated increase in histone acetylation-related accessibility of phosphorylated Smad3 (pSmad3) to the Foxp3 promoter (Xu et al. 2010). Furthermore, it appears that the presence of TNF inhibits the production of pTreg cells, as it selectively inhibits the TGF-β-Smad3 signaling pathway through TNFR2 receptor regulation (Zhang et al. 2013). In fact, some clones of T cells are able to develop into Treg cells by responding to non-self antigens, but the mechanism is not fully understood. However, the chronic exposure to antigens, like in a chronic *Leishmania major* infection, induces Treg cell capable to regulate the response to this pathogen, by IL-10-dependent and IL-10-independent suppressive mechanisms. It was also demonstrated that pTreg cells are responsible for the immunotolerance to microbiota and non-pathogenic substances from the environment. Disruption of the induction of these cells leads to an impaired immune system (Belkaid et al. 2002; Josefowicz et al. 2012).

12.3 Differentiation of Treg Cells Within the Thymus

Treg cells differentiate from single-positive CD4$^+$CD25$^+$Foxp3$^-$ precursors during negative selection in the thymus. Thymocytes interacting with some level of affinity to MHC class II in combination with costimulatory molecules (B7 molecules) and cytokines (like IL-2, TGF-β and IL-15), start a signaling program activating STAT5 and Foxp3, which leads to Treg cell development. In the following subsections, we briefly explain the differentiation process of tTreg cells.

12.3.1 Thymus and Development of T Lymphocytes

The thymus is a primary lymphoid organ covered by a dense connective capsule that projects septa dividing it in two lobes. This organ has an external cortical region, with a dense accumulation of lymphocytes, and a central medullary region, which is less populated by lymphoid cells. The thymic parenchyma is mainly composed of developing thymocytes and non-lymphoid stromal cells, such as cortical thymic epithelial cells (cTECs), medullary thymic epithelial cells (mTECs), fibroblasts, dendritic cells (DCs) and macrophages. The thymus has an extracellular matrix (ECM) network composed mainly by fibronectin, laminin and type-4 collagen that permeates all other cells. The microenvironmental cells and ECM emit multiple signals that promote the development of thymocytes. In response to these signals, developing thymocytes are induced to proliferate and migrate through the thymic parenchyma, ultimately differentiating into different subsets of mature T lymphocytes.

Briefly, the immature progenitors from the bone marrow enter the cortical-medullary junction. These cells migrate to the subcapsular zone in the cortex and are referred as double-negative (DN) cells for not expressing CD4 or CD8; at this stage, the DN cells also do not express CD25 but express CD44 ($CD4^-CD8^-CD44^+CD25^-$) (Ciofani and Zúñiga-Pflücker 2007; Petrie and Zúñiga-Pflücker 2007). In the cortex, cells start the recombination of their TCR, by the V-D-J recombination, and the thymocytes that pass through this stage present the pre-TCR. Cells that have an unproductive rearrangement suffer apoptosis and are phagocytosed by macrophages and dendritic cells (Dudley et al. 1994; Surh and Sprent 1994). If the production of a pre-TCR is successful, thymocytes continue their migration through the parenchyma and undergo the rearrangement of the TCR α chain. At the same time, CD4 and CD8 protein levels are also upregulated, so the cells express the two CD4 and CD8 receptors, defining the double-positive (DP) stage (Dudley et al. 1994; Petrie et al. 1993). During thymocyte migration, cortical thymic epithelial cells (cTECs) present peptides through the MHC to cells under development. Cells that interact with low affinity are able to survive, and recognizing MHC class I molecules become single-positive cells (SP) for the CD8 receptor and the cells that recognize the MHC class II molecules become SP CD4 T cells; such a selective mechanism is called positive selection (Morris and Allen 2012; Singer et al. 2009). In the medulla, cells that bind with high avidity to MHC undergoes negative selection (clonal deletion) or clonal deviation, mechanisms dependent on mTEC and DCs. The distinction in cell fate is made in terms of interaction duration and strength, therefore producing different levels of intracellular signaling (Hinterberger et al. 2010). mTECs express the *Aire* gene (named after "autoimmune regulator") that regulates the expression and presentation of several tissue-specific self-antigens (TSAs) to developing thymocytes in a dose-dependent manner, which prevents the emergence of autoreactive cells. Thymocytes that interact with high affinity and strong MHC-peptide complex signaling enter programmed cell death (apoptosis) through the activation of caspase 8. This process selects mature T cells that are ready to populate the periphery (Passos et al. 2015; Savino and Dardenne 2010; Xing and Hogquist 2012). Cells that recognize with high avidity the antigens presented by the mTECs trough MHC class II may undergo a clonal deviation, which comprises a series of stimuli that lead to the generation of Treg cells (Xing and Hogquist 2012).

12.3.2   Clonal Deviation

The clonal deviation is the process responsible for producing tTreg cells (Fig. 12.1). The difference between clonal deletion and clonal deviation is challenging, as both are related to the high intensity and duration of TCR interaction with the MHC (Xing and Hogquist 2012). In this way, it has been pointed out that the difference might be related to the interaction with different thymic antigen presenting cells (APCs) (Xing and Hogquist 2012). Although it has been reported that mTECs would be sufficient and DCs would be dispensable in Treg cell differentiation (Aschenbrenner et al. 2007), antigens expressed by DCs are involved in negative selection (Proietto et al. 2008). These studies showed that the mechanisms of deviation and deletion occur in a complex interaction between these cells, with signaling of chemokines like CCL19, CCL21, CCL17 and CCL22. Moreover, transgenic mice expressing lower levels of MHC class II in mTECs were less effective in clonal deletion, whereas supporting higher clonal deviation and consequently higher frequency of Treg cells in the thymus (Hinterberger et al. 2010). APCs, mainly DCs, B cells and macrophages influence the maturation of tTreg cells presenting antigens through MHC II. Other studies suggested that $CD8\alpha^+$ DCs play an important role in the differentiation of tTreg cells, since they are able to take up TSAs produced by mTECs and present them to differentiating thymocytes. About half of Aire-dependent deletion or Treg cell selection used a pathway dependent on antigen presentation by APCs, specifically $CD8\alpha^+$ DCs (Perry et al. 2014). This finding is at odds with Leventhal et al. (2016), which demonstrated that $CD8\alpha^+$ DCs are dispensable for Treg cell differentiation. However, a further study suggested that $CD8\alpha^+$ DCs uptake Aire-dependent peptides through the CD36 receptor, which recognizes and internalizes apoptotic bodies (Perry et al. 2018). Moreover, thymic B cells seem also to contribute to the maintenance and size of the tTreg compartment by promoting the generation of tTreg precursors (Lu et al. 2015).

Interestingly, age-associated involution seems to lead to the deregulation of clonal deletion mechanisms and to favor clonal deviation. Both conditional inactivation of *Foxn1* in postnatal TECs and natural aging produce no effect in the generation of tTreg cells, which raise the possibility that the atrophied thymus balances defective negative selection by sustaining tTreg cell numbers to maintain central T cell tolerance upon aging (Oh et al. 2017). This finding may be related to a decreased TCR signaling strength due to inefficient promiscuous expression of self-antigens by mTECs showing reduced levels of negative selection-related markers Nur77 and CD5 in $CD4^+$ SP thymocytes. Therefore, although there is an escape of autoreactive clones, homeostasis might be achieved as the thymus has increased production of tTreg cells (Oh et al. 2017).

12.3.3   Signaling to Treg Cell Differentiation

In addition to TCR signals, CD28 costimulation induces Treg cell differentiation. This was observed in CD28-deficient mice, which exhibited decreased numbers of Treg cells in the thymus and in peripheral tissues (Salomon et al. 2000; Tai et al.

2005). In this regard, CD28 appears to provide an intrinsic signal for the development of Treg cells by stimulating a higher production of Treg cell survival-related IL-2 (Burchill et al. 2008; Liu et al. 2008). Besides the interaction with the B7-1/B7r2 molecules, the intensity of interaction with ICAM-1 and CD40 contributes to the generation of tTreg cells, once their absence decreases cell numbers (Buhlmann et al. 2003; Sakaguchi 2005). In addition, TCR stimulation along with CD28 co-stimulation is fundamental for the initiation of Treg cell differentiation program, as CD28 participates in the induction of *Foxp3*, *Gitr*, and *Ctla* expression. The TCR-CD28 axis induces several signaling pathways, such as the activation of the NFκB family of transcription factors, where c-Rel is the main factor in tTreg cell development (Richards et al. 2015). Deficiency in c-Rel causes a severe reduction of tTreg cell numbers, mainly of the Helios⁺Foxp3⁺ Treg subset (Luu et al. 2017). Furthermore, murine thymic epithelium expresses high levels of IFN regulatory factor 4 (Irf4), which regulates mTEC maturation and is required to prime mTEC into efficient Treg cell inducers (Haljasorg et al. 2017).

As described above, soluble factors such as TGF-β and IL-2 are also centrally involved in the development of Treg cells. In particular, the signaling by IL-2 seems essential for tTreg cell differentiation, since mice double-mutant for IL-2 and IL-15, or for CD122 (IL-2 receptor subunit β) and IL-15, present almost complete depletion of tTreg cells. At molecular levels, IL-2 signaling is necessary for the translocation of phosphorylated STAT5 (pSTAT5) to the nucleus, subsequently leading to the induction of *Foxp3* expression. Indeed, STAT5 proteins are essential for the IL-2/IL-2R signaling pathway (Burchill et al. 2007; Soper et al. 2007). In particular, Snow et al. (2003) have showed that double-knockout mice for STAT-5a and STAT-5b have autoimmunity similar to IL-2 or IL-2R single knockout mice. In a broader perspective, tTreg precursors cells increase their competence to IL-2 during cell differentiation, which is crucial for the development of mature tTreg cells (Ross and Cantrell 2018). Moreover, OX40-deficient mice present reduced numbers of tTreg precursors and tTreg cells, whereas treatment with OX40L in wild-type mice or human cells induced cell proliferation in a TCR-independent manner (Kumar et al. 2018). Likewise, the absence of the transcription factor Thpok reduces the number of tTreg cells and the expression of the Treg cell marker OX40 (Carpenter et al. 2017).

12.3.4 Mechanisms for Regulating Gene Expression During tTreg Cell Development

The *Foxp3* promoter has less transcriptional activity than other cis-regulatory elements. Moreover, some epigenetics changes are independent of Foxp3. These studies proposed a model in which the organization of the chromatin prior to Foxp3 induction and the expression of cofactors interacting with Foxp3 are critical components for the regulation of Foxp3-dependent genes (Kitagawa et al. 2015; Ohkura et al. 2013). There are four conserved non-coding sequences (CNSs) on the *Foxp3* locus called CNS 0, 1, 2 and 3. The most recently discovered region, CNS0, contains the biding sites for Satb1, which can change the chromatin conformation and makes it more accessible for super-enhancers (Kitagawa et al. 2017). Specifically, the CNS3 is a very important sequence for tTreg cells, as its deletion leads to a severe reduction in the thymic production of Treg cells. Another sequence, the CNS2, is highly demethylated in functional Treg cells, and is a good marker of commitment to Treg cell lineage (Iizuka-Koga et al. 2017; Lee and Lee 2018). As described previously, c-Rel acts in Foxp3 binding to the transcription promoter, binding to and demethylation of CNS2 and binding to CNS3. Foxo family of transcription factors is also involved in regulating Foxp3 binding directly to CNS1 and CNS3. Moreover, the signaling triggered by TGF-β and TCR respectively prompt Smad3 and NFAT to modulate Foxp3 expression. NFAT can bind to CNS2, promote a chromatin loop between the promoter and CNS2, and transactivate Foxp3 induction (Iizuka-Koga et al. 2017; Lee and Lee 2018).

Besides, a recent study suggested the genome organizer and transcription factor Satb1 as critical to mediate the function of Treg cell-specific super-enhancers (Treg-SEs) in the Treg precursor cells. These super-enhancers act associated with *Foxp3* and other Treg cell signature genes like *Ctla4* and *Il2ra* (Kitagawa et al. 2017). Additionally, developing T cells requires the linear ubiquitin chain assembly complex (LUBAC) at the late stages of thymocyte differentiation maintaining Treg cell homeostasis (Teh et al. 2016). Such a complex is essential for innate immunity and is composed of at least three proteins: ring finger protein 31 (RNF31/HOIP), RanBP-type and C3HC4-type zinc finger containing 1 (RBCK1/HOIL-1) and SHANK-associated RH domain interacting protein (SHARPIN/SIPL1). In fact, SHARPIN-deficient mice have significantly reduced numbers of Foxp3[+] Treg cells and almost complete loss of tTreg development (Redecke et al. 2016).

The microRNA (*miRNA*) pathway is very important for the differentiation of tTreg cells. A pivotal study demonstrated that Dicer is crucial for the development of T cells, outlined a specific miRNA profile for Treg cells and also showed that FoxP3 reinforces this pattern (Cobb et al. 2006). The first mechanistic finding about the molecular role of a specific miRNA in the development of T cells revealed that miR181a controls the intensity of TCR signaling (Li et al. 2007). A further study showed a regulatory mechanism for the development of tTreg cells (Lu et al. 2009) by clarifying how miR155 acts over the signal transduction of IL-2 pathway. This miRNA targets SOCS1 and consequently increases the tTreg responsiveness to IL-2 stimulus in the thymus. Moreover, miR155-deficient Treg cells have markedly reduced STAT5 phosphorylation under a low induction of IL-2 (Lu et al. 2009). Interestingly, Foxp3 enhances miR155 expression by directly binding to its genomic locus (mir-155/Bic) (Marson et al. 2007) and generates a positive feedback for differentiation. The increase of miR155 during the process of differentiation is necessary for achieving the proper fitness to keep the population of tTreg cells (Lu et al. 2009).

Adding one more layer of control, the long noncoding RNAs (lncRNAs) are also involved in tTreg differentiation. Flicr (Foxp3 long intergenic noncoding RNA) was shown to negatively regulate Foxp3 activity at the post-transcriptional level (Zemmour et al. 2017). This lncRNA partially overlaps the *Foxp3* locus and is specifically expressed at the final stage of Treg cell maturation in the thymus, arising concomitantly to the Foxp3 expression within CD4$^+$CD25$^+$ SP thymocytes. Flicr is downmodulated by IL-2 signaling and seems to be associated to enhanced Treg cell activity, as Flicr-deficient NOD (nonobese diabetic) mice present less severe insulitis along with lower incidence of autoimmune diabetes (Zemmour et al. 2017).

Altogether, these findings point out noncoding transcripts, miRNAs and lncRNAs, as critical players adding up to the complex nature of transcriptional regulation of tTreg cell differentiation and function. Moreover, transcriptome studies aimed at building comprehensive database of both coding and noncoding transcripts might allow the search for expression patterns able to correlate Treg cell development and function with protective or aberrant immunity in human diseases. Such an approach may be pivotal to further target Treg cell signaling pathways for therapeutical purposes.

12.4   Learning About tTreg Development from Two Genetic Diseases

As defined above, the *FOXP3* gene is essential for the role of Treg cells in self-tolerance, as it regulates the intrathymic development and Treg cell function (Khattri et al. 2003). In fact, natural mutations in the *FOXP3* gene, located on chromosome Xp11.23 (Bennett et al. 2001), impair Treg cell development and function, leading to hyperactivation of effector T cells and the emergence of numerous spontaneous autoimmune pathologies (Fontenot et al. 2003; Sakaguchi 2005). Scurfy mice represent the clearest example how mutations in *Foxp3* cause multiorgan inflammation and enlargement of spleen and lymph nodes, due to the infiltration of lymphoid and myeloid cells (Bacchetta et al. 2018; Godfrey et al. 1991; Hadaschik et al. 2015). Interestingly, the transfer of CD4$^+$CD8$^-$CD25$^+$Foxp3$^+$ Treg cells into scurfy mice prevents the overproliferation of CD4$^+$CD8$^-$CD25$^-$ effector T lymphocytes, evidencing the importance of immunoregulatory mechanisms mediated by Treg cells to control the autoimmunity (Hori et al. 2003; Khattri et al. 2003).

In humans, mutations in the *FOXP3* gene results in a primary immunodeficiency known as IPEX (immune dysregulation, polyendocrinopathy, enteropathy X-linked). At least 70 distinct *FOXP3* mutations can occur in IPEX patients (Bacchetta et al. 2018), most of which result in a change in the amino acid sequence in the DNA-binding domain of the FOXP3 protein (Van Der Vliet and Nieuwenhuis 2007). Unlike Scurfy mice, which fail to translate Foxp3 protein, these mutations in IPEX patients may result in a protein totally translated, but functionally impaired. These mutations are related to the clinical outcomes; therefore, the more severe cases are associated with patients with a point mutation in the DNA-binding forkhead domain. In these patients, it is possible to find Treg cells in the circulation with suppressive activity, but not with the same level of the control cells (Bacchetta et al. 2006). Children with mutations outside of the forkhead domain show the milder disease but patients with mutation that results in non-expression of FOXP3 have severely reduced numbers of Treg cells (Ochs et al. 2005). IPEX syndrome is characterized by autoimmune endocrinopathy frequently observed in almost all patients and they suffer with Type 1 diabetes mellitus at the early months of life (Wildin et al. 2002), and the disease is also associated with thyroiditis, resulting in higher levels of thyroid-stimulating hormone or the presence of anti-thyroid antibodies (Ochs et al. 2007). Some patients develop inflammatory bowel disease, characterized by extensive lymphocytic infiltrates, which result in severe diarrhea and villous atrophy (Ochs et al. 2007), and severe atopy is also found, including eczema, food allergy, and eosinophilia (Chatila et al. 2000). IPEX is frequently fatal in infancy and the patients require nutritional support, immunosuppressive therapy, or allogeneic bone marrow transplantation, which can cure the disease (Barzaghi et al. 2018).

Interestingly, autoimmunity is also an important and very common complication in Wiskott-Aldrich Syndrome (WAS) patients (Sullivan et al. 1994), and about 70% of them are affected by at least one autoimmune disorder (Dupuis-Girod et al. 2003). This syndrome is a rare X-linked immunodeficiency, caused by mutations in *WAS* gene, located at Xp11.22-p11.23 region (Buchbinder et al. 2014). *WAS* encodes WASP (Wiskott-Aldrich Syndrome Protein), a key regulator of actin polymerization in hematopoietic cells (Symons et al. 1996), associated with receptor engagement and the signaling events, including cytoskeletal actin rearrangement (Massaad et al. 2013). The apparent incongruence between the presence of important immunodeficiency concomitant with autoimmunity events leads to the hypothesis that WASP deficiency can directly affect Treg cells (Adriani et al. 2007). In fact, WASP absence may imply disturbances in tissue localization and functional capacity of these cells (Humblet-baron et al. 2007; Maillard et al. 2007; Marangoni et al. 2007). WASP-deficient Treg cells are not able to expand normally in the periphery, have a reduced suppressor activity on effector T cells *in vitro*, defects in IL-10 secretion, and inability to mediate *in vivo* tolerance. Although FOXP3 expression seems not to be affected by the WASP deficiency (Humblet-baron et al. 2007; Marangoni et al. 2007), lower number of thymic Treg cells have been reported in Wasp-knockout mice (WKO) (Maillard et al. 2007). These mice exhibit many characteristics like the human disease, including defects in leukocyte migration, impaired T and B cell signaling, and the presence of autoimmunity as well (Snapper et al. 1998; Westerberg et al. 2005). In this context, studies of colitis induction, by CD45RB+ cell transfer in SCID mice, demonstrated that Treg cells from WKO mice are unable to prevent the development of this inflammatory process (Maillard et al. 2007).

It is important to point out that untreated patients may die in the first decade of life, unless hematologic and immune reconstitution is achieved by hematopoietic stem cell transplantation (HSCT) (Notarangelo et al. 2008; Symons et al. 1996). Gene therapy has also been reported to restore WASP expression and function in cells of children (Aiuti et al. 2013; Hacein-Bey Abina et al. 2015) and adults (Morris et al. 2017), including in CD4+CD25+FoxP3+ Tregs (Hacein-Bey Abina et al. 2015).

Ultimately, although autoreactivity in WAS patients might be ascribed to other changes in immune function (Adriani et al. 2011), these data point that autoimmunity in the context of WASP deficiency appears to reflect, at least in part, defects in Treg cell homeostasis and its function. This suggests an important role of WASP in regulating the major cellular activities for the control of autoreactive T and B cells and the maintenance of self-tolerance.

12.5 Concluding Hallmark: Treg Cell as Target of Therapies

Since Treg cells play prominent immunomodulatory roles, their adoptive transfer has been considered for clinical application in the treatment of immune disorders, such as autoimmunity, transplant rejection and graft-versus-host disease (GvHD) (Roncarolo and Battaglia 2007). The first evidence showing that Treg cell infusion could be used to treat both GvHD and autoimmunity in mice was shown by (Sakaguchi et al. 1995), who identified a population of CD4$^+$CD25$^+$ cells able to rescue peripheral tolerance. Since then, clinical studies revealed their efficacy (Bluestone et al. 2015; Brunstein et al. 2013; Chandran et al. 2017; Leventhal et al. 2012; Mathew et al. 2018; Todo et al. 2016; Trzonkowski et al. 2009) and numerous clinical trials are currently investigating ways to modulate the number of Treg cells in different organs and diseases (Gliwiński et al. 2017). Interestingly, human thymus obtained from children during cardiac surgery has been reported to be an efficient source for generating functional and long-lived Treg cells, with higher stability and suppressive capacity, as compared to peripheral blood-derived Treg cells (Dijke et al. 2016). Taken together, the intricate mechanisms involved in Treg cell development and the complexity of Treg cell biology in different disease settings point out to the necessity of further studies in order to support the clinical applications of Treg cells as a cellular medicinal product.

Acknowledgments This work was financially supported by the Oswaldo Cruz Institute/Fiocruz, the Brazilian National Institute of Science and Technology on Neuroimmunomodulation/INCT-NIM, MERCOSUL Fund for Structural Convergence/FOCEM, the Brazilian Research Council/CNPq, Estacio de Sá University (Productivity in Research Program) and the Rio de Janeiro State Research Council (FAPERJ).

References

Abbas AK, Benoist C, Bluestone JA, Campbell DJ, Ghosh S, Hori S et al (2013) Regulatory T cells: recommendations to simplify the nomenclature. Nat Immunol 14(4):307–308. https://doi.org/10.1038/ni.2554
Adriani M, Aoki J, Horai R, Thornton AM, Konno A, Kirby M et al (2007) Impaired in vitro regulatory T cell function associated with Wiskott-Aldrich syndrome. Clin Immunol 124(1):41–48

Adriani M, Jones KA, Uchiyama T, Kirby MR, Silvin C, Anderson SM et al (2011) Defective inhibition of B-cell proliferation by Wiskott-Aldrich syndrome protein-deficient regulatory T cells. Blood 117(24):6608–6611

Aiuti A, Biasco L, Scaramuzza S, Ferrua F, Cicalese MP, Baricordi C et al (2013) Lentivirus-based gene therapy of hematopoietic stem cells in Wiskott-Aldrich Syndrome. Science 341(6148):1233151

Akimova T, Beier UH, Wang L, Levine MH, Hancock WW (2011) Helios expression is a marker of T cell activation and proliferation. PLoS One 6(8):e24226

Aschenbrenner K, D'Cruz LM, Vollmann EH, Hinterberger M, Emmerich J, Swee LK et al (2007) Selection of Foxp3+ regulatory T cells specific for self antigen expressed and presented by Aire+ medullary thymic epithelial cells. Nat Immunol 8(4):351–358

Bacchetta R, Passerini L, Gambineri E, Dai M, Allan SE, Lawitschka A et al (2006) Defective regulatory and effector T cell functions in patients with FOXP3 mutations. J Clin Invest 116(6):1713–1722

Bacchetta R, Barzaghi F, Roncarolo MG (2018) From IPEX syndrome to FOXP3 mutation: a lesson on immune dysregulation. Ann N Y Acad Sci 1417(1):5–22

Barzaghi F, Amaya Hernandez LC, Neven B, Ricci S, Kucuk ZY, Bleesing JJ et al (2018) Long-term follow-up of IPEX syndrome patients after different therapeutic strategies: an international multicenter retrospective study. J Allergy Clin Immunol 141(3):1036–1049.e5

Belkaid Y, Piccirillo CA, Mendez S (2002) CD4+ CD25+ regulatory T cells control Leishmania major persistence and immunity. Nature 420(September):633–637

Bennett CL, Christie J, Ramsdell F, Brunkow ME, Ferguson PJ, Whitesell L et al (2001) The immune dysregulation, polyendocrinopathy, enteropathy, X-linked syndrome (IPEX) is caused by mutations of FOXP3. Nat Genet 27(1):20–21

Bluestone JA, Buckner JH, Fitch M, Gitelman SE, Gupta S, Hellerstein MK et al (2015) Type 1 diabetes immunotherapy using polyclonal regulatory T cells. Sci Transl Med 7(315):315ra189. http://stm.sciencemag.org/cgi/doi/10.1126/scitranslmed.aad4134

Brunstein CG, Miller JS, Cao Q, Mckenna DH, Hippen KL, Curtsinger J et al (2013) Infusion of ex vivo expanded T regulatory cells in adults transplanted with umbilical cord blood: safety profile and detection kinetics. Blood 117(3):1061–1070

Buchbinder D, Nugent DJ, Fillipovich AH (2014) Wiskott-Aldrich syndrome: diagnosis, current management, and emerging treatments. Appl Clin Genet 7:55–66. http://www.ncbi.nlm.nih.gov/pubmed/24817816

Buhlmann JE, Elkin SK, Sharpe AH (2003) A role for the B7-1/B7-2:CD28/CTLA-4 pathway during negative selection. J Immunol 170(11):5421–5428

Burchill MA, Yang J, Vogtenhuber C, Blazar BR, Farrar MA (2007) IL-2 receptor beta-dependent STAT5 activation is required for the development of Foxp3+ regulatory T cells. J Immunol 178(1):280–290. http://www.ncbi.nlm.nih.gov/pubmed/17182565. http://www.jimmunol.org/content/178/1/280.full.pdf

Burchill MA, Yang J, Vang KB, Moon JJ, Chu HH, Lio C-WJ et al (2008) Linked T cell receptor and cytokine signaling govern the development of the regulatory T cell repertoire. Immunity 28(1):112–121. http://www.sciencedirect.com/science/article/pii/S1074761307005870

Carpenter AC, Wohlfert E, Chopp LB, Vacchio MS, Nie J, Zhao Y et al (2017) Control of regulatory T cell differentiation by the transcription factors Thpok and LRF. J Immunol 199(5):1716–1728

Chandran S, Tang Q, Sarwal M, Laszik ZG, Putnam AL, Lee K et al (2017) Polyclonal regulatory T cell therapy for control of inflammation in kidney transplants. Am J Transplant 17(11):2945–2954

Chatila TA, Blaeser F, Ho N, Lederman HM, Voulgaropoulos C, Helms C et al (2000) JM2, encoding a fork head-related protein, is mutated in X-linked autoimmunity-allergic disregulation syndrome. J Clin Invest 106(12):75–81

Ciofani M, Zúñiga-Pflücker JC (2007) The thymus as an inductive site for T lymphopoiesis. Annu Rev Cell Dev Biol 23:463–493. http://www.ncbi.nlm.nih.gov/pubmed/17506693

Cobb BS, Hertweck A, Smith J, O'Connor E, Graf D, Cook T et al (2006) A role for Dicer in immune regulation. J Exp Med 203(11):2519–2527

Dijke IE, Hoeppli RE, Ellis T, Pearcey J, Huang Q, McMurchy AN et al (2016) Discarded human thymus is a novel source of stable and long-lived therapeutic regulatory T cells. Am J Transplant 16(1):58–71

Dudley EC, Petrie HT, Shah LM, Owen MJ, Hayday AC (1994) T cell receptor beta chain gene rearrangement and selection during thymocyte development in adult mice. Immunity 1(2):83–93

Dupuis-Girod S, Medioni J, Haddad E, Quartier P, Cavazzana-Calvo M, Le Deist F et al (2003) Autoimmunity in Wiskott-Aldrich syndrome: risk factors, clinical features, and outcome in a single-center cohort of 55 patients. Pediatrics 111(5 Pt 1):e622–e627

Fontenot JD, Gavin MA, Rudensky AY (2003) Foxp3 programs the development and function of CD4+CD25+ regulatory T cells. Nat Immunol 4(4):330–336

Furtado GC, Curotto de Lafaille MA, Kutchukhidze N, Lafaille JJ (2002) Interleukin 2 signaling is required for CD4(+) regulatory T cell function. J Exp Med 196(6):851–857

Gershon RK, Kondo K (1970) Cell interactions in the induction of tolerance: the role of thymic lymphocytes. Immunology 18(5):723–737

Getnet D, Grosso J, Goldberg M, Harris T, Yen H, Bruno T et al (2010) A role for the transcription factor Helios in human CD4+CD25+ regulatory T cells. Mol Immunol 47(7–18):1595–1600

Gliwiński M, Iwaszkiewicz-Grześ D, Trzonkowski P (2017) Cell-based therapies with T regulatory cells. BioDrugs 31(4):335–347

Godfrey VL, Wilkinson JE, Russell LB (1991) X-linked lymphoreticular disease in the scurfy (sf) mutant mouse. Am J Pathol 138(6):1379–1387. http://www.pubmedcentral.nih.gov/articlerender.fcgi?artid=1886400&tool=pmcentrez&rendertype=abstract

Hacein-Bey Abina S, Gaspar HB, Blondeau J, Caccavelli L, Charrier S, Buckland K et al (2015) Outcome following gene therapy in patients with severe Wiskott-Aldrich syndrome. HHS Public Access 33(4):395–401

Hadaschik EN, Wei X, Leiss H, Heckmann B, Niederreiter B, Steiner G et al (2015) Regulatory T cell-deficient scurfy mice develop systemic autoimmune features resembling lupus-like disease. Arthritis Res Ther 17(1):1–12

Haljasorg U, Dooley J, Laan M, Kisand K, Bichele R, Liston A et al (2017) Irf4 expression in thymic epithelium is critical for thymic regulatory T cell homeostasis. J Immunol 198(5):1952–1960

Hall BYBM, Pearce NW, Gurley KAYE, Dorschi SE (1990) Specific unresponsiveness in rats with prolonged cardiac allograft survival after treatment with cyclosporine. J Exp Med 171:141–157

Hinterberger M, Aichinger M, da Costa OP, Voehringer D, Hoffmann R, Klein L (2010) Autonomous role of medullary thymic epithelial cells in central CD4(+) T cell tolerance. Nat Immunol 11(6):512–519. https://doi.org/10.1038/ni.1874

Hori S, Nomura T, Sakaguchi S (2003) Control of regulatory T cell development by the transcription factor Foxp3. Science 299(5609):1057–1061

Humblet-baron S, Sather B, Anover S, Becker-herman S, Kasprowicz DJ, Khim S et al (2007) Wiskott-Aldrich syndrome protein is required for regulatory T cell homeostasis. J Clin Invest 117(2):407–418

Iizuka-Koga M, Nakatsukasa H, Ito M, Akanuma T, Lu Q, Yoshimura A (2017) Induction and maintenance of regulatory T cells by transcription factors and epigenetic modifications. J Autoimmun 83:113–121

Josefowicz SZ, Lu L-F, Rudensky AY (2012) Regulatory T cells: mechanisms of differentiation and function. Annu Rev Immunol 30(1):531–564

Kanamaru F, Youngnak P, Hashiguchi M, Nishioka T, Takahashi T, Sakaguchi S et al (2004) Costimulation via glucocorticoid-induced TNF receptor in both conventional and CD25+ regulatory CD4+ T cells. J Immunol 172(12):7306–7314. http://www.jimmunol.org/cgi/doi/10.4049/jimmunol.172.12.7306

Khattri R, Cox T, Yasayko S-A, Ramsdell F (2003) An essential role for Scurfin in CD4+CD25+ T regulatory cells. Nat Immunol 4(4):337–342

Kitagawa Y, Ohkura N, Sakaguchi S (2015) Epigenetic control of thymic Treg-cell development. Eur J Immunol 45(1):11–16. http://doi.wiley.com/10.1002/eji.201444577

Kitagawa Y, Ohkura N, Kidani Y, Vandenbon A, Hirota K, Kawakami R et al (2017) Guidance of regulatory T cell development by Satb1-dependent super-enhancer establishment. Nat Immunol 18(2):173–183

Kumar P, Marinelarena A, Raghunathan D, Ragothaman VK, Saini S, Bhattacharya P et al (2018) Critical role of OX40 signaling in the TCR-independent phase of human and murine thymic Treg generation. Cell Mol Immunol. https://doi.org/10.1038/cmi.2018.8

Lee W, Lee GR (2018) Transcriptional regulation and development of regulatory T cells. Exp Mol Med 50(3):e456

Leventhal J, Abecassis M, Miller J, Gallon L, Ravindra K, Tollerud DJ et al (2012) Chimerism and tolerance without GVHD or engraftment syndrome in HLA-mismatched combined kidney and hematopoietic stem cell transplantation. Sci Transl Med 4(124):124ra28

Leventhal DS, Gilmore DC, Berger JM, Nishi S, Lee V, Kline DE et al (2016) Dendritic cells coordinate the development and homeostasis of organ-specific regulatory T cells. Immunity 44(4):847–859

Li QJ, Chau J, Ebert PJR, Sylvester G, Min H, Liu G et al (2007) miR-181a is an intrinsic modulator of T cell sensitivity and selection. Cell 129(1):147–161

Lin W, Haribhai D, Relland LM, Truong N, Carlson MR, Williams CB et al (2007) Regulatory T cell development in the absence of functional Foxp3. Nat Immunol 8(4):359–368

Liu Y, Zhang P, Li J, Kulkarni AB, Perruche S, Chen W (2008) A critical function for TGF-beta signaling in the development of natural CD4+CD25+Foxp3+ regulatory T cells. Nat Immunol 9(6):632–640

Lu LF, Thai TH, Calado DP, Chaudhry A, Kubo M, Tanaka K et al (2009) Foxp3-dependent microRNA155 confers competitive fitness to regulatory T cells by targeting SOCS1 protein. Immunity 30(1):80–91

Lu F-T, Yang W, Wang Y-H, Ma H-D, Tang W, Yang J-B et al (2015) Thymic B cells promote thymus-derived regulatory T cell development and proliferation. J Autoimmun 61:62–72. http://linkinghub.elsevier.com/retrieve/pii/S0896841115000797

Luu M, Jenike E, Vachharajani N, Visekruna A (2017) Transcription factor c-Rel is indispensable for generation of thymic but not of peripheral Foxp3+ regulatory T cells. Oncotarget 8(32):52678–52689

Mahmud SA, Manlove LS, Schmitz HM, Xing Y, Wang Y, Owen DL et al (2014) Costimulation via the tumor-necrosis factor receptor superfamily couples TCR signal strength to the thymic differentiation of regulatory T cells. Nat Immunol 15(5):473–481. http://www.ncbi.nlm.nih.gov/pubmed/24633226

Maillard MH, Cotta-de-Almeida V, Takeshima F, Nguyen DD, Michetti P, Nagler C et al (2007) The Wiskott-Aldrich syndrome protein is required for the function of CD4(+)CD25(+) Foxp3(+) regulatory T cells. J Exp Med 204(2):381–391. http://www.pubmedcentral.nih.gov/articlerender.fcgi?artid=2118715&tool=pmcentrez&rendertype=abstract

Marangoni F, Trifari S, Scaramuzza S, Panaroni C, Martino S, Notarangelo LD et al (2007) WASP regulates suppressor activity of human and murine CD4(+)CD25(+)FOXP3(+) natural regulatory T cells. J Exp Med 204(2):369–380

Marson A, Kretschmer K, Frampton GM, Jacobsen ES, Polansky JK, MacIsaac KD et al (2007) Foxp3 occupancy and regulation of key target genes during T-cell stimulation. Nature 445(7130):931–935

Massaad MJ, Ramesh N, Geha RS (2013) Wiskott-Aldrich syndrome: a comprehensive review. Ann N Y Acad Sci 1285(1):26–43

Mathew JM, H-Voss J, LeFever A, Konieczna I, Stratton C, He J et al (2018) A phase I clinical trial with ex vivo expanded recipient regulatory T cells in living donor kidney transplants. Sci Rep 8(1):1–12

Miyara M, Yoshioka Y, Kitoh A, Shima T, Wing K, Niwa A et al (2009) Functional delineation and differentiation dynamics of human CD4+ T cells expressing the FoxP3 transcription factor. Immunity 30(6):899–911. https://doi.org/10.1016/j.immuni.2009.03.019

Morris GP, Allen PM (2012) How the TCR balances sensitivity and specificity for the recognition of self and pathogens. Nat Immunol 13(2):121–128

Morris EC, Fox T, Chakraverty R, Tendeiro R, Snell K, Rivat C et al (2017) Gene therapy for Wiskott-Aldrich syndrome in a severely affected adult. Blood 130(11):1327–1335

Nishizuka Y, Sakakura T (1969) Thymus and reproduction: sex-linked dysgenesia of the Gonad after neonatal thymectomy in mice. Science 166:753–755

Notarangelo L, Miao C, Ochs H (2008) Wiskott-Aldrich syndrome - UpToDate. Curr Opin Hematol 15:30–36. https://www.uptodate.com/contents/wiskott-aldrich-syndrome

Ochs HD, Ziegler SF, Torgerson TR (2005) FOXP3 acts as a rheostat of the immune response. Immunol Rev 203:156–164

Ochs HD, Gambineri E, Torgerson TR (2007) IPEX, FOXP3 and regulatory T-cells: a model for autoimmunity. Immunol Res 38(1–3):112–121

Oh J, Wang W, Thomas R, Su DM (2017) Capacity of tTreg generation is not impaired in the atrophied thymus. PLoS Biol 15(11):1–22

Ohkura N, Kitagawa Y, Sakaguchi S (2013) Development and maintenance of regulatory T cells. Immunity 38(3):414–423. http://www.ncbi.nlm.nih.gov/pubmed/23521883

Ouyang W, Beckett O, Ma Q, Li MO (2010) Transforming growth factor-beta signaling curbs thymic negative selection promoting regulatory T cell development. Immunity 32(5):642–653. https://doi.org/10.1016/j.immuni.2010.04.012

Passos GA, Mendes-da-Cruz DA, Oliveira EH (2015) The thymic orchestration involving Aire, miRNAs, and cell–cell interactions during the induction of central tolerance. Front Immunol 6(July):1–7. http://journal.frontiersin.org/Article/10.3389/fimmu.2015.00352/abstract

Perry JSA, Lio CJ, Kau AL, Nutsch K, Yang Z, Gordon JI et al (2014) Distinct contributions of Aire and antigen-presenting-cell subsets to the generation of self-tolerance in the thymus. Immunity 41(3):414–426. https://doi.org/10.1016/j.immuni.2014.08.007

Perry JSA, Russler-Germain EV, Zhou YW, Purtha W, Cooper ML, Choi J et al (2018) CD36 mediates cell-surface antigens to promote thymic development of the regulatory T cell receptor repertoire and allo-tolerance. Immunity 48(5):923–936.e4

Petrie HT, Zúñiga-Pflücker JC (2007) Zoned out: functional mapping of stromal signaling microenvironments in the thymus. Annu Rev Immunol 25:649–679. http://www.ncbi.nlm.nih.gov/pubmed/17291187

Petrie HT, Livak F, Schatz DG, Strasser A, Crispe IN, Shortman K (1993) Multiple rearrangements in T cell receptor alpha chain genes maximize the production of useful thymocytes. J Exp Med 178(2):615–622

Proietto AI, van Dommelen S, Zhou P, Rizzitelli A, D'Amico A, Steptoe RJ et al (2008) Dendritic cells in the thymus contribute to T-regulatory cell induction. Proc Natl Acad Sci U S A 105(50):19869–19874

Redecke V, Chaturvedi V, Kuriakose J, Häcker H (2016) SHARPIN controls the development of regulatory T cells. Immunology 148(2):216–226

Richards DM, Delacher M, Goldfarb Y, Kägebein D, Hofer A-C, Abramson J et al (2015) Treg cell differentiation: from thymus to peripheral tissue. Prog Mol Biol Transl Sci 136:175–205. http://linkinghub.elsevier.com/retrieve/pii/S1877117315001465

Roncarolo MG, Battaglia M (2007) Regulatory T-cell immunotherapy for tolerance to self antigens and alloantigens in humans. Nat Rev Immunol 7(8):585–598

Ross SH, Cantrell DA (2018) Signaling and function of interleukin-2 in T lymphocytes. Annu Rev Immunol 36:411–433

Sakaguchi S (2005) Naturally arising Foxp3-expressing CD25+CD4+ regulatory T cells in immunological tolerance to self and non-self. Nat Immunol 6(4):345–352

Sakaguchi S, Sakaguchi N, Asano M, Itoh M, Toda M (1995) Immunologic self-tolerance maintained by activated T cells expressing IL-2 receptor alpha-chains (CD25). Breakdown of a single mechanism of self-tolerance causes various autoimmune diseases. J Immunol 155(3):1151–1164

Salomon B, Lenschow DJ, Rhee L, Ashourian N, Singh B, Sharpe A et al (2000) B7/CD28 costimulation is essential for the homeostasis of the CD4+CD25+ immunoregulatory T cells that control autoimmune diabetes. Immunity 12(4):431–440

Savino W, Dardenne M (2010) Nutritional imbalances and infections affect the thymus: consequences on T-cell-mediated immune responses. Proc Nutr Soc 69(4):636–643. http://www.ncbi.nlm.nih.gov/pubmed/20860857

Shimizu J, Yamazaki S, Takahashi T, Ishida Y, Sakaguchi S (2002) Stimulation of CD25(+) CD4(+) regulatory T cells through GITR breaks immunological self-tolerance. Nat Immunol 3(2):135–142

Singer A, Adoro S, Park J (2009) Lineage fate and intense debate: myths, models and mechanisms of CD4/CD8 lineage choice. Nat Rev Immunol 8(10):788–801

Snapper SB, Rosen FS, Mizoguchi E, Cohen P, Khan W, Liu CH et al (1998) Wiskott-Aldrich syndrome protein-deficient mice reveal a role for WASP in T but not B cell activation. Immunity 9:81–91

Snow JW, Abraham N, Ma MC, Herndier BG, Pastuszak AW, Goldsmith MA (2003) Loss of tolerance and autoimmunity affecting multiple organs in STAT5A/5B-deficient mice. J Immunol 171(10):5042–5050

So T, Seung-Woo L, Croft M (2008) Immune regulation and control of regulatory T cells by OX40 and 41BB. Cytokine Growth Factor Rev 19:253–262

Soper DM, Kasprowicz DJ, Ziegler SF (2007) IL-2R?? links IL-2R signaling with Foxp3 expression. Eur J Immunol 37(7):1817–1826

Sugimoto N, Oida T, Hirota K, Nakamura K, Nomura T, Uchiyama T et al (2006) Foxp3-dependent and -independent molecules specific for CD25+CD4+ natural regulatory T cells revealed by DNA microarray analysis. Int Immunol 18(8):1197–1209

Sullivan KE, Mullen CA, Blaese RM, Winkelstein JA (1994) A multiinstitutional survey of the Wiskott-Aldrich syndrome. J Pediatr 125(6 Pt 1):876–885

Surh CD, Sprent J (1994) T-cell apoptosis detected in situ during positive and negative selection in the thymus. Nature 372(6501):100–103

Symons M, Derry JMJ, Karlak B, Jiang S, Lemahieu V, McCormick F et al (1996) Wiskott-Aldrich syndrome protein, a novel effector for the GTPase CDC42Hs, is implicated in actin polymerization. Cell 84(5):723–734

Tai X, Cowan M, Feigenbaum L, Singer A (2005) CD28 costimulation of developing thymocytes induces Foxp3 expression and regulatory T cell differentiation independently of interleukin 2. Nat Immunol 6(2):152–162

Takahashi T, Tagami T, Yamazaki S, Uede T, Shimizu J, Sakaguchi N et al (2000) Immunologic self-tolerance maintained by CD25(+)CD4(+) regulatory T cells constitutively expressing cytotoxic T lymphocyte-associated antigen 4. J Exp Med 192(2):303–310

Teh CE, Lalaoui N, Jain R, Policheni AN, Heinlein M, Alvarez-Diaz S et al (2016) Linear ubiquitin chain assembly complex coordinates late thymic T-cell differentiation and regulatory T-cell homeostasis. Nat Commun 7:13353

Thornton AM, Shevach EM (1998) CD4+CD25+ immunoregulatory T cells suppress polyclonal T cell activation in vitro by inhibiting interleukin 2 production. J Exp Med 188(2):287–296

Thornton AM, Donovan EE, Piccirillo CA, Shevach EM (2004) Cutting edge: IL-2 is critically required for the in vitro activation of CD4+CD25+ T cell suppressor function. J Immunol 172(11):6519–6523

Thornton AM, Korty PE, Tran DQ, Wohlfert EA, Murray PE, Belkaid Y et al (2010) Expression of Helios, an Ikaros transcription factor family member, differentiates thymic-derived from peripherally Induced Foxp3+ T regulatory cells. J Immunol 184(7):3433–3441. http://www.jimmunol.org/cgi/doi/10.4049/jimmunol.0904028

Todo S, Yamashita K, Goto R, Zaitsu M, Nagatsu A, Oura T et al (2016) A pilot study of operational tolerance with a regulatory T-cell-based cell therapy in living donor liver transplantation. Hepatology 64(2):632–643

272 L. Vasconcelos-Fontes et al.

Trzonkowski P, Bieniaszewska M, Juścińska J, Dobyszuk A, Krzystyniak A, Marek N et al (2009) First-in-man clinical results of the treatment of patients with graft versus host disease with human ex vivo expanded CD4+CD25+CD127- T regulatory cells. Clin Immunol 133(1):22–26

Van Der Vliet HJJ, Nieuwenhuis EE (2007) IPEX as a result of mutations in FOXP3. Clin Dev Immunol 2007:3–7

Verhagen J, Wraith D (2010) Comment on "Expression of Helios, an Ikaros transcription factor family member, differentiates thymic-derived from peripherally induced Foxp3+ T regulatory cells". J Immunol 184:3433–3441

Westerberg L, Larsson M, Hardy SJ, Fernández C, Thrasher AJ, Ferna C (2005) Wiskott-Aldrich syndrome protein deficiency leads to reduced B-cell adhesion, migration, and homing, and a delayed humoral immune response. Immunobiology 105(3):1144–1152

Wildin RS, Ramsdell F, Peake J, Faravelli F, Casanova JL, Buist N et al (2001) X-linked neonatal diabetes mellitus, enteropathy and endocrinopathy syndrome is the human equivalent of mouse scurfy. Nat Genet 27(1):18–20

Wildin RS, Smyk-Pearson S, Filipovich AH (2002) Clinical and molecular features of the immunodysregulation, polyendocrinopathy, enteropathy, X linked (IPEX) syndrome. J Med Genet 39(8):537–545. http://www.pubmedcentral.nih.gov/articlerender.fcgi?artid=1735203&tool=pmcentrez&rendertype=abstract

Xing Y, Hogquist KA (2012) T-Cell tolerance: central and peripheral. Cold Spring Harb Perspect Biol 4:1–15

Xu L, Kitani A, Stuelten C, McGrady G, Fuss I, Strober W (2010) Positive and negative transcriptional regulation of the Foxp3 gene is mediated by access and binding of the Smad3 protein to enhancer I. Immunity 33(3):313–325. https://doi.org/10.1016/j.immuni.2010.09.001

Zemmour D, Pratama A, Loughhead SM, Mathis D, Benoist C (2017) *Flicr*, a long noncoding RNA, modulates Foxp3 expression and autoimmunity. Proc Natl Acad Sci U S A 114(17):E3472–E3480

Zhang Q, Cui F, Fang L, Hong J, Zheng B, Zhang JZ (2013) TNF-alpha impairs differentiation and function of TGF-beta-induced Treg cells in autoimmune diseases through Akt and Smad3 signaling pathway. J Mol Cell Biol 5:85–98. http://www.ncbi.nlm.nih.gov/pubmed/23243069

Zheng SG, Wang JH, Stohl W, Kim KS, Gray JD, Horwitz DA (2006) TGF-beta requires CTLA-4 early after T cell activation to induce FoxP3 and generate adaptive CD4+CD25+ regulatory cells. J Immunol 176(6):3321–3329

Chapter 13
Adding Insult to Injury: Improving the Regenerative Capacity of the Aged Thymus Following Clinically Induced Damage

Jarrod Dudakov, Michael L. Hun, Kahlia Wong, Georg Holländer, and Ann P. Chidgey

Abstract T cell development in the thymus requires constant interactions between developing thymocytes and the surrounding resident stromal cells (TSCs). These functional, non-hematopoietic-derived cells are predominantly of epithelial origin and constitute 1% of total thymus cellularity. Thymic function and naïve T cell output are optimal in humans during the early years of life, but decline sharply after the first years of life and further after the onset of puberty, with progressive contraction of functional thymus tissue. The loss of thymic epithelial cells (TECs) eventually results in an altered microenvironment that affects normal T cell development, reducing naïve T cell output and contributing to the overall demise of immune functions. Other extrinsic factors that exacerbate the effects of thymic involution include cytoreductive conditioning regimes in the context of cancer treatments and hematopoietic stem cell transplantation (HSCT). In these scenarios, elderly patients are incapable of full immune recovery. With such complications present, we describe herein various preclinical approaches that could lead to potential new treatment strategies to restore thymic function and, more importantly, T-cell mediated immunity following clinically induced damage (Fig. 13.1).

J. Dudakov
Program in Immunology, Clinical Research Division, and Immunotherapy Integrated Research Center, Fred Hutchinson Cancer Research Center, Seattle, WA, USA

M. L. Hun · K. Wong · A. P. Chidgey (✉)
Department of Anatomy and Developmental Biology, Biomedicine Discovery Institute, Monash University, Melbourne, VIC, Australia
e-mail: ann.chidgey@monash.edu

G. Holländer
Department of Paediatrics and the Weatherall Institute of Molecular Medicine, University of Oxford, Oxford, UK

MRC Functional Genomics Unit, Department of Physiology, Anatomy and Genetics, University of Oxford, Oxford, UK

© Springer Nature Switzerland AG 2019
G. A. Passos (ed.), *Thymus Transcriptome and Cell Biology*,
https://doi.org/10.1007/978-3-030-12040-5_13

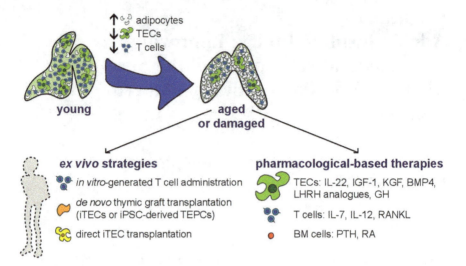

Fig. 13.1 Potential strategies for T cell reconstitution following clinically-induced damage. The effects of age-associated thymic damage (adipocyte infiltration, in conjunction with a loss of TECs and T cell output) negatively impact on T cell reconstitution post-chemotherapy or TBI-induced DNA damage. *Ex vivo* strategies designed to alleviate this phenomenon include the administration of *in vitro*-generated T cells, and transplantation of *de novo* thymic grafts (iTEC- or iPSC-derived TEPCs) or iTECs. Reversal may also be achieved by pharmacological-based therapies that target TECs (IL-22, IGF-1, KGF, BMP4, LHRH analogues, GH), T cells (IL-7, IL-12, RANKL), and BM cells (PTH, RA). *BM* bone marrow, *BMP4* bone morphogenic protein 4, *DNA* deoxyribose nucleic acid, *GH* growth hormone, *IGF-1* insulin-like growth factor-1, *IL* interleukin, *iPSC* induced pluripotent stem cell, *iTEC* forkhead box n1-induced TEC, *KGF* keratinocyte growth factor, *LHRH* luteinizing hormone-releasing hormone, *PTH* parathyroid hormone, *RANKL* receptor activator of nuclear factor Kappa-B, *RA* retinoid acid, *TBI* total body irradiation, *TEC* thymic epithelial cell, *TEPC* thymic epithelial progenitor cell

13.1 Thymic Involution

Age-associated thymic atrophy is a chronic condition portrayed by various morphological and cellular changes (Mackall and Gress 1997; Gruver et al. 2007). Extensive studies have demonstrated that increased sex steroid production from puberty is associated with thymic degeneration (reviewed in (Chaudhry et al. 2016; Lynch et al. 2009)). In mice, chronic thymus degeneration begins from approximately 4 weeks of age; a time where the onset of puberty is initiated, with further progressive loss in total thymic cellularity (Gray et al. 2006). The gradual contraction of the thymus organ is characterized by the widening of perivascular spaces, loss of thymocytes, increased adipose tissue deposition in these structures, decreased TEC cellularity and a disorganized stromal architecture at the cortico-medullary junction (CMJ) (Hirokawa and Makinodan 1975; Aspinall and Andrew 2000; Sutherland et al. 2005; Gray et al. 2006; Lynch et al. 2009; Heng et al. 2010) (Fig. 13.1). While naïve T cell output is dramatically decreased, the total peripheral T cell pool remains constant as aging progresses. This is due to homeostatic proliferation of peripheral memory T cell clones that have encountered their cognate antigens (Aspinall and Andrew 2000; Linton and Dorshkind 2004).

The reduction in thymic cellularity and naïve T cell generation by middle age is closely associated with a numerical loss and reduced turnover of TECs (Gray et al. 2006). The ratio of medullary TEC (mTEC) to cortical TEC (cTEC) decreases significantly, and the number of mature TECs expressing high levels of MHC II is also reduced. TEC proliferation assessed by Ki67, a nuclear antigen associated with proliferation (Scholzen and Gerdes 2000), is further reduced with age, suggesting a lower turnover (Gray et al. 2006). This is further confirmed by gene expression profiling of TECs between 1 and 6 months of age, which identifies decreased expression of cell-cycle-associated genes and a reduction in E2F3 activity, required for cell proliferation, constituting hallmarks of early thymic involution (Ki et al. 2014). The thymus, brain and testes-associated gene (*Tbata*; also known as *Spatial*) is present at increasing levels with age, and has been proposed to negatively impact on both TEC proliferation and thymic size; thymi from aged Tbata-deficient mice are larger and contain more dividing TECs than wild-type littermate controls (Flomerfelt et al. 2010). Other TEC intrinsic age-related abnormalities relate to the cell's thymopoietic function. For instance, interleukin-7 (IL-7) plays a key role in thymocyte expansion during early development, and reduced IL-7 levels in the aging thymus is associated with thymic involution (Alves et al. 2009; Henson et al. 2005), particularly within the cortex (Alves et al. 2010; Ribeiro et al. 2013).

A number of studies have implicated transcription factor forkhead-box N1 (Foxn1) in contributing to the mechanisms of thymic involution. Foxn1 is a master regulator of TEC lineage specification, and is indispensable for TEC lineage commitment during early thymus organogenesis. Given the increased number of Foxn1 transcripts in mature TECs, when compared to immature TECs, (Barsanti et al. 2017; Rode et al. 2015; O'Neill et al. 2016), it is conceivable that high Foxn1 levels are required for normal TEC differentiation and maturation. Thymopoiesis relies on continued Foxn1 expression (Zuklys et al. 2016) and thymic involution is accelerated by the engineered absence of *Foxn1* (Zuklys et al. 2016; Corbeaux et al. 2010). Indeed, reduced *Foxn1* levels have been found in the aged thymus (Barsanti et al. 2017; O'Neill et al. 2016; Ortman et al. 2002; Rode et al. 2015). In a transgenic mouse model where postnatal *Foxn1* expression is progressively lost, mirroring age-related *Foxn1* decline, the thymic microenvironment collapses, and consequently results in a reduced and eventually completely absent production of T cells (Chen et al. 2009; Sun et al. 2010; Zuklys et al. 2016). Senescence-related thymic involution is however partially reversible by several measures, including the engineered over- or re-expression of *Foxn1* (Zook et al. 2011; Sun et al. 2010). Importantly, the observed reversal of thymus size coincides with restoration of the epithelial architecture compared to controls, a return of effective thymopoiesis, an attenuation of naïve T cellularity in the periphery and an expansion of memory T cells therein (Zook et al. 2011; Sun et al. 2010). Molecules and signaling pathways downstream of Foxn1 were identified in extensive transcriptome studies in the postnatal thymus and include many factors essential for normal T cell development (Zuklys et al. 2016). Thus, understanding the molecular control of Foxn1 expression (Balciunaite et al. 2002; Garfin et al. 2013) will provide important knowledge in developing strategies to restore thymic function in aged individuals following tissue damage.

The Wnt signaling pathway is associated with many cellular processes during fetal development and adult homeostasis in several tissues (Akiyama 2000; Niehrs 2006; MacDonald et al. 2009). The binding of Wnt secreted glycoproteins (constituting in mice a family of 18 individual members) to their respective receptors triggers a signal transduction pathway involving several intermediary molecules, and culminating in gene transcription (MacDonald et al. 2009; Osada et al. 2010). Wnt signaling is linked to several aspects of thymic epithelial senescence in mice. Inhibition of Wnt signaling in TECs by the expression of Wnt antagonists or ablation of adapter molecules required for Wnt-targeted gene expression causes rapid thymic atrophy in adult transgenic mice (Osada et al. 2010; Liang et al. 2013). Wnt4, a regulator of Foxn1 expression in early thymic development (Balciunaite et al. 2002), decreases with age, and thus parallels the downregulation of Foxn1 (Kvell et al. 2010; Varecza et al. 2011). Reduced Wnt4 activity but increased expression of the thymopoietin, lamina-associated polypeptide 2α (LAP2α), were hypothesized to be involved in epithelial-mesenchymal transition (EMT) of TECs (Talaber et al. 2011; Kvell et al. 2010); a claim in line with a proceeding report that qualitatively demonstrates TEC transition into mesenchyme that expresses proadipogenic regulators (Youm et al. 2009). These observations previously offered an explanation for the increased frequency of adipocytes during involution that may arise from EMT of endogenous TECs. Quantitative analysis of EMT in β5t-Cre × loxP-EGFP mice (Ohigashi et al. 2013), which enable labeling of virtually all TECs, has since revealed that there is no significant increase in EMT from 2 weeks (young) to 11 months (aged) of age, with approximately 10% EGFP$^+$CD45$^-$PDGFRα^+ mesenchymal cells detected (Ohigashi et al. 2016).

Bone morphogenic protein-4 (Bmp4) is produced by fibroblasts and endothelial cells (ECs) in the thymus and is spatiotemporally involved in early thymus morphogenesis (Gordon et al. 2010). Bmp4 binds to the BMP receptors (BMPR) I and II, expressed mainly on immature TEC precursors (Barsanti et al. 2017) in the postnatal thymus; and acts on Smad4, the central mediator of canonical transforming growth factor-beta (TGF-β) signaling. Although loss of TEC Smad4 expression results in a progressively disorganized thymic microenvironment (Jeker et al. 2008), a hallmark of late stage thymic involution, loss of TEC TGF-β RII expression instead mitigates thymic senescence (Hauri-Hohl et al. 2008). Moreover, the increasing expression of TGF-β1 with age (Sempowski et al. 2000) is believed to be associated with thymus atrophy through the upregulation of inflammatory molecules, LIF and IL-6 (Schluns et al. 1997). Taken together, it is clear that an imbalance of TGF-β superfamily signaling molecules is contributing to thymic involution by reducing TEC capacity to support thymopoiesis.

Factors other than a decline in the naïve T cell pool, such as intrinsic age-related alterations in T cell function, can also severely compromise the immune system. In the periphery, defects in naïve T cells vary from poor responses to novel antigens following stimulation and reduced production of vital cytokines such as IL-2; to a diminished ability to activate other lymphoid subsets required for a normal immune response (Maue et al. 2009; Lynch et al. 2009). Transcriptional profiling of memory CD8[+] T cells from aged mice demonstrated a profile consistent with exhaustion marked by an increased expression of PD-1 and CTLA-4 (Decman et al. 2012). Spontaneous engagement of AMPK has also been shown to trigger p38 auto-phosphorylation in middle aged human T cells, with p38 MAP kinase (MAPK) signaling consequently inhibiting expression of TCR signalosome components (Lanna et al. 2014, 2017). Moreover, the primary CD8[+] T cell response is limited in aged mice, a finding in part due to an accumulation of antigen-naïve but incompletely differentiated and hence dysfunctional "virtual memory" T cells. These cells have a reduced proliferative capacity in both aged mice and humans and exhibit a RNA-Seq profile consistent with senescence (Quinn et al. 2018).

The molecular mechanisms that underpin thymus involution are still incompletely understood and the focus of ongoing research. If thymus regeneration can be achieved, even transiently, dislodging dysfunctionally aged T cells with newly generated naïve T cells may improve the immune response in the aged and elderly, enhance immune recovery following damage, and increase the success of immunotherapeutic approaches.

13.2 Thymic Response to Injury

13.2.1 Dysregulation of Thymic Function After Damage

The thymus is variably sensitive to different insults ranging from infection and stress. Severe and immediate injury is typically caused by cytoreductive chemotherapy and irradiation as part of cancer therapies. The decline in thymic function after these acute insults constitutes a particular problem as recovery of the adaptive immune system post-chemotherapy and the conditioning required for successful HSCT is slow, as reflected by both prolonged reduction of naïve T cells which in addition, have an impaired function (Weinberg et al. 1995; Roux et al. 1996, 2000; Dumont-Girard et al. 1998; Ault et al. 1985; Gratama et al. 1984; Roosnek et al. 1987; Soiffer et al. 1990). T cell recovery numbers can however only occur in two ways and both are limited in their efficiency; homeostatic expansion of peripheral (primarily memory) T cells and the new endogenous generation of T cells in the thymus. The former results in a narrowing of the T cell receptor repertoire and impaired adaptive immunity as only memory subsets are typically expanded (Mackall et al. 1994, 1995). Although recovered rapidly in both young and aged patients following HSCT (Mackall et al. 1997), CD8[+] T cells primarily expand to fill their peripheral niches via clonal expansion (Fagnoni et al. 2002; Heitger et al. 1997). Whether an effective reconstitution of the T cell antigen receptor repertoire diversity and capacity require a functional thymus remains however to be defined (Hakim et al. 2005). In striking contrast to children, adults with a senescent thymus are significantly impaired in their ability to recover age-related thymic cellularity and function following chemotherapy or HSCT (Mackall et al. 1995; Storek et al.

1995; Weinberg et al. 1995; Roux et al. 2000). Full immune recovery is likely in treated individuals until middle age, whereas in older patients, the naïve T cell receptor repertoire cannot be fully restored (Sfikakis et al. 2005), and has been linked to infectious morbidity (Storek et al. 1997). Thus, T lymphopenia and a limited T cell antigen receptor repertoire represent major clinical challenges for most patients following therapies that damage the thymus.

13.2.2 Endogenous Thymic Regeneration

Despite its high sensitivity to damage, the thymus also holds a remarkable capacity for reconstitution and repair; a phenomenon that has been noted for longer than the assignment of defined immunological function to this lymphoid organ (Jaffe 1924; Miller 1961). Recent studies have shown that thymic function can spontaneously regenerate in the young even after significant surgical resection of the tissue (van den Broek et al. 2016). This experimental observation maps to the contention that endogenous thymic regeneration plays a crucial role in restoring immune competence after clinically-induced damage (Chidgey et al. 2007; Dudakov et al. 2010), and thus has a profound impact on clinical outcomes. Indeed, thymic function prior to HSCT directly correlates to the patient's overall survival following cytoreductive therapy (Clave et al. 2005), which has been explained by the capacity of the young thymus to recover its thymopoietic function and thus contribute naïve T cells to anti-microbial and cancer responses. Although the mechanisms that governed thymic repair have been poorly understood for an extended time, recent studies could however identify at least two mechanistic pathways that appear critical for the potential of thymic regeneration.

Despite some circumstantial evidence that indicated a role for KGF in endogenous thymic regeneration (Alpdogan et al. 2006; Min et al. 2002; Rossi et al. 2002), the first pathway to be comprehensively described was centered on the production of interleukin-22 (IL-22) by innate lymphoid cells (Dudakov et al. 2012). In this pathway, thymic damage, and in particular the loss of CD4+CD8+ double positive (DP) thymocytes, triggers the upregulation of IL-23 by dendritic cells (DCs), which induces the production of IL-22 by innate lymphoid cells (ILCs). IL-22 acts on TECs, stimulating TEC proliferation, which ultimately leads to rejuvenation of thymopoiesis (Dudakov et al. 2012).

More recently, a second, IL-22-independent, pathway has been described revolving around the production of BMP4 by endothelial cells (ECs) in the thymus. Although the thymic vasculature condenses after damage commensurate to the wider damage to the thymus after total body irradiation (TBI), thymic ECs, similar to ILCs, are extremely resistant to damage (Wertheimer et al. 2018; Zhang et al. 2014). This radio-resistance enables the cells to be responsive to activation by hitherto unknown signals and to produce BMP4. This morphogen targets TECs to induce FOXN1 expression which controls the transcription of delta-like 4 (DLL4) and Kit ligand. Both of these factors are critical for thymopoiesis and can even control thymic size (Velardi et al. 2014).

13.3 Exogenous Regeneration of the Thymus

In view of the significant decline in thymic function with age, and the prolonged immune suppression observed after high dose radio-chemotherapy, there has been considerable interest in developing new evidence-based therapeutic strategies that improve thymic function and hence peripheral T cell reconstitution. Multiple approaches have been taken to achieve thymus reconstitution, including the use of cytokine and growth factors, modulation of sex hormones, cellular therapies with either hematopoietic or stromal cells, and even the *ex vivo* generation of a thymic microenvironment employing tissue engineering.

13.3.1 Cytokines and Growth Factors

One of the most promising strategies for thymic regeneration is the use of cytokines and growth factors that have been demonstrated to play an important role in native thymus function. The exogenous administration of KGF, IL-7, Flt-3 Ligand (Flt3L), IL-22, IL-21, and RANKL, for instance, has been documented to support the potential of the thymus to regenerate after injury. The experimental evidence for their utility will be detailed in the following paragraphs.

Interleukin-7 is a pro-lymphopoietic cytokine that acts on both developing B and T cells in the mouse (Kang and Der 2004; Sudo et al. 1993), and T cells in humans (Prieyl and LeBien 1996). Early clinical data demonstrated the effectiveness of IL-7 treatment on thymus recovery, making this approach one of the most promising thymus-boosting therapies to date. In aged mice and mice subjected to allogeneic-HSCT, IL-7 treatment promotes the proliferation of lymphoid precursors (Alpdogan et al. 2003a), thereby enhancing the generation and export of new naïve T cells, which directly contribute to the functionality of the peripheral T cell pool (Alpdogan et al. 2003a; Lu et al. 2005; Sempowski et al. 2002; Fry et al. 2003; Chu et al. 2004). Consequently, IL-7 increases the breadth of the peripheral T cell antigen receptor repertoire and, in addition, via its extra-thymic functions, fosters T cell proliferation in experimental animal models but also humans (Mackall et al. 2011). Clinical trials with recombinant human IL-7 (r-hIL-7) demonstrated an expansion of CD4$^+$ and CD8$^+$ T cells in patients with solid tumors or HIV infection (Rosenberg et al. 2006; Sportes et al. 2008, 2010; Levy et al. 2009, 2012; Sereti et al. 2009). More recently, a phase I trial of r-hIL-7 in recipients of T-cell depleted (TCD) allogeneic-HSCT showed that r-hIL-7 was well tolerated and induced a rapid increase in peripheral CD4$^+$ and CD8$^+$ T cells (Perales et al. 2012).

Administration of KGF has been shown to be effective at increasing thymic cellularity and function in aged mice, as well as in those receiving chemotherapy and HSCT (Alpdogan et al. 2006; Min et al. 2007; Rossi et al. 2002). KGF acts directly on TECs via a specific Fgf receptor variant, FgfR2IIIb, to promote their proliferation and expansion (Rossi et al. 2007b), but also protects TECs from GvHD-mediated thymic damage (Rossi et al. 2002). Although this growth factor has been approved for treatment of mucositis in recipients of high dose chemotherapy, there is currently no clinical data regarding the potential of KGF to enhance T cell reconstitution in patients, with studies in non-human primates demonstrating that KGF does not reverse age-related thymic involution (Wils et al. 2012).

Recent work has also concentrated on exploiting the endogenous mechanisms that guide thymic regeneration after injury. Following acute injury caused by sublethal TBI, IL-22 therapy enhances murine thymopoiesis via stimulation of TEC proliferation (Dudakov et al. 2012). Also, IL-22 can be given in the context of HSCT to improve reconstitution (Lopes et al. 2017), even in the face of fulminant GvHD (Dudakov et al. 2017). In addition to IL-22, ILCs also increase their production of RANKL after damage (Dudakov et al. 2012), which improves T cell reconstitution after HSCT (Lopes et al. 2017). The mechanisms by which these pathways operate are complex, with RANKL regulating IL-22 expression in an autocrine fashion (Bando et al. 2018).

Other cytokines and growth factors have also been evaluated for their beneficial role in regenerating the immune system following thymus injury and senescence of the organ. For example, IGF-1 promotes TEC expansion and enhances reconstitution following HSCT (Alpdogan et al. 2003b; Chu et al. 2008; Taguchi et al. 2006); IL-15 predominantly promotes proliferation of circulating NK and T cells (Alpdogan et al. 2005; Alpdogan and van den Brink 2005); growth hormone (GH) can reverse age-related thymic involution in mice (Dixit et al. 2007; Taub and Longo 2005) and promote thymus function in HIV patients (Herasimtschuk et al. 2008; Napolitano et al. 2002, 2008); and IL-12 stimulates thymic expression of IL-7 and then enhances hematopoietic engraftment after transplant (Chen et al. 2007, 2008a; Li et al. 2004). Preclinical studies have demonstrated that administration of Flt3L enhances thymus-dependent T cell reconstitution (Fry et al. 2004; Kenins et al. 2008), however, the effects of Flt3L are predominantly thymus-independent, and due to an expansion in Flt3$^+$ progenitors in the bone marrow (BM) (Wils et al. 2007). Recent studies have shown that IL-21 promotes thymic regeneration under different experimental conditions of acute and chronic involution including age, exogenous glucocorticoid administration and HSCT (Tormo et al. 2017; Al-Chami et al. 2016; Rafei et al. 2013). However, IL-21 signaling in alloreactive T cells can further contribute to GvHD-mediated thymus damage, limiting its therapeutic use (Dudakov et al. 2017).

13.3.2   Hormone Modulation

Sex hormones such as estrogen and androgen have been implicated in the natural, age-related and experimental reduction of thymopoiesis (Olsen and Kovacs 2001; Zoller and Kersh 2006) and B-lymphopoiesis (Grimaldi et al. 2006; Kincade et al. 2000; Viselli et al. 1997), as well as the senescence-related decrease in the frequency of early lymphoid precursors (Igarashi et al. 2001; Medina et al. 2001). For this reason, the use of sex steroid ablation (SSA) has been extensively evaluated to enhance immune reconstitution. SSA can be achieved pharmacologically by disrupting LHRH hormone signals, blocking the binding of sex steroid receptors, or by inactivating the hormones themselves (Conn and Crowley 1994; Velardi et al. 2013, 2015). These interventions of sex steroid activity result in transient restoration of thymic architecture and an enhanced ability to import circulating T cell progenitors (Williams et al. 2008). Taken together, these effects contribute to enhance thymopoiesis in aged mice and humans (Heng et al. 2005; Sutherland et al. 2005; Greenstein et al. 1987; Olsen et al. 1991; Roden et al. 2004; Dudakov et al. 2009b; Goldberg et al. 2009, 2010; Williams et al. 2008). The observed effects may be mediated via the induction of the Notch ligand DLL4, which is directly suppressed by androgen receptor binding (Velardi et al. 2014). Current investigations have also identified involvement of TGF-β family members in TEC aging and intrinsic TEC changes following SSA that contribute to thymus regeneration (Lepletier et al. 2019).

The beneficial effects of SSA extend beyond the thymus, to involve enhanced B-lymphopoiesis and an enhanced frequency of lymphoid progenitors (Erben et al. 1998, 2001; Ellis et al. 2001; Masuzawa et al. 1994; Dudakov et al. 2009a, 2010). Collectively, these changes improve the overall recovery of immunological competence following cytoablative therapy as well as autologous (Goldberg et al. 2005) and allogeneic (Goldberg et al. 2007) HSCT (Dudakov et al. 2009a, b; Goldberg et al. 2010). In the case of pharmacologic inhibition at the level of LHRH, the beneficial effects could also be partly attributed to the effects of LH on HSC function (Velardi et al. 2018). Retrospective analysis in patients receiving SSA as standard of care for prostate cancer demonstrate this treatment to be effective in improving thymic function (Sutherland et al. 2008) (NCT02406651), and current clinical trials are evaluating its effect in promoting T cell reconstitution following allogeneic-HSCT. Specifically, improved thymus function has been shown in prostate cancer patients who routinely receive treatment with an LHRH-agonist, as well as recipients of autologous- and allogeneic-HSCT in a dedicated Phase I trial.

13.3.3 Cellular Therapies

Given that the delay in T cell reconstitution is partly attributed to a limited supply of BM-derived progenitors (Zlotoff et al. 2011) and the time required to generate naïve T cell precursors derived from transplanted HSCs, several studies have explored the utility and effectiveness of BM-derived lymphoid precursors admixed with HSCs to accelerate and enhance immune rejuvenation (Arber et al. 2003). This approach however holds the challenge of the severely limited supply of these lymphoid progenitors. This limitation may now have been overcome with the recent development *of in vitro* systems that use Notch-1 stimulation to generate T lineage committed precursors (Holmes and Zuniga-Pflucker 2009; de Pooter and Zuniga-Pflucker 2007). The use of three-dimensional culture systems (Seet et al. 2017), and culture conditions that avoid the necessity for other cellular components have been most promising (Shukla et al. 2017). These approaches have proven successful at producing large numbers of T-lineage precursors effective at promoting immune reconstitution when used for adoptive therapy (Zakrzewski et al. 2006; Schmitt and Zuniga-Pflucker 2002; Awong et al. 2009, 2013; Reimann et al. 2010; Smith et al. 2017). Adoptive transfer of T cell precursors into lethally irradiated allogeneic-HSCT recipients significantly enhanced thymopoiesis and chimerism, effects that are expected to translate under clinical conditions to enhanced peripheral T cell reconstitution and competence (Zakrzewski et al. 2006; Holland et al. 2008). In addition to this significant benefit, *in vitro* generated pre-T cells also permit their genetic engineering to achieve bespoke tumor-specificity and thus their use for targeted tumor immunotherapy (Zakrzewski et al. 2008).

Given the senescence-related thymus involution and organ's damage by repeated cytoreductive treatments, efforts have also been undertaken to generate *ex vivo* a functionally complete thymic microenvironment that can eventually be grafted into patients (Chidgey et al. 2008; Seach et al. 2007; Tajima et al. 2016). While still in their preclinical infancy, some of these approaches target the use of endogenous thymic epithelial progenitor cells (TEPCs) that, in early development, give rise to all individual TEC lineages (Gill et al. 2002; Bennett et al. 2002; Depreter et al. 2008; Rossi et al. 2007a). Although the identity and existence of a similar population in the adult thymus remained elusive for many years (Jenkinson et al. 2008; Bleul et al. 2006), recent work suggests that several phenotypically distinct thymus endogenous precursor populations have the apparent ability in these mice to generate the different epithelial compartments (Ulyanchenko et al. 2016; Wong et al. 2014; Bleul et al. 2006). This advancement provides a promising framework for generating a *de novo* thymus. The generation of TECs from pluripotent stem cells or via induced cell reprogramming has also been actively pursued, with several groups demonstrating TEC progenitors can be differentiated from pluripotent stem cells (Sun et al. 2013; Parent et al. 2013; Soh et al. 2014). Furthermore, TEC-like cells derived from fibroblasts have recently been generated by targeted expression of FOXN1, and are able to support T cell development (Bredenkamp et al. 2014). These inducible TECs (iTECs) represent another promising source of TEC progenitor cells capable of supporting T cell development. Parallel to the efforts in generating an *ex vivo* thymic microenvironment, several studies have also suggested that TECs/TEC progenitors can also be transplanted directly into the thymus of mice, promoting T cell reconstitution after HSCT (Lai et al. 2011; Lai and Jin 2009).

The recent work on ECs and their production of BMP4 in endogenous regeneration also highlights their clinical potential as ECs could be propagated and expanded *ex vivo* and administered to promote thymic repair after acute injury (Wertheimer et al. 2018). This is consistent with other observations where ECs promote the regeneration of lung, liver, kidney and BM (Butler et al. 2010; Ding et al. 2010, 2011, 2014; Kobayashi et al. 2010; Rafii et al. 2016b). It is however of note that this capacity is tissue-specific (Nolan et al. 2013; Rafii et al. 2016a), and thus requires organ-specific EC cultures for cell expansion.

13.3.4 Combination Therapies

The extensive crosstalk that occurs between the thymic stromal compartment and developing hematopoietic cells (Anderson and Jenkinson 2001; Love and Bhandoola 2011; Hollander et al. 1995), suggests that a combination of strategies could best unlock the organ's potential for regeneration in lieu of a single approach. Preclinical studies using KGF have demonstrated that this cytokine can be effectively used in conjunction with other regenerative therapies such as the administration of pre-T cells (Zakrzewski et al. 2006), SSA (Kelly et al. 2008) or the temporary inhibition of p53 (Kelly et al. 2010) to enhance T cell reconstitution. However, it is unlikely that IL-7 could be used effectively in combination with KGF or SSA as both of these strategies promote intrathymic production of IL-7, and studies in knockout mice demonstrate that the biological effect of both of these regenerative strategies essentially depends on downstream IL-7 production and secretion (Goldberg et al. 2007; Alpdogan et al. 2006). Another promising strategy foresees the combination of therapies acting on TEC pre-thymic lymphoid progenitors. The rationale for this approach is given by studies demonstrating the supply of BM-derived lymphoid precursors to be limiting for thymic reconstitution after HSCT (Zlotoff et al. 2011). Other potentially rewarding strategies propose to target hematopoietic precursors, for example, with parathyroid hormone (PTH), which enhances HSC numbers (Adams et al. 2007) or retinoic acid (Chen et al. 2008b), which accelerates BM lymphopoiesis. Alternatively, there is evidence that modulation of the BM niche, specifically to enhance the mobilization of BM precursors, has a profound impact on thymic function (Rodrigues et al. 2018; Yu et al. 2015), making this an attractive secondary target for T cell reconstitution.

13.4 Conclusion

In conclusion, the quest for assisted thymus regeneration post-cytoablative conditioning regimes in an increasingly aged population is progressing in parallel with improvements in our understanding of thymic aging and its endogenous repair following clinically induced damage. Given the reliance of thymopoiesis on the supply of lymphoid precursors from the BM, it stands to reason that we must consider both primary immune organs in developing strategies for thymus and hence, immune regeneration. More recent advances in stem cell technologies in conjunction with biomedical engineering techniques, such as microfluidics for tissue and organ reconstruction, make this an exciting era in regenerative medicine that may enhance clinical strategies for thymus regeneration, which traditionally relied on pharmacological based interventions.

References

Adams GB, Martin RP, Alley IR, Chabner KT, Cohen KS, Calvi LM, Kronenberg HM, Scadden DT (2007) Therapeutic targeting of a stem cell niche. Nat Biotechnol 25(2):238–243

Akiyama T (2000) Wnt/beta-catenin signaling. Cytokine Growth Factor Rev 11(4):273–282

Al-Chami E, Tormo A, Pasquin S, Kanjarawi R, Ziouani S, Rafei M (2016) Interleukin-21 administration to aged mice rejuvenates their peripheral T-cell pool by triggering de novo thymopoiesis. Aging Cell 15(2):349–360

Alpdogan O, van den Brink MR (2005) IL-7 and IL-15: therapeutic cytokines for immunodeficiency. Trends Immunol 26(1):56–64

Alpdogan O, Muriglan SJ, Eng JM, Willis LM, Greenberg AS, Kappel BJ, van den Brink MR (2003a) IL-7 enhances peripheral T cell reconstitution after allogeneic hematopoietic stem cell transplantation. J Clin Invest 112(7):1095–1107

Alpdogan O, Muriglan SJ, Kappel BJ, Doubrovina E, Schmaltz C, Schiro R, Eng JM, Greenberg AS, Willis LM, Rotolo JA, O'Reilly RJ, van den Brink MR (2003b) Insulin-like growth factor-I enhances lymphoid and myeloid reconstitution after allogeneic bone marrow transplantation. Transplantation 75(12):1977–1983

Alpdogan O, Eng JM, Muriglan SJ, Willis LM, Hubbard VM, Tjoe KH, Terwey TH, Kochman A, van den Brink MR (2005) Interleukin-15 enhances immune reconstitution after allogeneic bone marrow transplantation. Blood 105(2):865–873

Alpdogan O, Hubbard VM, Smith OM, Patel N, Lu S, Goldberg GL, Gray DH, Feinman J, Kochman AA, Eng JM, Suh D, Muriglan SJ, Boyd RL, van den Brink MR (2006) Keratinocyte growth factor (KGF) is required for postnatal thymic regeneration. Blood 107(6):2453–2460

Alves NL, Richard-Le Goff O, Huntington ND, Sousa AP, Ribeiro VS, Bordack A, Vives FL, Peduto L, Chidgey A, Cumano A, Boyd R, Eberl G, Di Santo JP (2009) Characterization of the thymic IL-7 niche in vivo. Proc Natl Acad Sci U S A 106(5):1512–1517

Alves NL, Huntington ND, Mention JJ, Richard-Le Goff O, Di Santo JP (2010) Cutting Edge: a thymocyte-thymic epithelial cell cross-talk dynamically regulates intrathymic IL-7 expression in vivo. J Immunol 184(11):5949–5953

Anderson G, Jenkinson EJ (2001) Lymphostromal interactions in thymic development and function. Nat Rev Immunol 1(1):31–40

Arber C, BitMansour A, Sparer TE, Higgins JP, Mocarski ES, Weissman IL, Shizuru JA, Brown JM (2003) Common lymphoid progenitors rapidly engraft and protect against lethal murine cytomegalovirus infection after hematopoietic stem cell transplantation. Blood 102(2):421–428

Aspinall R, Andrew D (2000) Thymic atrophy in the mouse is a soluble problem of the thymic environment. Vaccine 18(16):1629–1637

Ault KA, Antin JH, Ginsburg D, Orkin SH, Rappeport JM, Keohan ML, Martin P, Smith BR (1985) Phenotype of recovering lymphoid cell populations after marrow transplantation. J Exp Med 161(6):1483–1502

Awong G, Herer E, Surh CD, Dick JE, La Motte-Mohs RN, Zuniga-Pflucker JC (2009) Characterization in vitro and engraftment potential in vivo of human progenitor T cells generated from hematopoietic stem cells. Blood 114(5):972–982

Awong G, Singh J, Mohtashami M, Malm M, La Motte-Mohs RN, Benveniste P, Serra P, Herer E, van den Brink M, Zúñiga-Pflücker JC (2013) Human proT-cells generated in vitro facilitate hematopoietic stem cell-derived T-lymphopoiesis in vivo and restore thymic architecture. Blood 122(26):4210–4219

Balciunaite G, Keller MP, Balciunaite E, Piali L, Zuklys S, Mathieu YD, Gill J, Boyd R, Sussman DJ, Hollander GA (2002) Wnt glycoproteins regulate the expression of FoxN1, the gene defective in nude mice. Nat Immunol 3(11):1102–1108

Bando JK, Gilfillan S, Song C, McDonald KG, Huang SCC, Newberry RD, Kobayashi Y, Allan DSJ, Carlyle JR, Cella M, Colonna M (2018) The tumor necrosis factor superfamily member RANKL suppresses effector cytokine production in group 3 innate lymphoid cells. Immunity 48(6):1208–1219

Barsanti M, Lim J, Hun ML, Lister N, Wong K, Hammett MV, Lepletier A, Boyd RL, Giudice A, Chidgey AP (2017) A novel Foxn1eGFP/+ mouse model identifies Bmp4-induced maintenance of Foxn1 expression and thymic epithelial progenitor populations. Eur J Immunol 47(2):291–304

Bennett AR, Farley A, Blair NF, Gordon J, Sharp L, Blackburn CC (2002) Identification and characterization of thymic epithelial progenitor cells. Immunity 16(6):803–814

Bleul CC, Corbeaux T, Reuter A, Fisch P, Monting JS, Boehm T (2006) Formation of a functional thymus initiated by a postnatal epithelial progenitor cell. Nature 441(7096):992–996

Bredenkamp N, Ulyanchenko S, O'Neill KE, Manley NR, Vaidya HJ, Blackburn CC (2014) An organized and functional thymus generated from FOXN1 reprogrammed fibroblasts. Nat Cell Biol 16(9).902–908

Butler JM, Nolan DJ, Vertes EL, Varnum-Finney B, Kobayashi H, Hooper AT, Seandel M, Shido K, White IA, Kobayashi M, Witte L, May C, Shawber C, Kimura Y, Kitajewski J, Rosenwaks Z, Bernstein ID, Rafii S (2010) Endothelial cells are essential for the self-renewal and repopulation of Notch-dependent hematopoietic stem cells. Cell Stem Cell 6(3):251–264

Chaudhry MS, Velardi E, Dudakov JA, Brink MR (2016) Thymus: the next (re)generation. Immunol Rev 271(1):56–71

Chen T, Burke KA, Zhan Y, Wang X, Shibata D, Zhao Y (2007) IL-12 facilitates both the recovery of endogenous hematopoiesis and the engraftment of stem cells after ionizing radiation. Exp Hematol 35(2):203–213

Chen J, Wang J, Li J, Wu Q, Chu Lim F, Yang P, Hsu HC, Curiel DT, Mountz JD (2008a) Enhancement of cytotoxic T-lymphocyte response in aged mice by a novel treatment with recombinant AdIL-12 and wild-type adenovirus in rapid succession. Mol Ther 16(8):1500–1506

Chen X, Esplin BL, Garrett KP, Welner RS, Webb CF, Kincade PW (2008b) Retinoids accelerate B lineage lymphoid differentiation. J Immunol 180(1):138–145

Chen L, Xiao S, Manley NR (2009) Foxn1 is required to maintain the postnatal thymic microenvironment in a dosage-sensitive manner. Blood 113(3):567–574

Chidgey A, Dudakov J, Seach N, Boyd R (2007) Impact of niche aging on thymic regeneration and immune reconstitution. Semin Immunol 19(5):331–340

Chidgey AP, Seach N, Dudakov J, Hammett MV, Boyd RL (2008) Strategies for reconstituting and boosting T cell-based immunity following haematopoietic stem cell transplantation: preclinical and clinical approaches. Semin Immunopathol 30:457–477

Chu YW, Memon SA, Sharrow SO, Hakim FT, Eckhaus M, Lucas PJ, Gress RE (2004) Exogenous IL-7 increases recent thymic emigrants in peripheral lymphoid tissue without enhanced thymic function. Blood 104(4):1110–1119

Chu YW, Schmitz S, Choudhury B, Telford W, Kapoor V, Garfield S, Howe D, Gress RE (2008) Exogenous insulin-like growth factor 1 enhances thymopoiesis predominantly through thymic epithelial cell expansion. Blood 112(7):2836–2846

Clave E, Rocha V, Talvensaari K, Busson M, Douay C, Appert ML, Rabian C, Carmagnat M, Garnier F, Filion A, Socie G, Gluckman E, Charron D, Toubert A (2005) Prognostic value of pretransplantation host thymic function in HLA-identical sibling hematopoietic stem cell transplantation. Blood 105(6):2608–2613

Conn PM, Crowley WF Jr (1994) Gonadotropin-releasing hormone and its analogs. Annu Rev Med 45:391–405

Corbeaux T, Hess I, Swann JB, Kanzler B, Haas-Assenbaum A, Boehm T (2010) Thymopoiesis in mice depends on a Foxn1-positive thymic epithelial cell lineage. Proc Natl Acad Sci U S A 107(38):16613–16618

de Pooter R, Zuniga-Pflucker JC (2007) T-cell potential and development in vitro: the OP9-DL1 approach. Curr Opin Immunol 19(2):163–168

Decman V, Laidlaw BJ, Doering TA, Leng J, Ertl HC, Goldstein DR, Wherry EJ (2012) Defective CD8 T cell responses in aged mice are due to quantitative and qualitative changes in virus-specific precursors. J Immunol 188(4):1933–1941

Depreter MG, Blair NF, Gaskell TL, Nowell CS, Davern K, Pagliocca A, Stenhouse FH, Farley AM, Fraser A, Vrana J, Robertson K, Morahan G, Tomlinson SR, Blackburn CC (2008)

Identification of Plet-1 as a specific marker of early thymic epithelial progenitor cells. Proc Natl Acad Sci U S A 105(3):961–966

Ding BS, Nolan DJ, Butler JM, James D, Babazadeh AO, Rosenwaks Z, Mittal V, Kobayashi H, Shido K, Lyden D, Sato TN, Rabbany SY, Rafii S (2010) Inductive angiocrine signals from sinusoidal endothelium are required for liver regeneration. Nature 468(7321):310–315

Ding BS, Nolan DJ, Guo P, Babazadeh AO, Cao Z, Rosenwaks Z, Crystal RG, Simons M, Sato TN, Worgall S, Shido K, Rabbany SY, Rafii S (2011) Endothelial-derived angiocrine signals induce and sustain regenerative lung alveolarization. Cell 147(3):539–553

Ding BS, Cao Z, Lis R, Nolan DJ, Guo P, Simons M, Penfold ME, Shido K, Rabbany SY, Rafii S (2014) Divergent angiocrine signals from vascular niche balance liver regeneration and fibrosis. Nature 505(7481):97–102

Dixit VD, Yang H, Sun Y, Weeraratna AT, Youm YH, Smith RG, Taub DD (2007) Ghrelin promotes thymopoiesis during aging. J Clin Invest 117(10):2778–2790

Dudakov JA, Goldberg GL, Reiseger JJ, Chidgey AP, Boyd RL (2009a) Withdrawal of sex steroids reverses age- and chemotherapy-related defects in bone marrow lymphopoiesis. J Immunol 182(10):6247–6260

Dudakov JA, Goldberg GL, Reiseger JJ, Vlahos K, Chidgey AP, Boyd RL (2009b) Sex steroid ablation enhances hematopoietic recovery following cytotoxic antineoplastic therapy in aged mice. J Immunol 183(11):7084–7094

Dudakov JA, Khong DMP, Boyd RL, Chidgey AP (2010) Feeding the fire: the role of defective bone marrow function in exacerbating thymic involution. Trends Immunol 31(5):191–198

Dudakov JA, Hanash AM, Jenq RR, Young LF, Ghosh A, Singer NV, West ML, Smith OM, Holland AM, Tsai JJ, Boyd RL, van den Brink MRM (2012) Interleukin-22 drives endogenous thymic regeneration in mice. Science 336(6077):91–95

Dudakov JA, Mertelsmann AM, O'Connor MH, Jenq RR, Velardi E, Young LF, Smith OM, Boyd RL, van den Brink MRM, Hanash AM (2017) Loss of thymic innate lymphoid cells leads to impaired thymopoiesis in experimental graft-versus-host disease. Blood 130(7):933–942

Dumont-Girard F, Roux E, van Lier RA, Hale G, Helg C, Chapuis B, Starobinski M, Roosnek E (1998) Reconstitution of the T-cell compartment after bone marrow transplantation: restoration of the repertoire by thymic emigrants. Blood 92(11):4464–4471

Ellis TM, Moser MT, Le PT, Flanigan RC, Kwon ED (2001) Alterations in peripheral B cells and B cell progenitors following androgen ablation in mice. Int Immunol 13(4):553–558

Erben RG, Raith S, Eberle J, Stangassinger M (1998) Ovariectomy augments B lymphopoiesis and generation of monocyte-macrophage precursors in rat bone marrow. Am J Physiol 274(3 Pt 1):E476–E483

Erben RG, Eberle J, Stangassinger M (2001) B lymphopoiesis is upregulated after orchiectomy and is correlated with estradiol but not testosterone serum levels in aged male rats. Horm Metab Res 33(8):491–498

Fagnoni FF, Lozza L, Zibera C, Zambelli A, Ponchio L, Gibelli N, Oliviero B, Pavesi L, Gennari R, Vescovini R, Sansoni P, Da Prada G, Robustelli Della Cuna G (2002) T-cell dynamics after high-dose chemotherapy in adults: elucidation of the elusive CD8+ subset reveals multiple homeostatic T-cell compartments with distinct implications for immune competence. Immunology 106(1):27–37

Flomerfelt FA, El Kassar N, Gurunathan C, Chua KS, League SC, Schmitz S, Gershon TR, Kapoor V, Yan XY, Schwartz RH, Gress RE (2010) Tbata modulates thymic stromal cell proliferation and thymus function. J Exp Med 207(11):2521–2532

Fry TJ, Moniuszko M, Creekmore S, Donohue SJ, Douek DC, Giardina S, Hecht TT, Hill BJ, Komschlies K, Tomaszewski J, Franchini G, Mackall CL (2003) IL-7 therapy dramatically alters peripheral T-cell homeostasis in normal and SIV-infected nonhuman primates. Blood 101(6):2294–2299

Fry TJ, Sinha M, Milliron M, Chu YW, Kapoor V, Gress RE, Thomas E, Mackall CL (2004) Flt3 ligand enhances thymic-dependent and thymic-independent immune reconstitution. Blood 104(9):2794–2800

Garfin PM, Min D, Bryson JL, Serwold T, Edris B, Blackburn CC, Richie ER, Weinberg KI, Manley NR, Sage J, Viatour P (2013) Inactivation of the RB family prevents thymus involution and promotes thymic function by direct control of Foxn1 expression. J Exp Med 210(6):1087–1097

Gill J, Malin M, Hollander GA, Boyd R (2002) Generation of a complete thymic microenvironment by MTS24(+) thymic epithelial cells. Nat Immunol 3(7):635–642

Goldberg GL, Sutherland JS, Hammet MV, Milton MK, Heng TS, Chidgey AP, Boyd RL (2005) Sex steroid ablation enhances lymphoid recovery following autologous hematopoietic stem cell transplantation. Transplantation 80(11):1604–1613

Goldberg GL, Alpdogan O, Muriglan SJ, Hammett MV, Milton MK, Eng JM, Hubbard VM, Kochman A, Willis LM, Greenberg AS, Tjoc KH, Sutherland JS, Chidgey A, van den Brink MRM, Boyd RL (2007) Enhanced immune reconstitution by sex steroid ablation following allogeneic hemopoietic stem cell transplantation. J Immunol 178(11):7473–7484

Goldberg GL, King CG, Nejat RA, Suh DY, Smith OM, Bretz JC, Samstein RM, Dudakov JA, Chidgey AP, Chen-Kiang S, Boyd RL, van den Brink MRM (2009) Luteinizing hormone-releasing hormone enhances T cell recovery following allogeneic bone marrow transplantation. J Immunol 182(9):5846–5854

Goldberg GL, Dudakov JA, Reiseger JJ, Seach N, Ueno T, Vlahos K, Hammett MV, Young LF, Heng TSP, Boyd RL, Chidgey AP (2010) Sex steroid ablation enhances immune reconstitution following cytotoxic antineoplastic therapy in young mice. J Immunol 184(11):6014–6024

Gordon J, Patel SR, Mishina Y, Manley NR (2010) Evidence for an early role for BMP4 signaling in thymus and parathyroid morphogenesis. Dev Biol 339(1):141–154

Gratama JW, Naipal A, Oljans P, Zwaan FE, Verdonck LF, de Witte T, Vossen JM, Bolhuis RL, de Gast GC, Jansen J (1984) T lymphocyte repopulation and differentiation after bone marrow transplantation. Early shifts in the ratio between T4+ and T8+ T lymphocytes correlate with the occurrence of acute graft-versus-host disease. Blood 63(6):1416–1423

Gray DH, Seach N, Ueno T, Milton MK, Liston A, Lew AM, Goodnow CC, Boyd RL (2006) Developmental kinetics, turnover, and stimulatory capacity of thymic epithelial cells. Blood 108(12):3777–3785

Greenstein BD, Fitzpatrick FT, Kendall MD, Wheeler MJ (1987) Regeneration of the thymus in old male rats treated with a stable analogue of LHRH. J Endocrinol 112(3):345–350

Grimaldi CM, Jeganathan V, Diamond B (2006) Hormonal regulation of B cell development: 17 beta-estradiol impairs negative selection of high-affinity DNA-reactive B cells at more than one developmental checkpoint. J Immunol 176(5):2703–2710

Gruver AL, Hudson LL, Sempowski GD (2007) Immunosenescence of ageing. J Pathol 211(2):144–156

Hakim FT, Memon SA, Cepeda R, Jones EC, Chow CK, Kasten-Sportes C, Odom J, Vance BA, Christensen BL, Mackall CL, Gress RE (2005) Age-dependent incidence, time course, and consequences of thymic renewal in adults. J Clin Invest 115(4):930–939

Hauri-Hohl MM, Zuklys S, Keller MP, Jeker LT, Barthlott T, Moon AM, Roes J, Hollander GA (2008) TGF-beta signaling in thymic epithelial cells regulates thymic involution and postirradiation reconstitution. Blood 112(3):626–634

Heitger A, Neu N, Kern H, Panzer-Grumayer ER, Greinix H, Nachbaur D, Niederwieser D, Fink FM (1997) Essential role of the thymus to reconstitute naive (CD45RA+) T-helper cells after human allogeneic bone marrow transplantation. Blood 90(2):850–857

Heng TS, Goldberg GL, Gray DH, Sutherland JS, Chidgey AP, Boyd RL (2005) Effects of castration on thymocyte development in two different models of thymic involution. J Immunol 175(5):2982–2993

Heng TSP, Chidgey AP, Boyd RL (2010) Getting back at nature: understanding thymic development and overcoming its atrophy. Curr Opin Pharmacol 10(4):425–433

Henson SM, Snelgrove R, Hussell T, Wells DJ, Aspinall R (2005) An IL-7 fusion protein that shows increased thymopoietic ability. J Immunol 175(6):4112–4118

Herasimtschuk AA, Westrop SJ, Moyle GJ, Downey JS, Imami N (2008) Effects of recombinant human growth hormone on HIV-1-specific T-cell responses, thymic output and proviral DNA in patients on HAART: 48-week follow-up. J Immune Based Ther Vaccines 6:7

Hirokawa K, Makinodan T (1975) Thymic involution: effect on T cell differentiation. J Immunol 114(6):1659–1664

Holland AM, Zakrzewski JL, Goldberg GL, Ghosh A, van den Brink MR (2008) Adoptive precursor cell therapy to enhance immune reconstitution after hematopoietic stem cell transplantation in mouse and man. Semin Immunopathol 30:479–487

Hollander GA, Wang B, Nichogiannopoulou A, Platenburg PP, van Ewijk W, Burakoff SJ, Gutierrez-Ramos JC, Terhorst C (1995) Developmental control point in induction of thymic cortex regulated by a subpopulation of prothymocytes. Nature 373(6512):350–353

Holmes R, Zuniga-Pflucker JC (2009) The OP9-DL1 system: generation of T-lymphocytes from embryonic or hematopoietic stem cells in vitro. Cold Spring Harb Protoc 2009(2):pdb.prot5156

Igarashi H, Kouro T, Yokota T, Comp PC, Kincade PW (2001) Age and stage dependency of estrogen receptor expression by lymphocyte precursors. Proc Natl Acad Sci U S A 98(26):15131–15136

Jaffe HL (1924) The influence of the suprarenal gland on the thymus: I. Regeneration of the thymus following double suprarenalectomy in the rat. J Exp Med 40(3):325–342

Jeker LT, Barthlott T, Keller MP, Zuklys S, Hauri-Hohl M, Deng C-X, Holländer GA (2008) Maintenance of a normal thymic microenvironment and T-cell homeostasis require Smad4-mediated signaling in thymic epithelial cells. Blood 112(9):3688–3695

Jenkinson WE, Bacon A, White AJ, Anderson G, Jenkinson EJ (2008) An epithelial progenitor pool regulates thymus growth. J Immunol 181(9):6101–6108

Kang J, Der SD (2004) Cytokine functions in the formative stages of a lymphocyte's life. Curr Opin Immunol 16(2):180–190

Kelly RM, Highfill SL, Panoskaltsis-Mortari A, Taylor PA, Boyd RL, Hollander GA, Blazar BR (2008) Keratinocyte growth factor and androgen blockade work in concert to protect against conditioning regimen-induced thymic epithelial damage and enhance T-cell reconstitution after murine bone marrow transplantation. Blood 111(12):5734–5744

Kelly RM, Goren EM, Taylor PA, Mueller SN, Stefanski HE, Osborn MJ, Scott HS, Komarova EA, Gudkov AV, Hollander GA, Blazar BR (2010) Short-term inhibition of p53 combined with keratinocyte growth factor improves thymic epithelial cell recovery and enhances T-cell reconstitution after murine bone marrow transplantation. Blood 115(5):1088–1097. https://doi.org/10.1182/blood-2009-05-223198

Kenins L, Gill JW, Boyd RL, Hollander GA, Wodnar-Filipowicz A (2008) Intrathymic expression of Flt3 ligand enhances thymic recovery after irradiation. J Exp Med 205(3):523–531

Ki S, Park D, Selden HJ, Seita J, Chung H, Kim J, Iyer VR, Ehrlich LIR (2014) Global transcriptional profiling reveals distinct functions of thymic stromal subsets and age-related changes during thymic involution. Cell Rep 9(1):402–415

Kincade PW, Medina KL, Payne KJ, Rossi MI, Tudor KS, Yamashita Y, Kouro T (2000) Early B-lymphocyte precursors and their regulation by sex steroids. Immunol Rev 175:128–137

Kobayashi H, Butler JM, O'Donnell R, Kobayashi M, Ding BS, Bonner B, Chiu VK, Nolan DJ, Shido K, Benjamin L, Rafii S (2010) Angiocrine factors from Akt-activated endothelial cells balance self-renewal and differentiation of haematopoietic stem cells. Nat Cell Biol 12(11):1046–1056

Kvell K, Varecza Z, Bartis D, Hesse S, Parnell S, Anderson G, Jenkinson EJ, Pongracz JE (2010) Wnt4 and LAP2alpha as pacemakers of thymic epithelial senescence. PLoS One 5(5):e10701

Lai L, Jin J (2009) Generation of thymic epithelial cell progenitors by mouse embryonic stem cells. Stem Cells 27(12):3012–3020

Lai L, Cui C, Jin J, Hao Z, Zheng Q, Ying M, Boyd R, Zhao Y (2011) Mouse embryonic stem cell-derived thymic epithelial cell progenitors enhance T-cell reconstitution after allogeneic bone marrow transplantation. Blood 118(12):3410–3418

Lanna A, Henson SM, Escors D, Akbar AN (2014) The kinase p38 activated by the metabolic regulator AMPK and scaffold TAB1 drives the senescence of human T cells. Nat Immunol 15(10):965–972

Lanna A, Gomes DC, Muller-Durovic B, McDonnell T, Escors D, Gilroy DW, Lee JH, Karin M, Akbar AN (2017) A sestrin-dependent Erk-Jnk-p38 MAPK activation complex inhibits immunity during aging. Nat Immunol 18(3):354–363

Lepletier A, Hun M, Hammett M, Wong K, Naeem H, Hedger M, Loveland K, Chidgey AP (2019) Interplay between activin bA follistatin and Bmp4 signalling regulates postnatal thymic epithelial progenitor self-renewal and differentiation. Cell Reports http://dx.doi.org/10.2139/ssrn.3305565

Levy Y, Lacabaratz C, Weiss L, Viard JP, Goujard C, Lelievre JD, Boue F, Molina JM, Rouzioux C, Avettand-Fenoel V, Croughs T, Beq S, Thiebaut R, Chene G, Morre M, Delfraissy JF (2009) Enhanced T cell recovery in HIV-1-infected adults through IL-7 treatment. J Clin Invest 119(4):997–1007

Levy Y, Sereti I, Tambussi G, Routy JP, Lelievre JD, Delfraissy JF, Molina JM, Fischl M, Goujard C, Rodriguez B, Rouzioux C, Avettand Fenoel V, Croughs T, Beq S, Morre M, Poulin JF, Sekaly RP, Thiebaut R, Lederman MM (2012) Effects of recombinant human interleukin 7 on T-cell recovery and thymic output in HIV-infected patients receiving antiretroviral therapy: results of a phase I/IIa randomized, placebo-controlled, multicenter study. Clin Infect Dis 55(2):291–300

Li L, Hsu HC, Stockard CR, Yang P, Zhou J, Wu Q, Grizzle WE, Mountz JD (2004) IL-12 inhibits thymic involution by enhancing IL-7- and IL-2-induced thymocyte proliferation. J Immunol 172(5):2909–2916

Liang CC, You LR, Yen JJ, Liao NS, Yang-Yen HF, Chen CM (2013) Thymic epithelial beta-catenin is required for adult thymic homeostasis and function. Immunol Cell Biol 91(8):511–523. https://doi.org/10.1038/icb.2013.34

Linton PJ, Dorshkind K (2004) Age-related changes in lymphocyte development and function. Nat Immunol 5(2):133–139

Lopes N, Vachon H, Marie J, Irla M (2017) Administration of RANKL boosts thymic regeneration upon bone marrow transplantation. EMBO Mol Med 9(6):835–851

Love PE, Bhandoola A (2011) Signal integration and crosstalk during thymocyte migration and emigration. Nat Rev Immunol 11(7):469–477

Lu H, Zhao Z, Kalina T, Gillespy T III, Liggitt D, Andrews RG, Maloney DG, Kiem HP, Storek J (2005) Interleukin-7 improves reconstitution of antiviral CD4 T cells. Clin Immunol 114(1):30–41

Lynch HE, Goldberg GL, Chidgey A, Van den Brink MR, Boyd R, Sempowski GD (2009) Thymic involution and immune reconstitution. Trends Immunol 30(7):366–373

MacDonald BT, Tamai K, He X (2009) Wnt/beta-catenin signaling: components, mechanisms, and diseases. Dev Cell 17(1):9–26

Mackall CL, Gress RE (1997) Thymic aging and T-cell regeneration. Immunol Rev 160:91–102

Mackall CL, Fleisher TA, Brown MR, Magrath IT, Shad AT, Horowitz ME, Wexler LH, Adde MA, McClure LL, Gress RE (1994) Lymphocyte depletion during treatment with intensive chemotherapy for cancer. Blood 84(7):2221–2228

Mackall CL, Fleisher TA, Brown MR, Andrich MP, Chen CC, Feuerstein IM, Horowitz ME, Magrath IT, Shad AT, Steinberg SM et al (1995) Age, thymopoiesis, and CD4+ T-lymphocyte regeneration after intensive chemotherapy. N Engl J Med 332(3):143–149

Mackall CL, Fleisher TA, Brown MR, Andrich MP, Chen CC, Feuerstein IM, Magrath IT, Wexler LH, Dimitrov DS, Gress RE (1997) Distinctions between CD8+ and CD4+ T-cell regenerative pathways result in prolonged T-cell subset imbalance after intensive chemotherapy. Blood 89(10):3700–3707

Mackall CL, Fry TJ, Gress RE (2011) Harnessing the biology of IL-7 for therapeutic application. Nat Rev Immunol 11(5):330–342

Masuzawa T, Miyaura C, Onoe Y, Kusano K, Ohta H, Nozawa S, Suda T (1994) Estrogen deficiency stimulates B lymphopoiesis in mouse bone marrow. J Clin Invest 94(3):1090–1097

Maue AC, Yager EJ, Swain SL, Woodland DL, Blackman MA, Haynes L (2009) T-cell immunosenescence: lessons learned from mouse models of aging. Trends Immunol 30(7):301–305

Medina KL, Garrett KP, Thompson LF, Rossi MI, Payne KJ, Kincade PW (2001) Identification of very early lymphoid precursors in bone marrow and their regulation by estrogen. Nat Immunol 2(8):718–724

Miller JF (1961) Immunological function of the thymus. Lancet 2(7205):748–749

Min D, Taylor PA, Panoskaltsis-Mortari A, Chung B, Danilenko DM, Farrell C, Lacey DL, Blazar BR, Weinberg KI (2002) Protection from thymic epithelial cell injury by keratinocyte growth factor: a new approach to improve thymic and peripheral T-cell reconstitution after bone marrow transplantation. Blood 99(12):4592–4600

Min D, Panoskaltsis-Mortari A, Kuro OM, Hollander GA, Blazar BR, Weinberg KI (2007) Sustained thymopoiesis and improvement in functional immunity induced by exogenous KGF administration in murine models of aging. Blood 109(6):2529–2537

Napolitano LA, Lo JC, Gotway MB, Mulligan K, Barbour JD, Schmidt D, Grant RM, Halvorsen RA, Schambelan M, McCune JM (2002) Increased thymic mass and circulating naive CD4 T cells in HIV-1-infected adults treated with growth hormone. AIDS 16(8):1103–1111

Napolitano LA, Schmidt D, Gotway MB, Ameli N, Filbert EL, Ng MM, Clor JL, Epling L, Sinclair E, Baum PD, Li K, Killian ML, Bacchetti P, McCune JM (2008) Growth hormone enhances thymic function in HIV-1-infected adults. J Clin Invest 118(3):1085–1098

Niehrs C (2006) Function and biological roles of the Dickkopf family of Wnt modulators. Oncogene 25(57):7469–7481

Nolan DJ, Ginsberg M, Israely E, Palikuqi B, Poulos MG, James D, Ding BS, Schachterle W, Liu Y, Rosenwaks Z, Butler JM, Xiang J, Rafii A, Shido K, Rabbany SY, Elemento O, Rafii S (2013) Molecular signatures of tissue-specific microvascular endothelial cell heterogeneity in organ maintenance and regeneration. Dev Cell 26(2):204–219

O'Neill KE, Bredenkamp N, Tischner C, Vaidya HJ, Stenhouse FH, Peddie CD, Nowell CS, Gaskell T, Blackburn CC (2016) Foxn1 is dynamically regulated in thymic epithelial cells during embryogenesis and at the onset of thymic involution. PLoS One 11(3):e0151666

Ohigashi I, Zuklys S, Sakata M, Mayer CE, Zhanybekova S, Murata S, Tanaka K, Hollander GA, Takahama Y (2013) Aire-expressing thymic medullary epithelial cells originate from beta5t-expressing progenitor cells. Proc Natl Acad Sci U S A 110(24):9885–9890

Ohigashi I, Kozai M, Takahama Y (2016) Development and developmental potential of cortical thymic epithelial cells. Immunol Rev 271(1):10–22

Olsen NJ, Kovacs WJ (2001) Effects of androgens on T and B lymphocyte development. Immunol Res 23(2–3):281–288

Olsen NJ, Watson MB, Henderson GS, Kovacs WJ (1991) Androgen deprivation induces phenotypic and functional changes in the thymus of adult male mice. Endocrinology 129(5):2471–2476

Ortman CL, Dittmar KA, Witte PL, Le PT (2002) Molecular characterization of the mouse involuted thymus: aberrations in expression of transcription regulators in thymocyte and epithelial compartments. Int Immunol 14(7):813–822

Osada M, Jardine L, Misir R, Andl T, Millar SE, Pezzano M (2010) DKK1 mediated inhibition of Wnt signaling in postnatal mice leads to loss of TEC progenitors and thymic degeneration. PLoS One 5(2):e9062

Parent AV, Russ HA, Khan IS, LaFlam TN, Metzger TC, Anderson MS, Hebrok M (2013) Generation of functional thymic epithelium from human embryonic stem cells that supports host T cell development. Cell Stem Cell 13(2):219–229

Perales MA, Goldberg JD, Yuan J, Koehne G, Lechner L, Papadopoulos EB, Young JW, Jakubowski AA, Zaidi B, Gallardo H, Liu C, Rasalan T, Wolchok JD, Croughs T, Morre M, Devlin SM, van den Brink MR (2012) Recombinant human interleukin-7 (CYT107) promotes T-cell recovery after allogeneic stem cell transplantation. Blood 120(24):4882–4891

Prieyl JA, LeBien TW (1996) Interleukin 7 independent development of human B cells. Proc Natl Acad Sci U S A 93(19):10348–10353

Quinn KM, Fox A, Harland KL, Russ BE, Li J, Nguyen THO, Loh L, Olshanksy M, Naeem H, Tsyganov K, Wiede F, Webster R, Blyth C, Sng XYX, Tiganis T, Powell D, Doherty PC, Turner SJ, Kedzierska K, La Gruta NL (2018) Age-related decline in primary CD8(+) T cell responses is associated with the development of senescence in virtual memory CD8(+) T cells. Cell Rep 23(12):3512–3524

Rafei M, Dumont-Lagace M, Rouette A, Perreault C (2013) Interleukin-21 accelerates thymic recovery from glucocorticoid-induced atrophy. PLoS One 8(9):e72801

Rafii S, Butler JM, Ding BS (2016a) Angiocrine functions of organ-specific endothelial cells. Nature 529(7586):316–325

Rafii S, Ginsberg M, Scandura J, Butler JM, Ding BS (2016b) Transplantation of endothelial cells to mitigate acute and chronic radiation injury to vital organs. Radiat Res 186(2):196–202

Reimann C, Dal Cortivo L, Hacein-Bey-Abina S, Fischer A, Andre-Schmutz I, Cavazzana-Calvo M (2010) Advances in adoptive immunotherapy to accelerate T-cellular immune reconstitution after HLA-incompatible hematopoietic stem cell transplantation. Immunotherapy 2(4):481–496

Ribeiro AR, Rodrigues PM, Meireles C, Di Santo JP, Alves NL (2013) Thymocyte selection regulates the homeostasis of IL-7–expressing thymic cortical epithelial cells in vivo. J Immunol 191(3):1200–1209

Rode I, Martins VC, Küblbeck G, Maltry N, Tessmer C, Rodewald H-R (2015) Foxn1 protein expression in the developing, aging, and regenerating thymus. J Immunol 195(12):5678–5687

Roden AC, Moser MT, Tri SD, Mercader M, Kuntz SM, Dong H, Hurwitz AA, McKean DJ, Celis E, Leibovich BC, Allison JP, Kwon ED (2004) Augmentation of T cell levels and responses induced by androgen deprivation. J Immunol 173(10):6098–6108

Rodrigues PM, Ribeiro AR, Serafini N, Meireles C, Di Santo JP, Alves NL (2018) Intrathymic deletion of IL-7 reveals a contribution of the bone marrow to thymic rebound induced by androgen blockade. J Immunol. https://doi.org/10.4049/jimmunol.1701112

Roosnek EE, Brouwer MC, Vossen JM, Roos MT, Schellekens PT, Zeijlemaker WP, Aarden LA (1987) The role of interleukin-2 in proliferative responses in vitro of T cells from patients after bone marrow transplantation. Evidence that minor defects can lead to in vitro unresponsiveness. Transplantation 43(6):855–860

Rosenberg SA, Sportes C, Ahmadzadeh M, Fry TJ, Ngo LT, Schwarz SL, Stetler-Stevenson M, Morton KE, Mavroukakis SA, Morre M, Buffet R, Mackall CL, Gress RE (2006) IL-7 administration to humans leads to expansion of CD8+ and CD4+ cells but a relative decrease of CD4+ T-regulatory cells. J Immunother 29(3):313–319

Rossi S, Blazar BR, Farrell CL, Danilenko DM, Lacey DL, Weinberg KI, Krenger W, Hollander GA (2002) Keratinocyte growth factor preserves normal thymopoiesis and thymic microenvironment during experimental graft-versus-host disease. Blood 100(2):682–691

Rossi SW, Chidgey AP, Parnell SM, Jenkinson WE, Scott HS, Boyd RL, Jenkinson EJ, Anderson G (2007a) Redefining epithelial progenitor potential in the developing thymus. Eur J Immunol 37(9):2411–2418

Rossi SW, Jeker LT, Ueno T, Kuse S, Keller MP, Zuklys S, Gudkov AV, Takahama Y, Krenger W, Blazar BR, Hollander GA (2007b) Keratinocyte growth factor (KGF) enhances postnatal T-cell development via enhancements in proliferation and function of thymic epithelial cells. Blood 109(9):3803–3811

Roux E, Helg C, Dumont-Girard F, Chapuis B, Jeannet M, Roosnek E (1996) Analysis of T-cell repopulation after allogeneic bone marrow transplantation: significant differences between recipients of T-cell depleted and unmanipulated grafts. Blood 87(9):3984–3992

Roux E, Dumont-Girard F, Starobinski M, Siegrist CA, Helg C, Chapuis B, Roosnek E (2000) Recovery of immune reactivity after T-cell-depleted bone marrow transplantation depends on thymic activity. Blood 96(6):2299–2303

Schluns KS, Cook JE, Le PT (1997) TGF-beta differentially modulates epidermal growth factor-mediated increases in leukemia-inhibitory factor, IL-6, IL-1 alpha, and IL-1 beta in human thymic epithelial cells. J Immunol 158(6):2704–2712

Schmitt TM, Zuniga-Pflucker JC (2002) Induction of T cell development from hematopoietic progenitor cells by delta-like-1 in vitro. Immunity 17(6):749–756

Scholzen T, Gerdes J (2000) The Ki-67 protein: from the known and the unknown. J Cell Physiol 182(3):311–322

Seach N, Layton D, Lim J, Chidgey A, Boyd R (2007) Thymic generation and regeneration: a new paradigm for establishing clinical tolerance of stem cell-based therapies. Curr Opin Biotechnol 18(5):441–447

Seet CS, He C, Bethune MT, Li S, Chick B, Gschweng EH, Zhu Y, Kim K, Kohn DB, Baltimore D, Crooks GM, Montel-Hagen A (2017) Generation of mature T cells from human hematopoietic stem and progenitor cells in artificial thymic organoids. Nat Methods 14(5):521–530

Sempowski GD, Hale LP, Sundy JS, Massey JM, Koup RA, Douek DC, Patel DD, Haynes BF (2000) Leukemia inhibitory factor, oncostatin M, IL-6, and stem cell factor mRNA expression in human thymus increases with age and is associated with thymic atrophy. J Immunol 164(4):2180–2187

Sempowski GD, Gooding ME, Liao HX, Le PT, Haynes BF (2002) T cell receptor excision circle assessment of thymopoiesis in aging mice. [erratum appears in Mol Immunol 2002 Oct;39(5–6):379–80]. Mol Immunol 38(11):841–848

Sereti I, Dunham RM, Spritzler J, Aga E, Proschan MA, Medvik K, Battaglia CA, Landay AL, Pahwa S, Fischl MA, Asmuth DM, Tenorio AR, Altman JD, Fox L, Moir S, Malaspina A, Morre M, Buffet R, Silvestri G, Lederman MM (2009) IL-7 administration drives T cell-cycle entry and expansion in HIV-1 infection. Blood 113(25):6304–6314

Sfikakis PP, Gourgoulis GM, Moulopoulos LA, Kouvatseas G, Theofilopoulos AN, Dimopoulos MA (2005) Age-related thymic activity in adults following chemotherapy-induced lymphopenia. Eur J Clin Investig 35(6):380–387

Shukla S, Langley MA, Singh J, Edgar JM, Mohtashami M, Zuniga-Pflucker JC, Zandstra PW (2017) Progenitor T-cell differentiation from hematopoietic stem cells using Delta-like-4 and VCAM-1. Nat Methods 14(5):531–538

Smith MJ, Reichenbach DK, Parker SL, Riddle MJ, Mitchell J, Osum KC, Mohtashami M, Stefanski HE, Fife BT, Bhandoola A, Hogquist KA, Hollander GA, Zuniga-Pflucker JC, Tolar J, Blazar BR (2017) T cell progenitor therapy-facilitated thymopoiesis depends upon thymic input and continued thymic microenvironment interaction. JCI Insight 2(10):92056

Soh C-L, Giudice A, Jenny RA, Elliott DA, Hatzistavrou T, Micallef SJ, Kianizad K, Seach N, Zúñiga-Pflücker JC, Chidgey AP, Trounson A, Nilsson SK, Haylock DN, Boyd RL, Elefanty AG, Stanley EG (2014) FOXN1GFP/w reporter hESCs enable identification of integrin-β4, HLA-DR, and EpCAM as markers of human PSC-derived FOXN1+ thymic epithelial progenitors. Stem Cell Rep 2(6):925–937

Soiffer RJ, Bosserman L, Murray C, Cochran K, Daley J, Ritz J (1990) Reconstitution of T-cell function after CD6-depleted allogeneic bone marrow transplantation. Blood 75(10):2076–2084

Sportes C, Hakim FT, Memon SA, Zhang H, Chua KS, Brown MR, Fleisher TA, Krumlauf MC, Babb RR, Chow CK, Fry TJ, Engels J, Buffet R, Morre M, Amato RJ, Venzon DJ, Korngold R, Pecora A, Gress RE, Mackall CL (2008) Administration of rhIL-7 in humans increases in vivo TCR repertoire diversity by preferential expansion of naive T cell subsets. J Exp Med 205(7):1701–1714

Sportes C, Babb RR, Krumlauf MC, Hakim FT, Steinberg SM, Chow CK, Brown MR, Fleisher TA, Noel P, Maric I, Stetler-Stevenson M, Engel J, Buffet R, Morre M, Amato RJ, Pecora A, Mackall CL, Gress RE (2010) Phase I study of recombinant human interleukin-7 administration in subjects with refractory malignancy. Clin Cancer Res 16(2):727–735

Storek J, Witherspoon RP, Storb R (1995) T cell reconstitution after bone marrow transplantation into adult patients does not resemble T cell development in early life. Bone Marrow Transplant 16(3):413–425

Storek J, Gooley T, Witherspoon RP, Sullivan KM, Storb R (1997) Infectious morbidity in long-term survivors of allogeneic marrow transplantation is associated with low CD4 T cell counts. Am J Hematol 54(2):131–138

Sudo T, Nishikawa S, Ohno N, Akiyama N, Tamakoshi M, Yoshida H (1993) Expression and function of the interleukin 7 receptor in murine lymphocytes. Proc Natl Acad Sci U S A 90(19):9125–9129

Sun L, Guo J, Brown R, Amagai T, Zhao Y, Su DM (2010) Declining expression of a single epithelial cell-autonomous gene accelerates age-related thymic involution. Aging Cell 9(3):347–357

Sun X, Xu J, Lu H, Liu W, Miao Z, Sui X, Liu H, Su L, Du W, He Q, Chen F, Shi Y, Deng H (2013) Directed differentiation of human embryonic stem cells into thymic epithelial progenitor-like cells reconstitutes the thymic microenvironment in vivo. Cell Stem Cell 13(2):230–236

Sutherland JS, Goldberg GL, Hammett MV, Uldrich AP, Berzins SP, Heng TS, Blazar BR, Millar JL, Malin MA, Chidgey AP, Boyd RL (2005) Activation of thymic regeneration in mice and humans following androgen blockade. J Immunol 175(4):2741–2753

Sutherland JS, Spyroglou L, Muirhead JL, Heng TS, Prieto-Hinojosa A, Prince HM, Chidgey AP, Schwarer AP, Boyd RL (2008) Enhanced immune system regeneration in humans following allogeneic or autologous hemopoietic stem cell transplantation by temporary sex steroid blockade. Clin Cancer Res 14(4):1138–1149

Taguchi T, Takenouchi H, Matsui J, Tang WR, Itagaki M, Shiozawa Y, Suzuki K, Sakaguchi S, Ktagiri YU, Takahashi T, Okita H, Fujimoto J, Kiyokawa N (2006) Involvement of insulin-like growth factor-I and insulin-like growth factor binding proteins in pro-B-cell development. Exp Hematol 34(4):508–518

Tajima A, Pradhan I, Trucco M, Fan Y (2016) Restoration of thymus function with bioengineered thymus organoids. Curr Stem Cell Rep 2(2):128–139

Talaber G, Kvell K, Varecza Z, Boldizsar F, Parnell SM, Jenkinson EJ, Anderson G, Berki T, Pongracz JE (2011) Wnt-4 protects thymic epithelial cells against dexamethasone-induced senescence. Rejuvenation Res 14(3):241–248

Taub DD, Longo DL (2005) Insights into thymic aging and regeneration. Immunol Rev 205:72–93

Tormo A, Khodayarian F, Cui Y, Al-Chami E, Kanjarawi R, Noe B, Wang H, Rafei M (2017) Interleukin-21 promotes thymopoiesis recovery following hematopoietic stem cell transplantation. J Hematol Oncol 10(1):120

Ulyanchenko S, O'Neill KE, Medley T, Farley AM, Vaidya HJ, Cook AM, Blair NF, Blackburn CC (2016) Identification of a bipotent epithelial progenitor population in the adult thymus. Cell Rep 14(12):2819–2832

van den Broek T, Delemarre EM, Janssen WJM, Nievelstein RAJ, Broen JC, Tesselaar K, Borghans JAM, Nieuwenhuis EES, Prakken BJ, Mokry M, Jansen NJG, van Wijk F (2016) Neonatal thymectomy reveals differentiation and plasticity within human naive T cells. J Clin Invest 126(3):1126–1136

Varecza Z, Kvell K, Talaber G, Miskei G, Csongei V, Bartis D, Anderson G, Jenkinson EJ, Pongracz JE (2011) Multiple suppression pathways of canonical Wnt signalling control thymic epithelial senescence. Mech Ageing Dev 132(5):249–256

Velardi E, Dudakov JA, Van den Brink MRM (2013) Clinical strategies to enhance thymic recovery after allogeneic hematopoietic stem cell transplantation. Immunol Lett 155(1–2):31–35

Velardi E, Tsai JJ, Holland AM, Wertheimer T, Yu VWC, Zakrzewski JL, Tuckett AZ, Singer NV, West ML, Smith OM, Young LF, Kreines FM, Levy ER, Boyd RL, Scadden DT, Dudakov JA, van den Brink MRM (2014) Sex steroid blockade enhances thymopoiesis by modulating Notch signaling. J Exp Med 211(12):2341–2349

Velardi E, Dudakov JA, van den Brink MR (2015) Sex steroid ablation: an immunoregenerative strategy for immunocompromised patients. Bone Marrow Transplant 50(Suppl 2):S77–S81

Velardi E, Tsai JJ, Radtke S, Cooper K, Argyropoulos KV, Jae-Hung S, Young LF, Lazrak A, Smith OM, Lieberman S, Kreines F, Shono Y, Wertheimer T, Jenq RR, Hanash AM, Narayan P, Lei Z, Moore MA, Kiem HP, van den Brink MRM, Dudakov JA (2018) Suppression of luteinizing hormone enhances HSC recovery after hematopoietic injury. Nat Med 24(2):239–246

Viselli SM, Reese KR, Fan J, Kovacs WJ, Olsen NJ (1997) Androgens alter B cell development in normal male mice. Cell Immunol 182(2):99–104

Weinberg K, Annett G, Kashyap A, Lenarsky C, Forman SJ, Parkman R (1995) The effect of thymic function on immunocompetence following bone marrow transplantation. Biol Blood Marrow Transplant 1(1):18–23

Wertheimer T, Velardi E, Tsai J, Cooper K, Xiao S, Kloss CC, Ottmüller KJ, Mokhtari Z, Brede C, deRoos P, Kinsella S, Palikuqi B, Ginsberg M, Young LF, Kreines F, Lieberman SR, Lazrak A, Guo P, Malard F, Smith OM, Shono Y, Jenq RR, Hanash AM, Nolan DJ, Butler JM, Beilhack A, Manley NR, Rafii S, Dudakov JA, van den Brink MRM (2018) Production of BMP4 by endothelial cells is crucial for endogenous thymic regeneration. Sci Immunol 3(19). https://doi.org/10.1126/sciimmunol.aal2736

Williams KM, Lucas PJ, Bare CV, Wang J, Chu YW, Tayler E, Kapoor V, Gress RE (2008) CCL25 increases thymopoiesis after androgen withdrawal. Blood 112(8):3255–3263

Wils EJ, Braakman E, Verjans GM, Rombouts EJ, Broers AE, Niesters HG, Wagemaker G, Staal FJ, Lowenberg B, Spits H, Cornelissen JJ (2007) Flt3 ligand expands lymphoid progenitors prior to recovery of thymopoiesis and accelerates T cell reconstitution after bone marrow transplantation. J Immunol 178(6):3551–3557

Wils EJ, Aerts-Kaya FS, Rombouts EJ, van Mourik I, Rijken-Schelen A, Visser TP, Braakman E, Wagemaker G, Cornelissen JJ (2012) Keratinocyte growth factor and stem cell factor to improve thymopoiesis after autologous CD34+ cell transplantation in rhesus macaques. Biol Blood Marrow Transplant 18(1):55–65

Wong K, Lister Natalie L, Barsanti M, Lim Joanna MC, Hammett Maree V, Khong Danika M, Siatskas C, Gray Daniel HD, Boyd Richard L, Chidgey AP (2014) Multilineage potential and self-renewal define an epithelial progenitor cell population in the adult thymus. Cell Rep 8(4):1198–1209

Youm YH, Yang H, Sun Y, Smith RG, Manley NR, Vandanmagsar B, Dixit VD (2009) Deficient ghrelin receptor-mediated signaling compromises thymic stromal cell microenvironment by accelerating thymic adiposity. J Biol Chem 284(11):7068–7077

Yu VWC, Saez B, Cook C, Lotinun S, Pardo-Saganta A, Wang Y-H, Lymperi S, Ferraro F, Raaijmakers MHGP, Wu JY, Zhou L, Rajagopal J, Kronenberg HM, Baron R, Scadden DT (2015) Specific bone cells produce DLL4 to generate thymus-seeding progenitors from bone marrow. J Exp Med 212(5):759–774

Zakrzewski JL, Kochman AA, Lu SX, Terwey TH, Kim TD, Hubbard VM, Muriglan SJ, Suh D, Smith OM, Grubin J, Patel N, Chow A, Cabrera-Perez J, Radhakrishnan R, Diab A, Perales MA, Rizzuto G, Menet E, Pamer EG, Heller G, Zuniga-Pflucker JC, Alpdogan O, van den Brink MR (2006) Adoptive transfer of T-cell precursors enhances T-cell reconstitution after allogeneic hematopoietic stem cell transplantation. Nat Med 12(9):1039–1047

Zakrzewski JL, Suh D, Markley JC, Smith OM, King C, Goldberg GL, Jenq R, Holland AM, Grubin J, Cabrera-Perez J, Brentjens RJ, Lu SX, Rizzuto G, Sant'Angelo DB, Riviere I, Sadelain M, Heller G, Zuniga-Pflucker JC, Lu C, van den Brink MR (2008) Tumor immunotherapy across MHC barriers using allogeneic T-cell precursors. Nat Biotechnol 26(4):453–461

Zhang SL, Wang X, Manna S, Zlotoff DA, Bryson JL, Blazar BR, Bhandoola A (2014) Chemokine treatment rescues profound T-lineage progenitor homing defect after bone marrow transplant conditioning in mice. Blood 124(2):296–304

Zlotoff DA, Zhang SL, De Obaldia ME, Hess PR, Todd SP, Logan TD, Bhandoola A (2011) Delivery of progenitors to the thymus limits T-lineage reconstitution after bone marrow transplantation. Blood 118(7):1962–1970

Zoller AL, Kersh GJ (2006) Estrogen induces thymic atrophy by eliminating early thymic progenitors and inhibiting proliferation of beta-selected thymocytes. J Immunol 176(12):7371–7378

Zook EC, Krishack PA, Zhang S, Zeleznik-Le NJ, Firulli AB, Witte PL, Le PT (2011) Overexpression of Foxn1 attenuates age-associated thymic involution and prevents the expansion of peripheral CD4 memory T cells. Blood 118(22):5723–5731

Zuklys S, Handel A, Zhanybekova S, Govani F, Keller M, Maio S, Mayer CE, Teh HY, Hafen K, Gallone G, Barthlott T, Ponting CP, Hollander GA (2016) Foxn1 regulates key target genes essential for T cell development in postnatal thymic epithelial cells. Nat Immunol 17(10):1206–1215

Chapter 14
Thymus Rejuvenation After Autologous Hematopoietic Stem Cell Transplantation in Patients with Autoimmune Diseases

João R. Lima-Júnior, Lucas C. M. Arruda, Maria Carolina de Oliveira, and Kelen C. R. Malmegrim

Abstract Autologous hematopoietic stem cell transplantation (AHSCT) has been established as an important therapeutic approach for patients with autoimmune diseases (AD) refractory to conventional treatment. This therapy is able to promote long-term remission in most patients without further use of immunosuppressive medication. High dose immunosuppression depletes autoreactive T and B cells and a new immune system re-emerges from the infused hematopoietic stem cells, a mechanism so-called "immune resetting". Thymic rejuvenation plays a crucial role

J. R. Lima-Júnior
Biosciences and Biotechnology Program, School of Pharmaceutical Sciences of Ribeirão Preto, University of São Paulo, Ribeirão Preto, SP, Brazil

Center for Cell-Based Therapy, Regional Hemotherapy Center of Ribeirão Preto Medical School, University of São Paulo, Ribeirão Preto, SP, Brazil

L. C. M. Arruda
Department of Clinical Science, Intervention and Technology, Karolinska Institutet, Stockholm, Sweden

M. C. de Oliveira
Center for Cell-Based Therapy, Regional Hemotherapy Center of Ribeirão Preto Medical School, University of São Paulo, Ribeirão Preto, SP, Brazil

Divison of Clinical Immunology, Department of Internal Medicine, Ribeirão Preto Medical School, University of São Paulo, Ribeirão Preto, SP, Brazil

Basic and Applied Immunology Program, Ribeirão Preto Medical School, University of São Paulo, Ribeirão Preto, SP, Brazil

K. C. R. Malmegrim (✉)
Biosciences and Biotechnology Program, School of Pharmaceutical Sciences of Ribeirão Preto, University of São Paulo, Ribeirão Preto, SP, Brazil

Center for Cell-Based Therapy, Regional Hemotherapy Center of Ribeirão Preto Medical School, University of São Paulo, Ribeirão Preto, SP, Brazil

Department of Clinical Analysis, Toxicology and Food Sciences, School of Pharmaceutical Sciences of Ribeirão Preto, University of São Paulo, Ribeirão Preto, SP, Brazil
e-mail: kelenfarias@fcfrp.usp.br

© Springer Nature Switzerland AG 2019
G. A. Passos (ed.), *Thymus Transcriptome and Cell Biology*,
https://doi.org/10.1007/978-3-030-12040-5_14

in the immune reconstitution and restoration of self-tolerance in AD patients treated with AHSCT. Indeed, recent thymic emigrants promote reestablishment of TCR diversity that is associated with favorable clinical outcomes. However, generally one third of the patients undergo disease reactivation after AHSCT and the involved mechanisms are not yet fully understood. Therefore, additional investigations should be made to improve the knowledge about immune mechanisms involved in AHSCT for AD and to unravel biomarkers of post-transplantation outcomes.

14.1 Introduction

Autoimmune diseases (AD) are characterized by specific tissue damage due to humoral and cellular self-reactivity. Their development is attributed to imbalances among genetic, environmental, hormonal and immunoregulatory factors, which culminate in loss of immune tolerance mechanisms (Theofilopoulos et al. 2017; Barturen et al. 2018).

The thymus is a primary lymphoid organ that plays a pivotal role in the development of immune tolerance and prevention of autoimmunity through the elimination of self-reactive T-cells and production of CD4+ Forkhead box *P3*+ (Foxp3+) natural regulatory T-cells (nTreg) (Boehm and Swann 2013; Cheng and Anderson 2018). In addition, a productive thymopoiesis sustains a polyclonal naïve T-cell repertoire (TCR) (Chung et al. 2014), which is characteristic of a healthy and self-tolerant immune system. However, thymus undergoes a progressive involution (or atrophy) associated with aging that can lead to dysfunction of negative selection mechanism, decreased output of naive T-cells and impaired nTreg development and function, thereby increasing the risk of AD development (Coder et al. 2015).

Although thymic involution is an inevitable process during life, thymic tissue plasticity allows this process to be stopped or even reversed (Toubert et al. 2012). Indeed, renewal of thymopoiesis has been reported in patients with AD that underwent immunoablation followed by autologous hematopoietic stem cell transplantation (AHSCT) (Fig. 14.1). Besides, thymic rejuvenation has been associated to clinical improvement MS patients after AHSCT (Muraro et al. 2005, 2014; Arruda et al. 2018a, b).

In this chapter, we provide the basic knowledge for understanding how thymus rejuvenation influences the clinical outcomes in AD patients treated with AHSCT. In addition, we detail different approaches available for human thymic function assessment.

14.2 A Brief Review of Thymus Physiology

14.2.1 Thymopoiesis

The thymus stroma and architecture provide a microenvironment that supports the differentiation of immature thymocytes (T-cell progenitors derived from bone marrow hematopoietic stem cells) into mature T-cell lymphocytes (Miller and Osoba 1967).

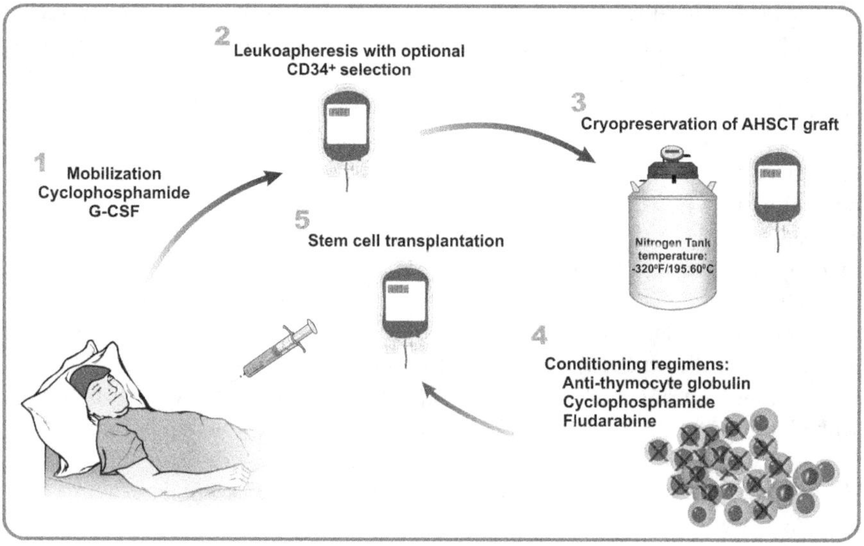

Fig. 14.1 Key steps of AHSCT. (1) The procedure starts with the mobilization of HSCs from bone marrow to peripheral blood through administration of low-dose cyclophosphamide and G-CSF. (2) HSCs are then harvest, with or without *ex vivo* manipulation (CD34+ HSCs selection), and (3) cryopreserved. (4) Prior to transplantation, the immune system is depleted by high-dose immunosuppression (conditioning regimen), followed by the (5) reinfusion of autologous HSCs. *HSCs* hematopoietic stem cells, *AHSCT* autologous hematopoietic stem cell transplantation, *G-CSF* granulocyte-colony stimulating factor

This microenvironment is composed of extracellular component matrix, chemokines (CXCR4-SDF-1, CCR7-CCL19/21, CCR9-CCL25, CCR4-CCL17/22), cytokines and growth factors (IL-7, IL-15, KGF, TSLP) and two thymic epithelial cell (TECs) types that can be subdivided into cortical TECs (cTECs) and medullary TECs (mTECs) based on their location in the thymus (Takahama 2006; Anderson et al. 2006; Schwarz et al. 2007; Trampont et al. 2010; Ribeiro et al. 2013; Meireles et al. 2017).

Thymocyte development occurs in different intrathymic sites and requires cell-to-cell interaction. T-cell progenitors enter the thymus close to the cortico-medullary junction. The earliest developmental stage of thymocytes is the CD4−CD8− double negative (DN) phenotype, further subcategorized into DN1, DN2, DN3, and DN4 based on CD44 and CD25 expressions (Zeng et al. 2007; Benz et al. 2008). During the development from DN to double positive (DP) stage, thymocytes migrate towards the sub-capsular zone of the cortex, where they interact with cTECs. At DN3 stage, the β-chain is rearranged and pre-TCR is formed. After several rounds of division, the TCR α-chain is rearranged and, upon successful rearrangement, the full TCR protein is expressed on thymocytes at the DP stage. DP cells undergo CD4 or CD8 lineage commitment, which leads to single positive (SP) thymocytes. SP cells migrate to thymic medulla, where they interact with mTECs and are scanned for self-antigen binding affinity. Thymocytes that interact strongly with self-peptides are finally deleted (Hoffman et al. 1997; Takahama 2006; Janas et al. 2010; Trampont et al. 2010; Kreslavsky et al. 2013).

It is important to reiterate that developing thymocytes are challenged to recognize self-antigens presented predominantly, but not exclusively, by cTECs (during positive selection process) and mTECs (during negative selection process) in the context of major histocompatibility complex (MHC), to ensure that no high-avidity self-reactive T-cells ultimately migrate to the periphery (Starr et al. 2003; Abramson and Anderson 2017). Those are key checkpoints to avoid autoimmunity and development of AD. However, currently is known that some degree of autoreactivity is physiologic and healthy individual have low-avidity autoreactive cells under control by peripheral immune regulatory mechanisms (Abramson and Anderson 2017).

mTECs plays a crucial role in negative selection due to their ability to express Autoimmune regulator (Aire) transcription factor. Aire controls the transcription of a range of peripheral tissue antigens in the thymus to promote the maintenance of self-tolerance. During thymic development, thymocytes that recognize self-antigens with high-avidity are eliminated (Oliveira et al. 2013; Speck-Hernandez et al. 2018; Passos et al. 2018). Noteworthy, Coder et al. (2015) observed that mice lacking Foxn1 (a transcription factor crucial to development of TECs) showed thymic atrophy, decreased Aire expression, negative selection impairment and increased susceptibility to autoimmunity.

14.2.2 Consequences of Thymic Involution for Immune Tolerance Mechanisms

Thymic involution is a physiological phenomenon mainly (but not exclusively) associated with aging (Boehm and Swann 2013; Ayasoufi et al. 2017). The thymus increases in size at birth until puberty and thereafter starts to atrophy. During this process, thymic microenvironment undergoes structural changes that lead to a loss of cellularity, including a decrease of mTECs and cTECs numbers. As consequence, thymocyte differentiation is decreased and stromal tissue is replaced by adipocytes (Dixit 2010; Aw et al. 2008). It has been suggested that thymic architecture alterations are related to development of AD (Abramson and Anderson 2017).

In parallel, as already mentioned, nTreg cell generation is an essential mechanism for self-tolerance. These cells comprise a subset of CD4+ T-cells (Foxp3+CD25high) that are rescued from the programmed cell death process after recognizing self-antigens with high/intermediate-avidity in the thymus and thereafter acquired suppressive capacity (Josefowicz et al. 2012; Ohkura et al. 2013). This process is predominantly AIRE-dependent and, therefore, sensitive to changes in the thymic microenvironment (Lin et al. 2016; Malchow et al. 2016). Indeed, nTreg cells play an important role in the maintenance of peripheral self-tolerance for being able to inhibit activation and/or proliferation of effector cells, predominantly by IL-10 production. Nevertheless, their frequency and suppressive capacity may be decreased in several AD (Bonelli et al. 2008; Dominguez-Villar et al. 2011; Haseda et al. 2013; Arruda et al. 2018a, b), probably as consequence of thymic involution or thymopoiesis deregulation.

14.3 AHSCT: Restoration of Thymic Function and Immune Tolerance in Patients with AD

To understand how AHSCT drives thymic rejuvenation, and consequently the TCR repertoire renewal and immune tolerance reestablishment, it is important to highlight certain aspects of this cellular therapy.

14.3.1 Basics Concepts of AHSCT

AHSCT has been used in the last 20 years for treatment of severe and refractory patients with AD. Several translational studies have provided a broad knowledge for clinical management and protocol improvement, increasing the safety of AHSCT through the years (Burt et al. 2009, 2011; Farge et al. 2010; van Laar et al. 2014; Muraro et al. 2014; Malmegrim et al. 2017; Snowden et al. 2017, 2018; Arruda et al. 2018a, b).

The AHSCT procedure can be divided into four main steps according to the guidelines of the European Society for Blood and Marrow Transplantation (Snowden et al. 2017). Initially, HSCs (CD34+) are recruited from the bone marrow to peripheral blood through a process called mobilization. This procedure consists on the administration of low-dose cyclophosphamide and granulocyte colony-stimulating factor (G-CSF). After mobilization, HSC-enriched peripheral blood is harvested via leukapheresis (with or without CD34+ *ex vivo* selection) and cryopreserved. Subsequently, high-dose immunosuppressive drugs (such as anti-thymocyte globulin, cyclophosphamide or fludarabine) are administrated into patients promoting immune ablation. This step is named conditioning regimen. Finally, at the transplant day, previously harvested autologous HSC are reinfused into the patient to promote bone marrow rescue and immune reconstitution (Swart et al. 2017; Muraro et al. 2017) (Fig. 14.1).

The rationale of AHSCT for AD is the elimination of autoreactive T and B-cells by the conditioning regimen, followed by regeneration of a new, diverse and self-tolerant immune system ontogeny from HSCs, the so-called "immune resetting" (Fig. 14.2) mechanism (Muraro et al. 2005; Abrahamsson and Muraro 2008). In addition, newly generated nTreg cells also contribute to modulate effector T-cell activation and function (Delemarre et al. 2014, 2016) (Fig. 14.2).

14.3.2 T-Cell Reconstitution After AHSCT

Early after transplantation (within 6 months), the lymphopenic environment induces homeostatic proliferation of peripheral residual non-depleted and/or re-infused mature T-cells (mostly T CD8+CD45RA−CD45RO+ memory cells), by self or

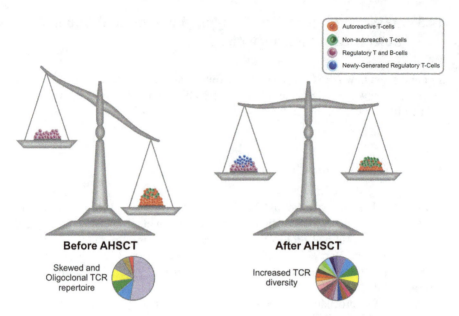

Fig. 14.2 Immune resetting after AHSCT for autoimmune diseases. Before AHSCT, there is an imbalance between effector and regulatory cells, high frequency of autoreactive cells, skewed TCR repertoire distribution, and reduced number/function of regulatory cells. The AHSCT shifts this "immune imbalance state" towards an "immune homeostasis state" by regenerating a new, diverse and self-tolerant immune system. *AHSCT* autologous hematopoietic stem cell transplantation, *TCR* T-cell receptor

foreign antigens and/or IL-7/IL-15 stimulation, leading to restricted or skewed repertoire expansions (Tchao and Turka 2012; Arruda et al. 2018a, b). This process is physiologically important to avoid profound lymphopenia and to control infections, but it does not reconstitute a diverse T-cell compartment.

Subsequently, the second phase of immune reconstitution (Fig. 14.3) comprises *de novo* thymic production of naive T-cells, shifting the T-cell repertoire pattern (from oligoclonal to polyclonal) after 1–2 years post-transplantation (Muraro et al. 2005). Thymic rebound (or thymic reactivation, or thymic rejuvenation) is the most important immune mechanism of the AHSCT leading to long-term maintenance of the T-cell compartment, renewal of TCR specificities and broader T-cell repertoire diversity, necessary to patient immunity.

Thymic function can be evaluated by three different approaches: (1) Measurement by flow cytometry of CD4+CD31+CD45RA+ recent thymic emigrants (RTEs) cell numbers in the peripheral blood (Thiel et al. 2008; Douaisi et al. 2017). After AHSCT, increased numbers of RTEs are associated to thymic rebound; (2) Evaluation of TCR repertoire diversity through complementarity determining region 3 (CDR3) analyses, which can be achieved by next-generation sequencing (NGS) and spectratyping/immunoscope methodologies (Rosati et al. 2017). After AHSCT,

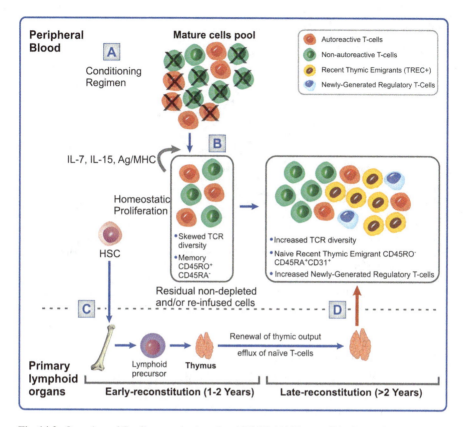

Fig. 14.3 Overview of T-cell reconstitution after AHSCT. (A) The conditioning regimen promotes indistinct depletion of autoreactive and non-autoreactive mature T-cells in the peripheral blood. (B) Early after transplantation, the lymphopenic environment induces homeostatic proliferation of residual non-depleted or re-infused T-cells through self-antigen or foreign antigen presentation and/or cytokine stimulation, leading to a restricted TCR diversity and memory T cell phenotype acquisition. (C) Transplanted CD34+ HSCs migrate and graft at the bone marrow, where lymphoid precursors are generated. Thereafter, these T cell precursors migrate to the thymus where they undergo positive and negative selection, emerging as naive self-tolerant T-cells. (D) Finally, thymic reactivation promotes efflux of newly generated T-cells and regulatory T-cells with new TCR specificities. *TCR* T-cell receptor, *HSC* hematopoietic stem cells, *Ag* antigen, *MHC* major histocompatibility complex

thymic rejuvenation leads to renewal of TCR specificities and increased T cell repertoire diversity; (3) Analysis of TREC (T cell receptor excision circles) levels in peripheral blood. TRECs are episomal, non-replicative, DNA excised from the TCRB locus excision during β-chain and recombination (βTREC and sjTREC, respectively) in developing T cells within the thymus. The sjTREC/βTREC ratio indicates the thymocyte intrathymic proliferation rate (Toubert et al. 2012; Mensen et al. 2013) (Fig. 14.4).

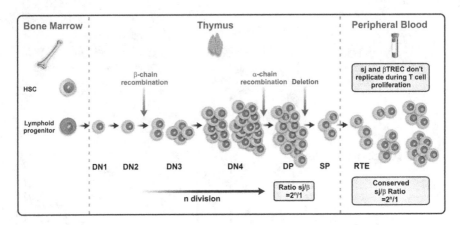

Fig. 14.4 The basic principle of βTREC/sjTREC analysis. βTREC comprises an episomal DNA which arises from TCRB gene locus excision during beta-chain recombination and remains stable and do not duplicated upon cell division. Thereafter, another excision occurs at the TCRA/D locus, also generating another episomal DNA, the sjTREC. The sjTREC/βTREC ratio indicates the thymocyte proliferative within the thymus and reflect the thymic function. *TREC* T-cell receptor excision circles, *HSC* hematopoietic stem cells, *DN* double negative, *DP* double positive, *SP* single positive, *RTE* recent thymic emigrants, *TCR* T-cell receptor

14.4 Biomarkers of Thymic Rejuvenation and Favorable Clinical Outcomes in Patients with AD After AHSCT: TCR Repertoire Renewal, Increased TREC Levels and Treg Cell Expansion

Since 2004 several research groups have reported different immune mechanisms of AHSCT for different AD, which are involved in early and/or late phases of immune reconstitution.

14.4.1 Immune Mechanisms of AHSCT in Multiple Sclerosis

Multiple Sclerosis (MS) is a demyelinating, neurodegenerative and inflammatory T-cell mediated disease of the central nervous system, characterized by axon injuries and neurological disabilities (Milo and Miller 2014; Yadav et al. 2015). It has been demonstrated that MS patients treated with AHSCT showed increased numbers of RTEs, significant TCR repertoire changes and decreased frequency of memory CD4+ T-cells after AHSCT as compared to baseline levels (Sun et al. 2004; Muraro et al. 2005; Dubinsky et al. 2010).

MS patients exhibited resolution of brain inflammation over AHSCT follow-up, which may be related to the immune resetting mechanism (Muraro et al. 2005). In addition, deep sequencing TCR analysis demonstrated substantial renewal of the

CD4+ T-cell repertoire mainly. Several CD4+ T-cell clones present at baseline were undetectable post-transplantation, while a large number of CD8+ T-cell clones have resisted the conditioning regimen and were detected after transplantation. Therefore, after AHSCT there is a mix of clonally expanded residual CD8+ T-cells and newly generated CD4+ T-cells with different TCR specificities. Furthermore, MS patients who failed to respond to AHSCT showed lower TCR repertoire diversity at 2 months after transplantation when compared to responder patients, suggesting that thymic reactivation and increased TCR diversity may be associated with favorable clinical response (Muraro et al. 2014).

As aforementioned, Treg cells play a crucial role in the maintenance of immune homeostasis. Therefore, decreased numbers and/or function of Treg cells may be associated with loss of tolerance and autoimmunity development (Coder et al. 2015). Indeed, some studies have demonstrated Treg cell expansion in MS patients following AHSCT (Darlington et al. 2013; Abrahamsson et al. 2013; Arruda et al. 2016).

Additionally, Arruda et al. (2015) demonstrated decreased expressions of FOXP3, FOXO1, PDCD1 and IRF2BP2 immunoregulatory genes in MS patients before AHSCT, compared to controls. These genes are target of miR-16, miR-155 and miR-142-3p, which were up-regulated in the patients at the same period. However, 2-years after AHSCT the expressions of the miRs and their target genes were normalized (Arruda et al. 2015).

Taken together these studies in MS patients indicate that thymic rejuvenation after AHSCT reestablishes the T-cell repertoire diversity and the balance between regulatory and effector T-cells.

14.4.2 Immune Mechanisms of AHSCT in Systemic Sclerosis

Systemic Sclerosis (SSc) is an autoimmune connective tissue disease characterized by microvascular damage as consequence of T-cell activation, autoantibodies production and abnormal secretion pro-fibrotic cytokines. This milieu leads to fibroblast activation, excessive production and deposition of collagen in the skin and internal organs, and ultimately to fibrosis (Allanore et al. 2015; Katsumoto et al. 2011).

Defective Treg compartment and reduced thymic function have been reported in SSc patients (Reiff et al. 2009; Papp et al. 2011). Noteworthy, absolute numbers of Treg increased in SSc patients 1-year after AHSCT and positively correlated with sjTREC and RTE counts, suggesting thymic reactivation in these patients. Reconstituted Tregs also produces high amounts of IL-10 when compared to baseline (Baraut et al. 2014; Arruda et al. 2018a, b).

In addition, responder SSc patients reconstituted a polyclonal repertoire after AHSCT, in contrast to non-responder patients that maintained a skewed TCR distribution (Farge et al. 2017; Arruda et al. 2018a, b). Around 25% of the TCR clonotypes present at baseline were found at 6 months after AHSCT in responder SSc patients, as opposed to 60% observed in non-responder patients. These shared clonotypes consisted around 10–60% of the entire TCR repertoire at 2-years post-transplantation in responder and non-responder patients, respectively (Arruda et al. 2018a, b). These data prove that thymic rejuvenation and consequent renewal of TCR specificities correlates with favorable clinical outcomes after AHSCT in SSc patients.

Altogether, these data suggested that newly generated T-cells and Treg cells might control autoimmunity and are associated with clinical remission of SSc patients after AHSCT.

14.4.3   Immune Mechanisms of AHSCT in Juvenile Idiopathic Arthritis

Juvenile idiopathic arthritis (JIA) is an autoimmune rheumatic disease characterized by T-cell dependent, chronic synovial inflammation, resulting in progressive joint damage (Mellins et al. 2011). Decreased thymic output has been reported in patients with JIA (Mayerl and Prelog 2012). Furthermore, Treg cells from peripheral blood and synovial fluid of JIA patients displayed reduced TCR diversity and impaired suppressive function (Wehrens et al. 2011; Haufe et al. 2011; Rossetti et al. 2017; Henderson et al. 2016).

In this context, Delemarre et al. (2016) suggested that restricted TCR diversity of CD4+ Treg cells might be associated with chronic inflammation in JIA. Notably, JIA patients that achieved clinical remission after AHSCT presented renewal of the Treg-TCR repertoire and thymic reactivation (Delemarre et al. 2016).

Collectively, these studies demonstrated the importance of thymic reactivation for Treg-TCR renewal after AHSCT in patients with JIA.

14.4.4   Immune Mechanisms of AHSCT in Type 1 Diabetes

Type 1 diabetes (T1D) is an AD characterized by destruction of pancreatic β cells predominantly by autoreactive CD8+ T-cells (Roep and Tree 2014). Recently, Malmegrim et al. (2017) reported that T1D patients became insulin-free for 43 months on average after undergoing AHSCT. Metabolic improvement was associated with high levels of C-peptide levels. Post-transplantation TREC levels increased from 18 to 60 months, indicating also thymic rebound in T1D patients after AHSCT. Of note, Treg cell counts were higher in prolonged-remission patients compared to short-remission ones (Malmegrim et al. 2017).

Immune reconstitution studies in T1D patients also demonstrated the pivotal role of thymic reactivation and production of new naive T-cells and Treg cells (with high TREC levels) in determining the clinical outcome after AHSCT.

14.4.5 Immune Mechanisms of AHSCT in Systemic Lupus Erythematosus

Systemic lupus erythematosus (SLE) is a chronic, systemic autoimmune disease characterized by autoantibody production against a wide spectrum of self-antigens and excessive immune complex deposition in specific tissues (Suárez-Fueyo et al. 2016; Moulton et al. 2017). SLE patients display decreased T cell repertoire diversity (Thapa et al. 2015). In this context, Alexander et al. (2009) demonstrated that long-term remission in SLE patients treated with AHSCT was also associated with *de novo* generation of T-cells and renewal of TCR repertoire. Furthermore, the frequency and absolute counts of nTreg cell achieved control levels in SLE patients after transplantation (Alexander et al. 2009).

Increased frequencies of CD8+FoxP3+ Treg cells, whose function was predominantly TGF-β dependent, have been associated with clinical remission of SLE patients after AHSCT (Churlaud et al. 2015; Zhang et al. 2009).

Taken together, these studies demonstrated increased frequencies of regulatory T subsets consists an important immune mechanism of AHSCT in SLE patients.

14.5 Conclusion

Currently, AHSCT is the only therapy able to promote long-term clinical remissions in AD patients without additional medication. The rational of this therapy is the depletion of autoreactive cells by high-dose immunosuppression and the resetting of a new, diverse and self-tolerant immune system. In this context, thymus rejuvenation gives rise to a highly diverse peripheral TCR repertoire by production of new naive T-cells and nTreg cells. Furthermore, several studies have reported correlation between clinical remission and effective thymic reactivation. Although the impact of thymic rejuvenation following AHSCT for AD has been elucidated in recent years, this topic needs to be further and deeply explored for harnessing its clinical potential.

Acknowledgments The authors thank Sandra Navarro Bresciani for the illustration and graphic design, Center for Cell-Based Therapy of Ribeirão Preto Medical School, Coordination for the Improvement of Higher Education Personnel (CAPES), and São Paulo Research Foundation (FAPESP) through grant No. 2016/24443-7 to João R. Lima-Júnior.

References

Abrahamsson S, Muraro PA (2008) Immune re-education following autologous hematopoietic stem cell transplantation. Autoimmunity 41(8):577–584

Abrahamsson SV et al (2013) Non-myeloablative autologous haematopoietic stem cell transplantation expands regulatory cells and depletes IL-17 producing mucosal-associated invariant T cells in multiple sclerosis. Brain 136(Pt 9):2888–2903

Abramson J, Anderson G (2017) Thymic epithelial cells. Annu Rev Immunol 35(1):85–118

Alexander T et al (2009) Depletion of autoreactive immunologic memory followed by autologous hematopoietic stem cell transplantation in patients with refractory SLE induces long-term remission through de novo generation of a juvenile and tolerant immune system. Blood 113(1):214–223

Allanore Y et al (2015) Systemic sclerosis. Nat Rev Dis Primers 1:15002

Anderson G et al (2006) Establishment and functioning of intrathymic microenvironments. Immunol Rev 209:10–27

Arruda LCMM et al (2015) Autologous hematopoietic SCT normalizes miR-16, -155 and -142-3p expression in multiple sclerosis patients. Bone Marrow Transplant 50(3):380–389

Arruda LCMM et al (2016) Immunological correlates of favorable long-term clinical outcome in multiple sclerosis patients after autologous hematopoietic stem cell transplantation. Clin Immunol 169:47–57

Arruda LCM et al (2018a) Homeostatic proliferation leads to telomere attrition and increased PD-1 expression after autologous hematopoietic SCT for systemic sclerosis. Bone Marrow Transplant 53:1319–1327

Arruda LCM et al (2018b) Immune rebound associates with a favorable clinical response to autologous HSCT in systemic sclerosis patients. Blood Adv 2(2):126–141

Aw D et al (2008) Architectural changes in the thymus of aging mice. Aging Cell 7(2):158–167

Ayasoufi K, Fan R, Valujskikh A (2017) Depletion-resistant CD4 T cells enhance thymopoiesis during lymphopenia. Am J Transplant 17(8):2008–2019

Baraut J et al (2014) Peripheral blood regulatory T cells in patients with diffuse systemic sclerosis (SSc) before and after autologous hematopoietic SCT: a pilot study. Bone Marrow Transplant 49(3):349–354

Barturen G et al (2018) Moving towards a molecular taxonomy of autoimmune rheumatic diseases. Nat Rev Rheumatol 14(2):75–93

Benz C et al (2008) The stream of precursors that colonizes the thymus proceeds selectively through the early T lineage precursor stage of T cell development. J Exp Med 205(5):1187–1199

Boehm T, Swann JB (2013) Thymus involution and regeneration: two sides of the same coin? Nat Rev Immunol 13(11):831–838

Bonelli M et al (2008) Quantitative and qualitative deficiencies of regulatory T cells in patients with systemic lupus erythematosus (SLE). Int Immunol 20(7):861–868

Burt RK et al (2009) Autologous non-myeloablative haemopoietic stem cell transplantation in relapsing-remitting multiple sclerosis: a phase I/II study. Lancet Neurol 8(3):244–253

Burt RK et al (2011) Autologous non-myeloablative haemopoietic stem-cell transplantation compared with pulse cyclophosphamide once per month for systemic sclerosis (ASSIST): an open-label, randomised phase 2 trial. Lancet 378(9790):498–506

Cheng M, Anderson MS (2018) Thymic tolerance as a key brake on autoimmunity. Nat Immunol 19(July):1

Chung B et al (2014) Engineering the human thymic microenvironment to support thymopoiesis in vivo. Stem Cells 32(9):2386–2396

Churlaud G et al (2015) Human and mouse CD8+CD25+FOXP3+ regulatory T cells at steady state and during interleukin-2 therapy. Front Immunol 6(April):2–11

Coder BD et al (2015) Thymic involution perturbs negative selection leading to autoreactive T cells that induce chronic inflammation. J Immunol 194(12):5825–5837

Darlington PJ et al (2013) Diminished Th17 (not Th1) responses underlie multiple sclerosis disease abrogation after hematopoietic stem cell transplantation. Ann Neurol 73(3):341–354

Delemarre EM et al (2014) Autologous stem cell transplantation restores immune tolerance in experimental arthritis by renewal and modulation of the Teff cell compartment. Arthritis Rheumatol 66(2):350–356

Delemarre EM et al (2016) Autologous stem cell transplantation aids autoimmune patients by functional renewal and TCR diversification of regulatory T cells. Blood 127(1):91–102

Dixit VD (2010) Thymic fatness and approaches to enhance thymopoietic fitness in aging. Curr Opin Immunol 22(4):521–528

Dominguez-Villar M, Baecher-Allan CM, Hafler DA (2011) Identification of T helper type 1-like, Foxp3+ regulatory T cells in human autoimmune disease. Nat Med 17(6):673–675

Douaisi M et al (2017) CD31, a Valuable Marker to Identify Early and Late Stages of T Cell Differentiation in the Human Thymus. J Immunol 198(6):2310–2319. https://doi.org/10.4049/jimmunol.1500350

Dubinsky AN et al (2010) T-cell clones persisting in the circulation after autologous hematopoietic SCT are undetectable in the peripheral CD34+ selected graft. Bone Marrow Transplant 45(2):325–331

Farge D et al (2010) Autologous hematopoietic stem cell transplantation for autoimmune diseases: an observational study on 12 years' experience from the European Group for Blood and Marrow Transplantation Working Party on Autoimmune Diseases. Haematologica 95(2):284–292

Farge D et al (2017) Long-term immune reconstitution and T cell repertoire analysis after autologous hematopoietic stem cell transplantation in systemic sclerosis patients. J Hematol Oncol 10(1):21

Haseda F et al (2013) CD4+CD45RA-FoxP3highactivated regulatory T cells are functionally impaired and related to residual insulin-secreting capacity in patients with type 1 diabetes. Clin Exp Immunol 173(2):207–216

Haufe S et al (2011) Impaired suppression of synovial fluid CD4+CD25− T cells from patients with juvenile idiopathic arthritis by CD4+CD25+ Treg cells. Arthritis Rheum 63(10):3153–3162

Henderson LA et al (2016) Next-generation sequencing reveals restriction and clonotypic expansion of Treg cells in juvenile idiopathic arthritis. Arthritis Rheumatol 68(7):1758–1768

Hoffman ES et al (1997) Productive T-cell receptor-chain gene rearrangement: coincident regulation of cell cycle and clonality during development in vivo. Genes Dev 10:948–962

Janas ML et al (2010) Thymic development beyond β-selection requires phosphatidylinositol 3-kinase activation by CXCR4. J Exp Med 207(1):247–261

Josefowicz SZ, Lu L-F, Rudensky AY (2012) Regulatory T cells: mechanisms of differentiation and function. Annu Rev Immunol 30(1):531–564

Katsumoto TR, Whitfield ML, Connolly MK (2011) The pathogenesis of systemic sclerosis. Annu Rev Pathol 6:509–537

Kreslavsky T et al (2013) Negative selection, not receptor editing, is a physiological response of autoreactive thymocytes. J Exp Med 210(10):1911–1918

Lin J et al (2016) Increased generation of Foxp3+ regulatory T cells by manipulating antigen presentation in the thymus. Nat Commun 7:1–12

Malchow S et al (2016) Aire enforces immune tolerance by directing autoreactive T cells into the regulatory T cell lineage. Immunity 44(5):1102–1113

Malmegrim KCR et al (2017) Immunological balance is associated with clinical outcome after autologous hematopoietic stem cell transplantation in Type 1 diabetes. Front Immunol 8:167

Mayerl C, Prelog M (2012) Immunosenescence and juvenile idiopathic arthritis. Autoimmun Rev 11(5):297–300

Meireles C et al (2017) Thymic crosstalk restrains the pool of cortical thymic epithelial cells with progenitor properties. Eur J Immunol 47(6):958–969

Mellins ED, MacAubas C, Grom AA (2011) Pathogenesis of systemic juvenile idiopathic arthritis: some answers, more questions. Nat Rev Rheumatol 7(7):416–426

Mensen A et al (2013) Utilization of TREC and KREC quantification for the monitoring of early T- and B-cell neogenesis in adult patients after allogeneic hematopoietic stem cell transplantation. J Transl Med 11:1

Miller JF, Osoba D (1967) Current concepts of the immunological function of the thymus. Physiol Rev 47(3):437–520

Milo R, Miller A (2014) Revised diagnostic criteria of multiple sclerosis. Autoimmun Rev 13(4–5):518–524

Moulton VR et al (2017) Pathogenesis of human systemic lupus erythematosus: a cellular perspective. Trends Mol Med 23(7):615–635

Muraro PA et al (2005) Thymic output generates a new and diverse TCR repertoire after autologous stem cell transplantation in multiple sclerosis patients. J Exp Med 201(5):805–816

Muraro PA et al (2014) Brief report T cell repertoire following autologous stem cell transplantation for multiple sclerosis. J Clin Invest 124(3):1168–1172

Muraro PA et al (2017) Autologous haematopoietic stem cell transplantation for treatment of multiple sclerosis. Nat Rev Neurol 13(7):391–405

Ohkura N, Kitagawa Y, Sakaguchi S (2013) Development and maintenance of regulatory T cells. Immunity 38(3):414–423

Oliveira EH et al (2013) Expression profile of peripheral tissue antigen genes in medullary thymic epithelial cells (mTECs) is dependent on mRNA levels of autoimmune regulator (Aire). Immunobiology 218(1):96–104

Papp G et al (2011) Altered T-cell and regulatory cell repertoire in patients with diffuse cutaneous systemic sclerosis. Scand J Rheumatol 40(3):205–210

Passos GA et al (2018) Update on Aire and thymic negative selection. Immunology 153(1):10–20

Reiff A et al (2009) Study of thymic size and function in children and adolescents with treatment refractory systemic sclerosis eligible for immunoablative therapy. Clin Immunol 133(3):295–302

Ribeiro AR et al (2013) Thymocyte selection regulates the homeostasis of IL-7-expressing thymic cortical epithelial cells in vivo. J Immunol 191(3):1200–1209

Roep BO, Tree TIM (2014) Immune modulation in humans: implications for type 1 diabetes mellitus. Nat Rev Endocrinol 10(4):229–242

Rosati E et al (2017) Overview of methodologies for T-cell receptor repertoire analysis. BMC Biotechnol 17(1):1–16

Rossetti M et al (2017) TCR repertoire sequencing identifies synovial Treg cell clonotypes in the bloodstream during active inflammation in human arthritis. Ann Rheum Dis 76(2):435–441

Schwarz BA et al (2007) Selective thymus settling regulated by cytokine and chemokine receptors. J Immunol 178(4):2008–2017

Snowden JA et al (2017) Evolution, trends, outcomes, and economics of hematopoietic stem cell transplantation in severe autoimmune diseases. Blood Adv 1(27):2742–2755

Snowden JA et al (2018) Autologous haematopoietic stem cell transplantation (AHSCT) in severe Crohn's Disease: a review on behalf of ECCO and EBMT. J Crohns Colitis 12(4):476–488

Speck-Hernandez CA et al (2018) Aire disruption influences the medullary thymic epithelial cell transcriptome and interaction with thymocytes. Front Immunol 9:1–15

Starr TK, Jameson SC, Hogquist KA (2003) Positive and negative selection of T cells. Annu Rev Immunol 21:139–176

Suárez-Fueyo A, Bradley SJ, Tsokos GC (2016) T cells in systemic lupus erythematosus. Curr Opin Immunol 43:32–38

Sun W et al (2004) Characteristics of T-cell receptor repertoire and myelin-reactive T cells reconstituted from autologous haematopoietic stem-cell grafts in multiple sclerosis. Brain 127(5):996–1008

Swart JF et al (2017) Haematopoietic stem cell transplantation for autoimmune diseases. Nat Rev Rheumatol 13(4):244–256

Takahama Y (2006) Journey through the thymus: stromal guides for T-cell development and selection. Nat Rev Immunol 6(2):127–135

Tchao NK, Turka LA (2012) Lymphodepletion and homeostatic proliferation: implications for transplantation. Am J Transplant 12(5):1079–1090

Thapa DR et al (2015) Longitudinal analysis of peripheral blood T cell receptor diversity in patients with systemic lupus erythematosus by next-generation sequencing. Arthritis Res Ther 17(1):132

Theofilopoulos AN, Kono DH, Baccala R (2017) The multiple pathways to autoimmunity. Nat Immunol 18(7):716–724

Thiel A et al (2008) Direct assessment of thymic reactivation after autologous stem cell transplantation. Acta Haematol 119(1):22–27

Toubert A et al (2012) Thymus and immune reconstitution after allogeneic hematopoietic stem cell transplantation in humans: never say never again. Tissue Antigens 79(2):83–89

Trampont PC et al (2010) CXCR4 acts as a costimulator during thymic B-selection. Nat Immunol 11(2):162–170

van Laar JM et al (2014) Autologous hematopoietic stem cell transplantation vs intravenous pulse cyclophosphamide in diffuse cutaneous systemic sclerosis. JAMA 311(24):2490

Wehrens EJ et al (2011) Functional human regulatory T cells fail to control autoimmune inflammation due to PKB/c-akt hyperactivation in effector cells. Blood 118(13):3538–3548

Yadav SK et al (2015) Advances in the immunopathogenesis of multiple sclerosis. Curr Opin Neurol 28(3):206–219

Zeng L, Dalheimer SL, Yankee TM (2007) Gads−/− mice reveal functionally distinct subsets of TCR+ CD4−CD8− double-negative thymocytes. J Immunol 179(2):1013–1021

Zhang L et al (2009) Regulatory T cell (Treg) subsets return in patients with refractory lupus following stem cell transplantation, and TGF-beta-producing CD8+ Treg cells are associated with immunological remission of lupus. J Immunol 183(10):6346–6358

Concluding Remarks and Perspectives

I hope this book has served as an overview of the thymus gland, from fundamental aspects, historical background, its ontogeny, maturation of T cells and thymic cross-talk, control of transcriptional gene expression and gene networking to the implication of *Aire* gene mutations in autoimmunity to preclinical and clinical approaches that could restore thymic function following clinically induced damage. Interestingly, the major functions of the thymus are well established and effector T lymphocytes correspond to a cell type most studied in immunology. Nevertheless, the thymus as an organ is still neglected. However, two of the fundamental properties of the immune system occur within this organ, i.e. the generation of diversity of TCR receptors through V (D) J recombination and the self-non-self distinction. In fact, immunologists have made significant advances in understanding of these fundamental properties. Still, research of the organ itself, i.e. its ontogeny, developmental biology, its ageing and involution, effect of stress or infections and control of thymic gene expression, emerged only recently. Important perspectives involving multidisciplinary disciplines are opened from now. Understanding biology of autoimmunity is a challenge for twenty-first century and this emphasis places research on thymus in priority. We have better understand the control of promiscuous gene expression (PGE), since it is the basis of the diversity of autoantigens and the induction of immunological tolerance and prevention of autoimmune diseases. More than one hundred of *Aire* gene mutations were described and associated to clinical manifestations of APECED (APS-1) syndrome. However, nothing is still known about the possible effect of non-lethal mutations in the *Fezf2* gene, which also is a controller of PGE in the thymus, on aggressive autoimmunity in man. More recently, a new genome-editing technology called Crispr-Cas9 system has emerged. As with other strategies for the therapeutic use of Crispr-Cas9 as in cancer cases, this opened up perspectives for its use in an eventual thymus intervention of patients seeking to correct *Aire* mutations. In addition, interventions pursuing the regeneration of the aged thymus or even the thymus of patients who have had infections or autoimmune diseases or who have undergone chemotherapy or radiotherapy are currently being investigated. Therefore, the thymus gland offers us a vast possibility for studies

© Springer Nature Switzerland AG 2019
G. A. Passos (ed.), *Thymus Transcriptome and Cell Biology*,
https://doi.org/10.1007/978-3-030-12040-5

ranging from fundamental research up to its therapeutic intervention whose aspects represent challenges for the next years.

Ribeirão Preto, November 2018   Geraldo A. Passos

Index

CPSIA information can be obtained
at www.ICGtesting.com
Printed in the USA
LVHW081059150519
617153LV00001BA/3/P